AF805618

Al-Si Alloys

Francisco C. Robles Hernandez
Jose Martin Herrera Ramírez • Robert Mackay

Al-Si Alloys

Automotive, Aeronautical, and Aerospace Applications

Springer

Francisco C. Robles Hernandez
College of Technology
University of Houston
Houston, Texas, USA

Jose Martin Herrera Ramírez
Centro de Investigación en Materiales Avanzados
Chihuahua, Mexico

Robert Mackay
Metallurgical & Heat Treatment
Nemak US/Canada Business Unit
Windsor, Ontario, Canada

ISBN 978-3-319-58379-2 ISBN 978-3-319-58380-8 (eBook)
DOI 10.1007/978-3-319-58380-8

Library of Congress Control Number: 2017943091

© Springer International Publishing AG 2017
This work is subject to copyright. All rights are reserved by the Publisher, whether the whole or part of the material is concerned, specifically the rights of translation, reprinting, reuse of illustrations, recitation, broadcasting, reproduction on microfilms or in any other physical way, and transmission or information storage and retrieval, electronic adaptation, computer software, or by similar or dissimilar methodology now known or hereafter developed.
The use of general descriptive names, registered names, trademarks, service marks, etc. in this publication does not imply, even in the absence of a specific statement, that such names are exempt from the relevant protective laws and regulations and therefore free for general use.
The publisher, the authors and the editors are safe to assume that the advice and information in this book are believed to be true and accurate at the date of publication. Neither the publisher nor the authors or the editors give a warranty, express or implied, with respect to the material contained herein or for any errors or omissions that may have been made. The publisher remains neutral with regard to jurisdictional claims in published maps and institutional affiliations.

Printed on acid-free paper

This Springer imprint is published by Springer Nature
The registered company is Springer International Publishing AG
The registered company address is: Gewerbestrasse 11, 6330 Cham, Switzerland

To the loves of my life: my wife, my daughters, and my mother…

Dr. Francisco C. Robles Hernandez

I dedicate this book especially to my wife, Ana Maria Hernandez Rojas, and my son, Martin Alejandro Herrera Hernandez, for all their love and support not only in the edition of this book but throughout my professional life. I apologize for my absence both physically and mentally.

I am grateful to Centro de Investigacion en Materiales Avanzados (CIMAV) for its support in my life as a researcher. My thanks go also to Consejo Nacional de Ciencia y Tecnologia (CONACyT) in Mexico and the University of Houston in the United States for supporting me in my sabbatical stay.

Dr. J.M. Herrera Ramirez

I dedicate my contribution to this textbook to my late father, Dr. Alward Murry Charles Mackay, and my late mother, Yvonne Marilyn Taylor Mackay. Their support during my life in academia and in my early professional career has allowed me to make the meaningful contribution found in this text.

Dr. R.I. Mackay

Authorship Statement

The authors claim the following authorship in this work:
Dr. Francisco C. Robles Hernandez is the lead author of this book. Dr. J. Martin Herrera Ramirez and Dr. Robert Mackay are contributing authors. Chapters 1, 2, 5, and 9 were written by Dr. Robles. Chapters 3, 6, 7, and 8 were written by Dr. Mackay Chapter 4 was written collaboratively by Dr. Herrera and Mr. Armando Tejeda Ochoa. Both Dr. Robles and Dr. Herrera proofread the entire book. The main editing of all chapters, including most figures, was conducted by Dr. Herrera. Dr. Robles helped partially with the editorial work.

Bio-sketches

Dr. Francisco C. Robles Hernandez has a BSc, an MSc, and a PhD in materials science and engineering. He did his BSc and MSc at the Instituto Politecnico Nacional in Mexico City. After that, he went to the University of Windsor to do his PhD. His PhD work is on the influence of electromagnetic stirring and vibration on Al-Si alloys, particularly hypereutectic, as a means of soundness improvement and microstructure refinement to demonstrate its positive influence on mechanical properties and service characteristics. His work on Al-Si alloys had a direct focus on automotive applications. After completion of his PhD, he moved to the Transportation Technology Center, Inc., as a principal investigator and a chief metallurgist. Later he relocated to the University of Houston as an assistant professor and he is currently an associate professor with tenure. He has authored more than 200 works, including conference presentations, proceedings, books, patents, journal articles, and industrial reports.

Dr. Martin Herrera conducted his BSc in Chemical Engineering at Mexican Army (Escuela Militar de Ingenieros-Universidad del Ejercito y Fuerza Aerea) in Mexico City. Then he made his MSc in Metallurgy Engineering Science at the Instituto Politecnico Nacional also in Mexico City. Later he studied his PhD in Materials Science and Engineering at Ecole Nationale Superieure des Mines de Paris in Paris, France. After graduating from his BSc, he worked for the Army at the Direccion General de Industria Militar where he held various positions including the Head and project leader of the Applied Research Center and Technology Development for the Mexican Military Industry. Once he retired from the Army he moved to Centro de Investigacion en Materiales Avanzados in Chihuahua Mexico as a full-time researcher. His current research consists on powder metallurgy of metallic alloys and composites, especially aluminum and magnesium base. Dr. Herrera has authored or co-authored around 120 publications including journal papers, proceedings, conference presentations, books, and basic science and technological projects. He has been thesis advisor of doctoral, master and bachelor students.

Dr. Robert Mackay began his formal education in the area of physics, completing a BSc at the University of Prince Edward Island, and then an MSc in geophysics at Memorial University of Newfoundland. From there he went to McGill University to complete a master of engineering degree in metallurgical engineering. Dr. Mackay began his career in the metal casting industry 20 years ago at Oberdorfer Industries Limited, Syracuse, New York, in product engineering of cast aluminum components. Following that, he moved to Haley Industries Limited, Haley, Ontario, as a process engineer with a focus on aerospace magnesium and aluminum castings. After that, Dr. Mackay began an internship with the Ford Motor Company's Casting Process Development Centre (CPDC) in 1999 while working on a doctorate of philosophy in materials engineering at the University of Windsor. In 2003, Dr. Mackay joined Nemak of Canada after completing his doctorate, and his current title is Metallurgical and Heat Treatment Specialist. Robert is also a former adjunct professor in the Mechanical, Automotive and Materials Engineering Department at the University of Windsor. To date, Dr. Mackay has authored or coauthored about 60 publications on work stemming mainly from metal casting science and its application in product development and/or manufacturing. Dr. Mackay has served extensively in the Detroit-Windsor Chapter of the American Foundry Society as a three-term chairman, one term as vice chairman, and finally as a two-time member of the board of directors.

Acknowledgments

Dr. Francisco C. Robles Hernandez acknowledges and appreciates the outstanding support of the government of Mexico by providing a fellowship through the Consejo Nacional de Ciencia y Technologia (CONACyT) to conduct his PhD studies. Part of the experimental work included in this book was completed during his PhD studies.

Dr. Martin Herrera Ramirez and MSc. Armando Tejada Ochoa greatly appreciate the support of Dr. Roberto Martinez-Sanchez and Dr. Ivanovich Estrada-Guel for facilitating the use of the mechanical alloying facilities at CIMAV-Chihuahua as well as their technical guidance for the mechanical alloying experiments. Special thanks to Mrs. Adriana Ramos-Palomino and Mr. Jose Ernesto Ledezma-Sillas for editing some of the images and for designing others. Martin Herrera Ramirez thanks CONACyT for the support during his sabbatical year (grant number 255695). Armando Tejeda Ochoa also thanks CONACyT for his doctoral scholarship (grant number 445419).

Dr. Robert Mackay acknowledges the Nemak of Canada Corporation and his staff and coworkers for supporting his objectives in both forensic analysis of casting defects and the pursuit of aluminum alloy and process development for the metal casting industry. Dr. Mackay is also grateful to his former graduate supervisors and his current employer for encouraging him to pursue the extensive publication record he has achieved to date, all of which has instilled the confidence to participate in the coauthorship of this textbook.

Contents

The original version of this book was revised. An erratum to this book can be found at DOI 10.1007/978-3-319-58380-8_10

Chapter 1
Al-Si Alloys, Minor, Major, and Impurity Elements

1.1 General Overview: Aluminum Silicon Alloys

To facilitate the identification of Al alloys, the Aluminum Association (AA) has classified these alloys based on their chemical composition. The classification differentiates the cast alloys with three digits, and four digits are used to identify the wrought alloys. The first digit (e.g. 1xxx) indicates the purity or the family (second most abundant element) of the alloy. The rest of the elements can be found in the alloys with a range of compositions. For instance, Al-Si alloys constitute the 3xx.x family [1–5]. The most common Al-Si hypoeutectic alloys are the 319, 356 and 357 with 6-7 wt%Si. In the hypereutectic alloys typical compositions are the 390 and 393 containing between 13–18 wt% Si and 25–30 wt% Si, respectively, and other alloying elements. The digit after the decimal point is usually 1 or 0 and it is used to identify if the alloy is a cast or an ingot, respectively. A more detailed explanation will be given in Sect. 1.2.

Aluminum Silicon (Al-Si) alloys are the most important among cast alloys and have widespread applications, especially in the aerospace and automotive industries [1, 2]. The dominant group of alloys is the Al-Si foundry alloys which contain between 5 and 25 wt% Si. Other additions include: Mg, Ni, and Cu. The microstructure of the Al-Si alloys consists of a primary phase (α-Al or primary Si for hypo- and hypereutectic alloys, respectively), Al-Si eutectic, and Mg, Fe, Ni, and Cu-rich phases [2, 3]. Figure 1.1 shows microstructures of typical Al-Si alloys.

Al-Si castings constitute 80–90% of the total Al castings produced worldwide [1, 5, 6]. Fe is the main impurity and the most undesirable element for Al-Si alloys because it forms detrimental brittle crystals (intermetallics) [1, 2]. Mn is added to transform the detrimental Fe intermetallics into Chinese script phases, which is a less harmful phase [2, 5]. The Al-Si binary system is composed of a eutectic reac-

© Springer International Publishing AG 2017

F.C. Robles Hernandez et al., *Al-Si Alloys*, DOI 10.1007/978-3-319-58380-8_1

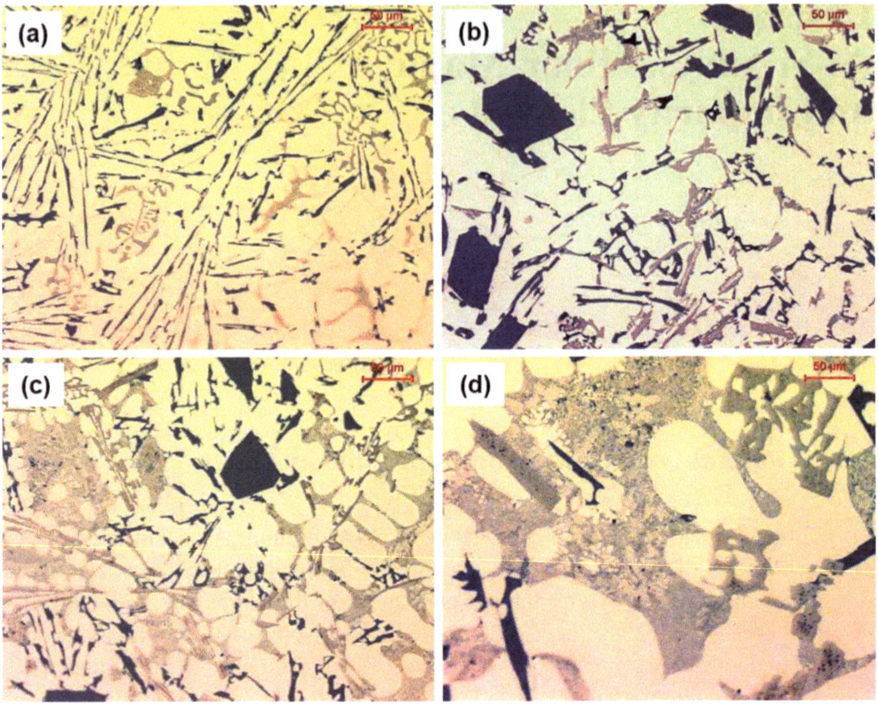

Fig. 1.1 Optical microscopy of Al-Si alloys for (**a**) Al-Si eutectic or hypoeutectic, (**b**) Al-Si hypereutectic, (**c**) Al-Si hypo- or hypereutectic showing the dendritic structure, and (**d**) Al-Cu eutectic

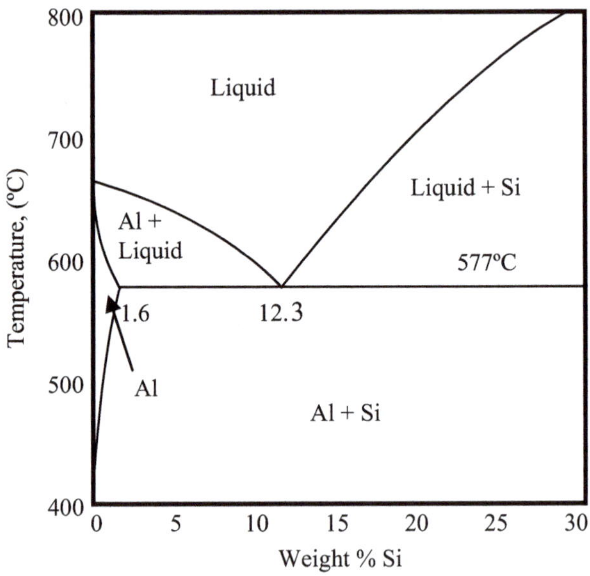

Fig. 1.2 Section of the binary Al-Si phase diagram [7] that covers the most common Al-Si compositions for the 3xx.x series of alloys

tion, which takes place at approximately 12.3 wt% Si at a temperature of 577 °C (Fig. 1.2). All of the Al-Si compositions with less than 12.3 wt% Si are known as hypoeutectic, with more than 12.3 wt% Si, hypereutectic, and 12.3 wt% Si, eutectic [7]. Figure 1.2 shows the Al-Si phase diagram.

Al-Si alloys are ideal to replace cast iron and other high-density alloys and composites due to their mechanical properties, wear resistance, and physical characteristics (e.g., fluidity, density, thermal expansion coefficient, etc.). For example, Al-Si hypereutectic alloys are the ideal candidates to substitute cast iron cylinder liners. Currently Al-Si engine blocks use cast iron cylinder liners that are subjected to excessive wear, dynamic loading, and high temperatures.

Al-Si hypereutectic alloys are mainly used for die casting in applications such as pistons, cylinder liners, and limited number of engine blocks [7–15]. So far, the successful development of the Al-Si hypereutectic engine blocks is still questionable, but has been used for selected vehicles [6, 8–14]. The lack of success of Al-Si hypereutectic alloys is presumably the presence of coarse and brittle primary Si particles that crack easily in the presence of high/cyclic loading. Once a primary Si particle cracks, it exposes the soft Al-matrix to extreme wear conditions. Some authors propose that the solution is to refine primary Si particles before they can be successfully applied for dynamic loading [16–21]. Alpas' group from the University of Windsor has published quite a comprehensive tribological work on Al-Si hypereutectic alloys. They recommend using these alloys for automotive applications under ultra-mild wear conditions [22–28].

1.2 Alloy Designations

There is not an international designation for aluminum alloys. However, the designation system of the Aluminum Association is the most common. Cast alloys are identified by a three-digit numerical system with a decimal point: xxx.x. The nomenclature reads as follows: the first digit is the principal alloying constituent. The second and third digits are arbitrary numbers to identify alloys in a specific family. The digit following the decimal point, "fourth digit", is an indicator if the alloy was manufactured as a casting (0) or an ingot (1 or 2). A capital preceding letter (A, B, C, etc.) indicates too small variations in the composition limits. Table 1.1 presents the designation system for casting alloys. The wrought alloy designation system is also listed for comparative purposes, where a four-digit numerical system is used.

1.3 ASTM Standards for Testing AL-SI Castings

Table 1.2 summarizes the main ASTM standards used for testing automotive and aerospace casting components. This information is provided so the reader familiarizes with the types of tests available for Al-Si castings.

Table 1.1 Aluminum alloy designation system – Aluminum Association

Casting alloys		Wrought alloys	
Alloy series	Principal alloying element	Alloy Series	Principal alloying element
1xx.x	≥99.000% aluminum	1xxx	≥99.000% aluminum
2xx.x	Aluminum-copper	2xxx	Aluminum-copper
3xx.x	Aluminum-silicon plus copper and/or magnesium	3xxx	Aluminum-manganese
4xx.x	Aluminum-silicon	4xxx	Aluminum-silicon
5xx.x	Aluminum-magnesium	5xxx	Aluminum-magnesium
6xx.x	Unused series	6xxx	Aluminum-magnesium and silicon
7xx.x	Aluminum-zinc	7xxx	Aluminum-zinc
8xx.x	Aluminum-tin	8xxx	Other elements
9xx.x	Other elements	9xxx	Unused series

1.4 Effect of Main Chemical Constituents of AL-SI Hypereutectic Alloys

Al-Si alloys are mainly divided into primary and secondary (recycled). Both usually have various other elements that can be divided into major (Mg, Ni, Cu, Zn), minor (Sr, Ti, B, P, rare elements), and impurity elements (Fe, Pb, Sb, Sn) [3, 4, 29–68]. Major alloying elements are added to improve the mechanical, physical, and service characteristics. These elements have strengthening effects. Some of them have the potential to be enhanced following a heat treatment, microstructure refinement, or phase modification [1–5, 69]. Minor alloying elements are added to improve the alloy's mechanical, service characteristics, and integrity (e.g., porosity, wear, soundness, etc.) [1–5, 53–68]. The presence of impurity elements is most common for secondary alloys or recycled. These impurities are usually undesirable, except for Cu; perhaps this lowers their price, keeping reasonably good mechanical properties. Yet, these alloys are attractive for various applications [2]. Due to the importance of the chemical modification of Al-Si hypereutectic alloys, it is discussed in detail in Chap. 2.

1.5 Major Alloying Elements

1.5.1 Silicon

Silicon (Si) improves the metal castability and fluidity and reduces metal shrinkage during solidification. Si requires large amounts of energy to melt mainly due to its highly ordered structure that requires a large amount of entropy to transform from solid to liquid. This change is a direct function of the enthalpy of transformation

Table 1.2 Commonly listed ASTM standards for the testing of Al-Si castings

Test	Standard	Description
Corrosion	ASTM G31	Standard Practice for Laboratory Immersion Corrosion Testing of Metals
	ASTM G111	Standard Guide for Corrosion Tests in High Temperature or High Pressure Environment, or Both
	ASTM B117	Standard Practice for Operating Salt Spray (Fog) Apparatus
Mechanical	ASTM E466	Standard Practice for Conducting Force Controlled Constant Amplitude Axial Fatigue Tests of Metallic Materials
	ASTM B557	Standard Test Methods for Tension Testing Wrought and Cast Aluminum- and Magnesium-Alloy Products
	ASTM E8/E8M	Standard Test Methods for Tension Testing of Metallic Materials
	ASTM B769	Standard Test Method for Shear Testing of Aluminum Alloys
	ASTM E9	Standard Test Methods of Compression Testing of Metallic Materials at Room Temperature
Hardness	ASTM E10	Standard Test Method for Brinell Hardness of Metallic Materials
	ASTM E384	Standard Test Method for Knoop and Vickers Hardness of Materials
Radiography	ASTM E155	Standard Reference Radiographs for Inspection of Aluminum and Magnesium Castings
	ASTM E94	Standard Guide for Radiographic Examination
	ASTM E2422	Standard Digital Reference Images for Inspection of Aluminum Castings
	ASTM E2973	Standard Digital Reference Images for Inspection of Aluminum and Magnesium Die Castings
	ASTM E1030	Standard Test Method for Radiographic Examination of Metallic Castings
Metallography	ASTM E3-11	Standard Guide for Preparation of Metallographic Specimens
	ASTM E112	Standard Test Methods for Determining Average Grain Size
	ASTM E340	Standard Test Method for Macroetching Metals and Alloys
	ASTM E407	Standard Practice for Microetching Metals and Alloys
	ASTM E883	Standard Guide for Reflected–Light Photomicrography

($\Delta H^T_{S\text{-}L}$), thus heat is required for melting. This energy is stored during melting and then released during solidification giving the Al-Si hypereutectic alloys good fluidity that improves with the amount of Si. It means that as the amount of Si increases, the required energy for melting increases. During solidification, alloys rich in Si require longer solidification times, that helps in their improved fluidity, castability, and interdendritic feeding of the alloy. However, this can be detrimental to die casting equipment. In addition [70], amounts of more than 14–18 wt% Si can cause problems during continuous casting [1–3, 7, 71].

The first phase to nucleate during solidification in Al-Si hypereutectic alloys is the primary Si particles that are polyhedral crystals and due to their nondendritic nature, they can freely flow together with the remaining liquid metal pool. Primary Si particles nucleate at higher temperatures than the Al-Si eutectic (next phase that solidifies). The solidification of primary Si has negligible effects on the viscosity of the alloy. This is because the primary Si particles can be dragged with the molten alloy because the solubility of Si in aluminum and other alloying elements is almost negligible [7]. The liquidus temperature can be precisely determined for binary and multi-component alloys using the Si equivalent (Si_{EQ}) methodology as a function of the alloy's chemical composition [41].

Primary Si particles are identified not only in Al-Si hypereutectic, but also in eutectic and near eutectic alloys [50]. This is because a nonequilibrium solidification may result in a solidification known as the skewed couple zone [72, 73]. In conventional casting procedures, chemical and solution treatments are widely used to modify the Si morphologies into a less harmful geometry [22].

In Al-Si hypo and eutectic alloys the first phase to solidify is the α-Al. The Al-Si eutectic is fibrous and the Si particles are long and needle-like [74, 75]. Additions of Si below 0.01wt% are usually dissolved in the aluminum matrix, increasing the mechanical properties. Further additions result in the precipitation of the Al-Si eutectic. Si additions of up to approximately 11–13 wt% have an almost direct increase in the alloy's strength. Below this limit the strength drops considerably. The drop is the result of primary Si precipitates [74, 75]. Further improvements in mechanical properties can be accomplished by modifying the Al-Si eutectic [76–79]. The most common Al-Si modifiers are Sr, Na, and Ba; however, Sr is the most popular modifier since it is not as reactive as other modifiers and it is more stable [80]. Following the precipitation of the α-Al phase (that is a soft matrix), the next phase to solidify is the Al-Si eutectic.

1.5.2 Copper

Copper (Cu) in Al-Si alloys improves as-cast and high temperature strength properties, but it reduces ductility. Cu is added as the main hardening strengtheners. In general the bulk hardness of Al-Si alloys is acceptable, but heat treatments (i.e., solution and aging) in the presence of Cu promote precipitation hardening allowing the full development of the hardness and strength of Al-Si alloys [22, 52]. Often Al-Si castings are overaged to provide a compromise between strength, hardness, and dimensional stability in service characteristics [10]. Cu-rich phases take several morphologies depending on the chemical composition, solidification, and heat treatment conditions [22]. Figure 1.3 shows some of these phases. The approximate compositions and a brief description of the Cu-rich eutectics identified by M. Djurjevic et al. [69] are given in Table 1.3.

Cu has a maximum solid solubility in aluminum at approximately 6 wt% and a temperature of 546 °C. In order to have good precipitation hardening results, it is

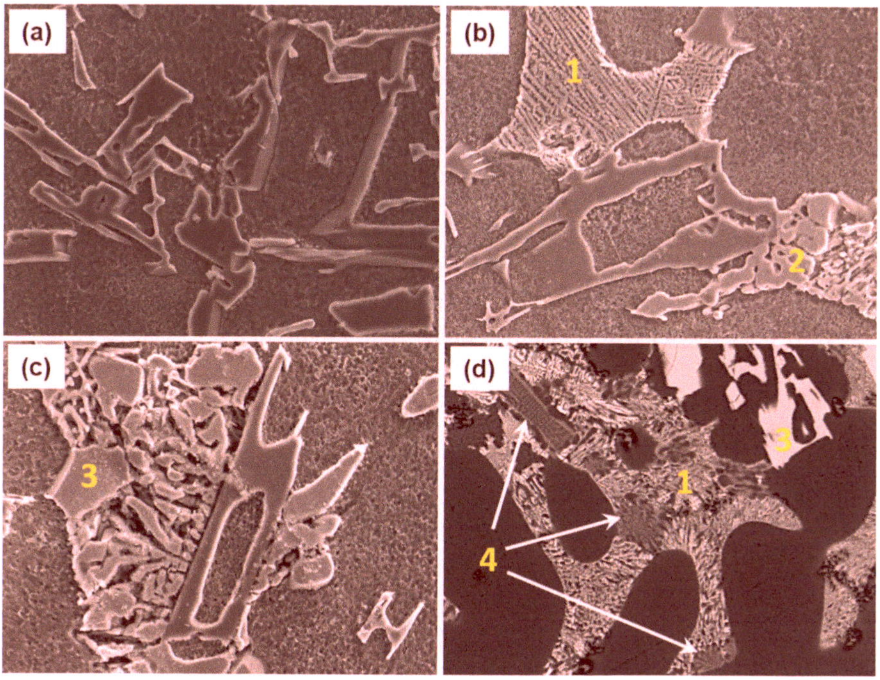

Fig. 1.3 Scanning electron microscopy of Al-Si alloys showing different eutectic phases such as (**a**) Al-Si, (**b**) Al-Cu, (**c**) Al-Cu-Ni, and (**d**) mix. The numbers stand for: Al-Cu fine (*1*), blocky Al-Cu (*2*), Al_3CuNi (*3*), and Mg_2Si (*4*)

Table 1.3 Stoichiometric composition of the Cu-enriched phases found in Al-Si alloys [69]

Morphology of the Cu-Enriched Phases	Stoichiometric Composition
"Blocky"	$Cu_{10.59}Al_{4.09}Si_{2.82}Mg_{2.48}$
"Eutectic"	$Cu_{7.27}Al_{12.25}Si_{0.47}$
"Fine Eutectic"	$Cu_{4.14}Al_{13.32}Si_{1.72}Mg_{0.81}$

recommended that Cu additions do not exceed 5 wt% Cu on Al-Si alloys, but preferably between 2 and 4 wt% Cu. It is suggested that after solidification casting with Cu additions, the alloy should be subjected to solubilization heat treatment at temperatures as close as possible, but below their respective solvus temperature (in most cases above 546 °C). This accelerates the dissolution rate of Cu in the Al-matrix that is usually followed by a water- or air-quenching process and natural or artificial aging (T4 or T6 heat treatments, respectively). The supersaturation of Cu in Al causes an instability forcing the Cu atoms to precipitate in the fashion described by Guinier and Preston and forming the so-called GP, GP-I and GP-II zones [51, 81, 82]. The final phase to form is $CuAl_2$ (θ) that is incoherent with Al-matrix [51, 83, 84].

Cu forms four types of precipitates that grow progressively starting with plates of Cu atoms one or two atoms thick. The plates are usually composed of approximately 25 atoms in diameter, oriented parallel to the {100} planes in the Al-rich matrix. The presence of such precipitates coarsens and distorts the lattice, increasing the hardness of the Al-matrix. Such precipitates are stable at room temperature, but when the temperature is increased to approximately 200 °C, the precipitates transform into the GP-I followed by the GP-II. The highest hardness is reached when the size of the GP-I precipitates is maximum or right before they transform into the GP-II precipitates. The GP-II are semicoherent with the Al-matrix losing some of the residual stress and strength in the lattice. These precipitates are parallel to {100} plane because the elastic modulus is minimum in this direction. A sketch of the GP zones' mechanisms is given in Fig. 1.4.

The GP-II are thicker (10 atoms) and of larger diameter (75 atoms), but they are not just big GP-I precipitates. The GP-II zones have different crystalline structure and composition when compared to those of the GP-I. The average composition and stoichiometry of the GP-II precipitates is about Cu_2Al_5. The θ' is a third metastable

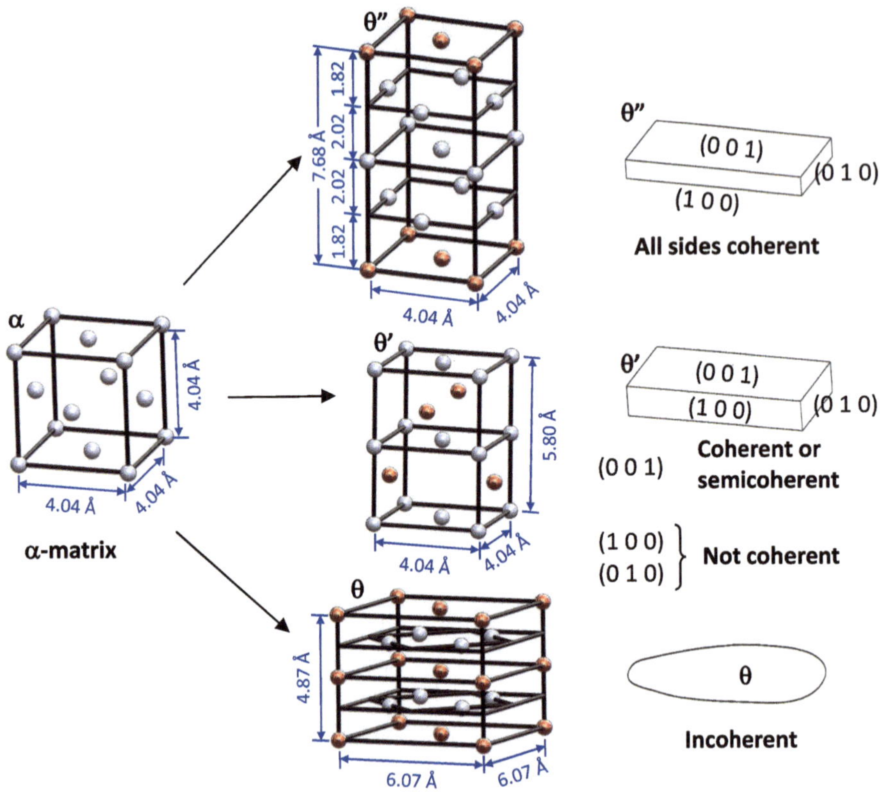

Fig. 1.4 Mechanisms of precipitation of the GP zones. Silver spheres represent Al and brown spheres, Cu

precipitate with a nominal composition of $CuAl_2$ and grow epitaxially with the Al-matrix creating distortions on it. Once these precipitates start nucleating, the material loses hardness and the properties, as a whole, fade. Here it is important to mention that this is what is called over aged, and it is desirable for components that require high dimensional stability.

1.5.3 Nickel

Nickel is added to some Al-Si alloys (e.g. 339 and 390) to increase tensile strength, yield strength at temperatures in the range 250–375 °C. Ni reduces elongation at temperatures below 50 °C. The increase in strength is mainly due to the precipitation of the Al_3Ni intermetallic [3, 30] . Additions of less than 0.3 wt% Ni have a negligible effect on mechanical properties. Additions between 0.3–0.5 wt% Ni can result in yield strength improvements of up to 50 MPa [3]. Ni has similar strengthening effects as Cu, but in comparison, Ni is more expensive. However, for pistons that are only artificially aged (in order to maximize their thermal stability) additions of up to 2 wt% are common. Ni additions are complementary to Cu. The main mechanisms that improve the mechanical properties is based on the increase in the amount of strengthening particles (e.g., Al_3Ni) in the Al-matrix [10]. However, additions above 4 wt% Ni are not suggested because they affect casting soundness. The main drawback is the presence of a large number of brittle precipitates (Al_3Ni intermetallics) that are susceptible to cracking [71].

Usually, in Al-Si hypereutectic alloys Ni and Cu are identified together by Energy-dispersive X-ray spectroscopy (EDX) because these elements are alike (atomic numbers 28 and 29). Ni and Cu present complete solubility in solid, semisolid, and liquid states. They have the same crystalline structure (FCC), similar electronegativity (1.91 and 1.90 respectively), valence and lattice parameters (a_{Cu} = 3.61 Å and a_{Ni} = 3.55 Å). Based on the Hume-Rothery Empirical Rules, Cu and Ni form an ideal solution [85]. In a solid solution, the atoms can freely interchange within the lattice due to their free energy of mixture ($\Delta G^{MIX} = 0$) [85].

1.5.4 Magnesium

Magnesium is added as a precipitation hardener. The application of solution treatments (e.g., T5 or T6) promotes the formation of a supersaturate solid solution with silicon. After solution treatment, the components are usually quenched to precipitate the Mg_xSi_y particles. The Mg_xSi_y particles are then transformed into Mg_2Si by a second transformation (aging) that can be natural or artificial. Typical artificial aging temperatures are approximately 200 °C [52, 57]. Under controlled precipitation conditions, Mg_2Si is homogeneously distributed improving the homogeneity of the alloy and thus improving its mechanical properties [52]. A microstructure showing the Mg_2Si precipitates is given in Fig. 1.3. Armstrong [53] reported that strontium can have certain advantages when combined with large amounts of magnesium. However, additions below 0.7 wt% Mg are not recommended since there is no apparent or measurable improvement in the mechanical properties [54, 55]. More than 3 wt% Mg promotes microstructure coarsening [56]. Ideally the Mg additions should be kept below 10 wt% Mg. To maximize its effects on ultimate tensile strength (UTS), it is recommended to keep it between 0.6 and 3 wt% Mg [22]. Mg additions are detrimental to ductility; therefore, there is a compromise between the benefits and disadvantages.

1.5.5 Zinc

Additions of zinc (Zn) in the range 1 wt%–4 wt% slightly improve the tensile properties in heat-treated (T5 or T6) Al-Si castings. The advantage of Zn is that it does not affect ductility. The effects of Zn are negligible when it is added in amounts below 1 wt% Zn [3, 31]. Mackay et al. [34] reported that additions of up to 15 wt% Zn can be used to attain superior mechanical properties.

1.6 Impurity Elements

1.6.1 Iron

Iron (Fe) is considered the principal impurity element of secondary Al-Si alloys. Fe is undesirable because it forms brittle Al-Fe intermetallic crystals [35–39]. Iron is usually present in Al-Si alloys as a byproduct of recycling. Al-Si castings (e.g., engine blocks) have cast-in or press-in cast iron components, such as cylinder liners. Technically the cylinder liners are removed prior to recycling. However, in most cases, not all the iron is taken away, resulting in a contaminant for secondary or recycled alloys.

In Al-Si castings, the intermetallic Al_5FeSi (β-phase) that has a plate-like or needle-like morphology. This morphology is detrimental due to its effect as a stress concentrator. Additionally, this phase is brittle and susceptible to cracking. This in turn jeopardizes the integrity of cast components and increases machining costs [30]. The amount of Fe should be kept as low as possible. Primary alloys (e.g., 356 and 357) have less than 0.2 wt% Fe. On recycled alloys (e.g., 319 and 390) the Fe concentration is up to 1 wt% Fe [30]. Fe intermetallics are not susceptible to modification by heat treatments or cooling rates. However, they can be chemically modified to a less hazardous morphology with Mn additions [30, 31, 35]. Figure 1.5a gives an example of the unmodified Fe-rich intermetallic.

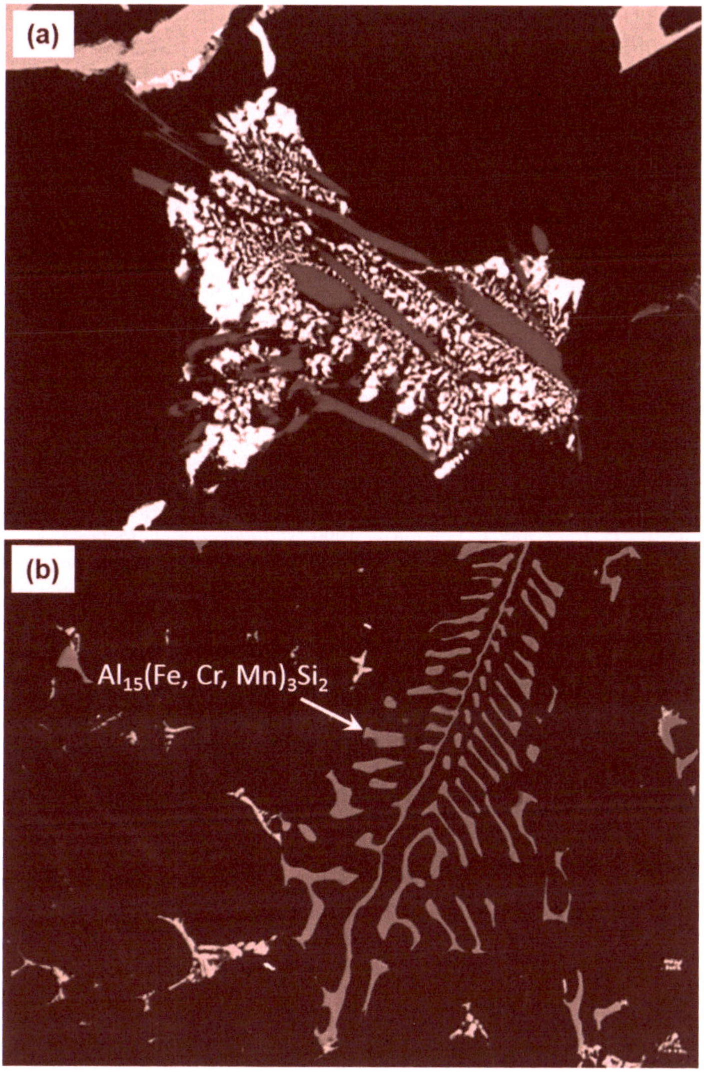

Fig. 1.5 (**a**) Unmodified Fe-rich intermetallic and (**b**) modified structure of the Chinese script intermetallic

1.6.2 Manganese

Manganese (Mn) has very low solubility in Al, but good susceptibility for Fe and Si. The combination of these four elements form the $Al_{15}(Mn, Fe)_3Si_2$ intermetallic [30, 31], as shown in Fig. 1.5b. This is also known as the Chinese script phase. The Chinese script phase is either blocky or dendritic [30]. A ratio of Mn:2 Fe is usually recommended to transform the Fe-rich phases (β-phase) into an α-phase

(Chinese script). The Chinese script phase is preferred because it is less detrimental to mechanical properties and integrity of castings [35]. Wang et al. [36] have reported positive effects of using up to 5 wt% Fe and 3 wt% Mn additions to Al-20 wt% Si. Yet, the application for these alloys is in coatings by means of spray atomization. Using spray deposition methods the wear resistance of Al–20 wt% Si alloy can be considerably affected by the presence of Fe (up to 5 wt%). Nevertheless, a combination 5 Fe:3 Mn significantly improves the wear resistance of the Al-20 wt% Si alloy [36]. Figure 1.5b shows the modified structure of the Chinese script intermetallic.

References

1. Gruzleski, J.E., and B.M. Closset. 1990. *The treatment of liquid aluminum-silicon alloys*, 256. Des Plaines: American Foundrymen's Society, Inc.
2. Campbell, J. 2003. *Castings*. 2nd ed., Oxford, UK: Butterworth Heinemann.
3. Bäckerud, L., G. Chai, and J. Tamminen. 1990. *Solidification characteristics of aluminum alloys: Foundry alloys*. Vol. 2. 1st ed., 256. Stockholm: AFS/Skan Aluminium.
4. Bäckerud, L., G. Chai, and J. Tamminen. 1990. *Solidification characteristics of aluminum alloys: Wrought alloys*. Vol. 1. 1st ed. Oslo: AFS/Skan Aluminium.
5. Adams, J.H. et al. 1990. *ASM metals handbook Vol. 2: Properties and selection: Nonferrous alloys and special-purpose materials*. Metals Park: ASM International.
6. Jorstad, J.L. 1971. The hypereutectic aluminum-silicon alloy used to cast the Vega engine block. *Modern Casting* 60 (4): 59–64.
7. Robles Hernandez, F.C. 2004. *Improvement in functional characteristics of aluminum-silicon cast components through the utilization of a novel electromagnetic treatment of liquid melts*. Ph.D Thesis, University of Windsor (Canada).
8. Gadou, R. 2001. *Thermal spray coatings for cylinder liners*. Shrewsbury: MPR.
9. Jorstad, J.L. 1970. *Paper no. 105*, 5–11. Detroit: SDCE.
10. Kolbemschmidt, A.G. 1985. *75 Years Kolbemschmidt, Brochure*, 1–51.
11. Jorstad, J.L. 1986. Understanding sludge. *Die Casting Engineer* 30 (6): 30.
12. Norbye, J.P. 1970. Chevrolet vega 2300 engine. *Automobile Engineering*: 320–326.
13. Green, R.E. 1970. Die casting the Vega engine block. *Die Casting Engineer* 14: 12–26.
14. Jorstad, J.L. 1970. *6th SDCE international die casting congress*, 1–6. Cleveland.
15. Kisielowski, C., P. Specht, S.M. Gygax, B. Barton, H.A. Calderon, J.H. Kang, and R. Cieslinski. 2015. Instrumental requirements for the detection of electron beam-induced object excitations at the single atom level in high-resolution transmission electron microscopy. *Micron* 68: 186–193.
16. Hernandez, F.C.R., J.H. Sokolowski, and J.D.C. Rivera. 2007. Micro-Raman analysis of the Si particles present in Al-Si hypereutectic alloys in liquid and semi-solid states. *Advanced Engineering Materials* 9 (1–2): 46–51.
17. Robles Hernandez, F.C., and J.H. Sokolowski. 2006. Comparison among chemical and electromagnetic stirring and vibration melt treatments for Al-Si hypereutectic alloys. *Journal of Alloys and Compounds* 426 (1–2): 205–212.
18. Robles Hernandez, F.C., and J.H. Sokolowski. 2006. Thermal analysis and microscopical characterization of Al-Si hypereutectic alloys. *Journal of Alloys and Compounds* 419 (1–2): 180–190.
19. Robles Hernandez, F.C., and J.H. Sokolowski. 2009. Effects and on-line prediction of electromagnetic stirring on microstructure refinement of the 319 Al-Si hypoeutectic alloy. *Journal of Alloys and Compounds* 480 (2): 416–421.

20. Robles Hernández, F.C., and J.H. Sokolowski. 2005. Identification of silicon agglomerates in quenched Al-Si hypereutectic alloys from liquid state. *Advanced Engineering Materials* 7 (11): 1037–1043.
21. Robles Hernandez, F.C., J.H. Sokolowski, and J.J. De Cruz Rivera. 2007. Micro-Raman analysis of the Si particles present in Al-Si hypereutectic alloys in liquid and semi-solid states. *Advanced Engineering Materials* 9 (1–2): 46–51.
22. Riahi, A.R., and A.T. Alpas. 2006. Fracture of silicon-rich particles during sliding contact of Al–Si alloys. *Materials Science and Engineering A* 441 (1–2): 326–330.
23. Li, J., et al. 2006. FIB and TEM characterization of subsurfaces of an Al–Si alloy (A390) subjected to sliding wear. *Materials Science and Engineering A* 421 (1–2): 317–327.
24. Riahi, A.R., T. Perry, and A.T. Alpas. 2003. Scuffing resistances of Al–Si alloys: Effects of etching condition, surface roughness and particle morphology. *Materials Science and Engineering A* 343 (1–2): 76–81.
25. Elmadagli, M., T. Perry, and A.T. Alpas. 2007. A parametric study of the relationship between microstructure and wear resistance of Al–Si alloys. *Wear* 262 (1–2): 79–92.
26. Elmadagli, M., and A.T. Alpas. 2006. Sliding wear of an Al–18.5 wt.% Si alloy tested in an argon atmosphere and against DLC coated counterfaces. *Wear* 261 (7–8): 823–834.
27. Chen, M., T. Perry, and A.T. Alpas. 2007. Ultra-mild wear in eutectic Al–Si alloys. *Wear* 263 (1–6): 552–561.
28. Elmadagli, M., and A.T. Alpas. 2006. Progression of wear in the mild wear regime of an Al-18.5% Si (A390) alloy. *Wear* 261 (3–4): 367–381.
29. Gustafsson, G., T. Thorvaldsson, and G.L. Dunlop. 1986. The influence of Fe and Cr on the microstructure of cast Al-Si-Mg alloys. *Metallurgical Transactions A* 17 (1): 45–52.
30. Davis, J.R. 1993. *ASM specialty handbook: Aluminum and aluminum alloys*, 784. Materials Park: ASM International.
31. Heusler, L., F.J. Feijus, and M.O. Otte. 2001. Alloy and casting process optimization for engine block application. *AFS Transactions* 109: 443–451.
32. Li, B.Q. 1998. Solidification processing of materials in magnetic fields. *JOM* 50 (2): 1–10.
33. Król, J. 1997. The precipitation strengthening of directionally solidified Al Si Cu alloys. *Materials Science and Engineering A* 234–236: 169–172.
34. Mackay, R. and J.H. Sokolowski. 2001. Investigation of Zn in an Al-Si alloy. In *40th Conference of Metallurgists (COM), Toronto*, 467.
35. Shabestari, S.G., and J.E. Gruzleski. 1995. Modification of iron intermetallics by strontium in 413 aluminum alloys. *AFS Transactions* 103: 285–293.
36. Wang, F., et al. 2004. Effect of Fe and Mn additions on microstructure and wear properties of spray-deposited Al–20Si alloy. *Materials Letters* 58 (19): 2442–2446.
37. Bian, X., and W. Wang. 2000. Thermal-rate treatment and structure transformation of Al–13 wt.% Si alloy melt. *Materials Letters* 44 (1): 54–58.
38. Wang, Q.G., and C.J. Davidson. 2001. Solidification and precipitation behaviour of Al-Si-Mg casting alloys. *Journal of Materials Science* 36 (3): 739–750.
39. Cullity, B.D., and S.R. Stock. 2001. Elements of x-ray diffraction. In *Addison-Wesley series in metallurgy and materials*, 3rd ed., 680. Reading: Addison-Wesley Pub. Co.
40. Malaczynski, G.W., et al. 1997. Diamond-like carbon coating for aluminum 390 alloy — automotive applications. *Surface and Coatings Technology* 93 (2–3): 280–286.
41. Robles Hernandez, F.C., et al. 2005. Calculation of the liquidus temperature for hypo and hypereutectic aluminum silicon alloys. *Materials Science and Engineering A* 396 (1–2): 271–276.
42. Kim, E.S., K.H. Lee, and Y.H. Moon. 2000. A feasibility study of the partial squeeze and vacuum die casting process. *Journal of Materials Processing Technology* 105 (1–2): 42–48.
43. Lee, E.S. 2000. A study on the economics of hypereutectic aluminium-silicon (Al-Si) alloy machining. *The International Journal of Advanced Manufacturing Technology* 16 (10): 700–708.

44. Spigarelli, S., E. Evangelista, and S. Cucchieri. 2004. Analysis of the creep response of an Al–17Si–4Cu–0.55Mg alloy. *Materials Science and Engineering A* 387–389: 702–705.
45. Chen, C.M., C.C. Yang, and C.G. Chao. 2005. A novel method for net-shape forming of hypereutectic Al–Si alloys by thixocasting with powder preforms. *Journal of Materials Processing Technology* 167 (1): 103–109.
46. Xu, C.L., et al. 2007. Effect of Al–P–Ti–TiC–Nd2O3 modifier on the microstructure and mechanical properties of hypereutectic Al–20 wt.%Si alloy. *Materials Science and Engineering A* 452–453: 341–346.
47. Jiang, Q.C., et al. 2007. Estimation of the shifting distance of the eutectic point in hypereutectic Al–Si alloys by the lever rule. *Scripta Materialia* 56 (5): 329–332.
48. Xu, C.L., et al. 2006. Effect of La2O3 in the Al–P–Ti–TiC–La2O3 modifier on primary silicon in hypereutectic Al–Si alloys. *Journal of Alloys and Compounds* 421 (1–2): 128–132.
49. Rea, K.E., et al. 2005. FIB cross-sectioning of a single rapidly solidified hypereutectic Al-Si powder particle for HRTEM. *Microscopy Research and Technique* 66 (1): 10–16.
50. Nave, M.D., A.K. Dahle, and D.H. StJohn. 2002. Halo formation in directional solidification. *Acta Materialia* 50 (11): 2837–2849.
51. Dieter, G.E. 1986. *Mechanical metallurgy*. New York: McGraw-Hill.
52. Haque, M.M., and M.A. Maleque. 1998. Effect of process variables on structure and properties of aluminium–silicon piston alloy. *Journal of Materials Processing Technology* 77 (1–3): 122–128.
53. Armstrong, G.L. 1978. *Alloy selections for automotive aluminium castings*. SAE Technical Paper No. 780249.
54. Dasgupta, R., C.C. Brown, and S. Marek. 1988. Optimization of properties in strontium modified 319 alloy castings. *AFS Transactions* 96: 297–310.
55. ———. 1989. Effect of increased magnesium content on the mechanical properties of sand-cast 319 aluminum alloy. *AFS Transactions* 97: 245–254.
56. Nishi, S., T. Shinoda, and E. Kato. 1968. Effects of iron contents, beryllium addition, and cooling velocity on mechanical properties of Al-Si-Mg-Zn cast alloys. *Journal of Japan Institute of Light Metals* 18 (12): 627–635.
57. Kanicki, D.P., and W.M. Rasmussen. 1990. Processing molten aluminum - part 1: Understanding silicon modification. *Modern Casting* 1: 24.
58. Lidman, W.G. 1984. Master alloys improve aluminum casting properties. *Foundry M&T* 112 (8): 46–47.
59. Mohanty, P.S., F.H. Samuel, and J.E. Gruzleski. 1995. Experimental study on pore nucleation by inclusions in aluminum castings. *AFS Transactions* 103: 555–564.
60. Kyffin, W.J., W.M. Rainforth, and H. Jones. 2001. Effect of phosphorus additions on the spacing between primary silicon particles in a Bridgman solidified hypereutectic Al-Si alloy. *Journal of Materials Science* 36 (11): 2667–2672.
61. Liu, X., et al. 2002. Application of Al-P master alloy to Al-Si piston alloy. *Special Casting & Nonferrous Alloys* 6: 43–45.
62. Müller, K. 1997. Advanced light alloys and composites. In *NATO Advanced Study Institute*. Zakopane: Kluwer Academic Publishers.
63. Kim, H.J. 2003. Effect of calcium on primary silicon particle size in hypereutectic Al–Si alloys. *Materials Science and Technology* 19 (7): 915–918.
64. Kyffin, W.J., W.M. Rainforth, and H. Jones. 2001. Effect of treatment variables on size refinement by phosphide inoculants of primary silicon in hypereutectic Al–Si alloys. *Materials Science and Technology* 17 (8): 901–905.
65. Kattoh, H., et al. 2002. Critical temperature for grain refining of primary Si in hyper-eutectic Al–Si alloy with phosphorus addition. *Journal of Japan Institute of Light Metals* 52 (1): 18–23.
66. Chang, J., I. Moon, and C. Choi. 1998. Refinement of cast microstructure of hypereutectic Al-Si alloys through the addition of rare earth metals. *Journal of Materials Science* 33 (20): 5015–5023.
67. Yi, H., and D. Zhang. 2003. Morphologies of Si phase and La-rich phase in as-cast hypereutectic Al–Si–xLa alloys. *Materials Letters* 57 (16–17): 2523–2529.

68. Yi, H., et al. 2003. Microstructures and La-rich compounds in a Cu-containing hypereutectic Al–Si alloy. *Journal of Alloys and Compounds* 354 (1–2): 159–164.
69. Djurdjevic, M.B., et al. 2001. Quantification of Cu enriched phases in synthetic 3XX aluminum alloys using the thermal analysis technique. *AFS Transactions* 16: 1–12.
70. Kamiya, M., et al. 2008. Effect of Si content on turning machinability of Al-Si binary alloy castings. *Materials Transactions* 49 (3): 587–592.
71. Kapranos, P., et al. 2003. Thixoforming of an automotive part in A390 hypereutectic Al–Si alloy. *Journal of Materials Processing Technology* 135 (2–3): 271–277.
72. Gruzleski, J.E. 2000. *Microstructure development during metal casting*, 99–116. Des Plaines: American Foundrymen's Society.
73. McDonald, S.D., K. Nogita, and A.K. Dahle. 2004. Eutectic nucleation in Al–Si alloys. *Acta Materialia* 52 (14): 4273–4280.
74. Saheb, N., et al. 2001. Influence of Ti addition on wear properties of Al-Si eutectic alloys. *Wear* 249 (8): 656–662.
75. Warmuzek, M. 2004. *Aluminum-silicon casting alloys: Atlas of microfractographs*, 124. Materials Park: ASM International.
76. Dahle, A.K., et al. 2005. Eutectic modification and microstructure development in Al-Si alloys. *Materials Science and Engineering A-Structural Materials Properties Microstructure and Processing* 413: 243–248.
77. Flood, S.C., and J.D. Hunt. 1981. Modification of Al-Si eutectic alloys with Na. *Metal Science* 15 (7): 287–294.
78. Hegde, S., and K.N. Prabhu. 2008. Modification of eutectic silicon in Al-Si alloys. *Journal of Materials Science* 43 (9): 3009–3027.
79. Heshmatpour, B. 1997. Modification of silicon in eutectic and hyper-eutectic Al-Si alloys. *Light Metals* 1997: 801–808.
80. Tiedje, N.S., et al. 2012. A solidification model for unmodified, Na-modified and Sr-modified Al-Si alloys. IOP Conference Series: Materials Science and Engineering 27 (1): 012033. IOP Publishing.
81. Preston, G.D. 1938. Structure of age-hardened aluminium-copper alloys. *Nature* 142 (3595): 570–570.
82. Guinier, A. 1938. Structure of age-hardened aluminium-copper alloys. *Nature* 142 (3595): 569–570.
83. Callister, W.D. 2007. *Materials science and engineering: An introduction*. 7th ed. New York: Wiley.
84. Porter, D.A., and K.E. Easterling. 2001. *Phase transformations in metals and alloys*. 2nd ed. Boca Raton: CRC Press.
85. Evans, E.L., C.B. Alcock, and O. Kubaschewski, eds. 1967. *Metallurgical thermochemistry*. 4th ed. Oxford, UK/New York: Pergamon Press.

Chapter 2
Thermal Analysis

2.1 Introduction

Thermal analysis is a powerful characterization tool used to determine transformation temperatures, fraction solids or fraction transforming, latent heats, among other characteristics on castings. It is usually applied during heating or cooling of materials, particularly melting or solidification, respectively. In ferrous and other alloys the method has been extensively used in solid-solid transformations. The most common and market-ready technique is differential scanning calorimetry, also known as DSC, that can be accompanied with thermogravimetry (TGA), and dilatometry.

Thermal analysis mainly consists of the precise determination of the historical data of the changes in temperature. The analysis is mainly based on the understanding of the cooling and/or heating curves as a function of slope changes. The slope curve is also known as the first derivative, and every change is the result of a phase transformation. The quantification of the energy stored during the changes in the slope is directly related to the energy required for a transformation to take place. This amount of energy is also known as the latent heat (for melting, transformation, solidification, etc.). In general terms we can interpret the first derivative curve as an enhanced plot of the slop changes observed in the cooling curve. The most common practice during cooling is carried out under natural heat exchange conditions. This means that the samples are allowed to cool naturally or normalized [1–4].

Thermal analysis has been effectively used for Al-Si alloys including hypo- and hypereutectic alloys [1–4]. Several groups have made tremendous efforts and have been dedicated to determine the solidification characteristics of Al-Si alloys, which in some cases are related to the properties of the alloy. This practice is arguable; nevertheless, the empirical results have good correlation. Some examples include the effects of strontium or boron to modify eutectic silicon [5–10], dendrite coherency point [11, 12], and microstructure refining [13–17]. The main difference among the

© Springer International Publishing AG 2017

F.C. Robles Hernandez et al., *Al-Si Alloys*, DOI 10.1007/978-3-319-58380-8_2

Table 2.1 Chemical Composition for the Investigated 319 Alloy (wt%)

Alloy	Al	Si	Cu	Ni	Fe	Mg	Zn	Sr (ppm)
319	Bal.	7.55	3.45	0.008	0.40	0.25	0.009	6
390	Bal.	15	5.5	0.01	0.3	0.65	0.065	

Al-Si hypoeutectic and hypereutectic is the amount of Si. For hypereutectic alloys the first phase to solidify is primary Si that is usually modified with P [16, 18] by the precipitation of AlP particles and is well documented by thermal analysis [19–21].

Thermal analysis is well supported by unique algorithms that help in determining the main characteristics of the heating and/or cooling curves. One of those algorithms has been developed to determine the baseline that can be complementary to the determination of the fraction solid. The first derivative can be calculated using a Newtonian approach. On the other hand, the following are the most common methods to determine the baseline:

1. Linear,
2. Polynomial,
3. Fourier.

The use of each of these methods is rather based on preference, and in the following chapter, we will focus on the polynomial method [13, 14, 22]. The analysis carried out herein is centralized on two of the most common Al-Si alloys that are the 319 and 390 Al-Si alloys (hypo and hypereutectic, respectively). The respective chemical compositions for the alloys are given in Table 2.1. The aim of investigating those Al-Si alloy systems is to expand their understanding and to share some of the knowledge of this family of alloys. The Al-Si hypoeutectic alloys have been widely investigated. In contrast, Al-Si hypereutectic alloys are behind. This is primarily attributed to the level of consumption of each of those alloys.

Bäkerud et al. have reported some of the most thorough characterization coupling thermal analysis and microscopy including optical and SEM/EDX on a large number of Al-Si alloys [23, 24]. In this chapter, a descriptive methodology to convey the understanding of those alloys is provided. The methodology proposed herein is complementary and it is proposed for any alloy. The algorithms that are provided herein include the first derivative, baseline, fraction transforming, dendrite coherency point, among others [25–27]. This method is flexible enough to be used for a wide variety of materials, including, but not limited to, metals [28–32], biomaterials [33–35], polymers, [36–38], and food [39–46].

Nafisi et al. [47] used thermal analysis to determine the optimal amount of eutectic and primary Si modifiers (Sr and/or P, respectively). In this method, the idea is to determine the required level of microstructure modification. The analysis is based on the changes in liquidus temperature by the effect of the Si modifiers. The findings are in agreement with those reported by Djurdjevic et al. [48]. Addition of modifiers above the required level causes a drop in the liquidus temperatures. The results conclusively show that the modifiers increase the liquidus temperature the modification level. In order to reach the optimum modification level one add the modifier to reach the maximum achievable liquidus temperature [49–51]. As an example of this approach Apelian et al. identified a depression of ≥5.5 °C [51] corresponding to a chemical modification of >90% of Si in the 356 Al-Si alloy.

2.2 Thermal Analysis Procedure

The thermal analysis can be carried out using sensitive equipment that allows high speed and high resolution of the thermal heating and/or cooling history of a material. For Al-Si alloys we recommend K-type thermocouple(s) that can be connected to data acquisition systems. The K thermocouple is selected for its working temperature range, accuracy, and affordability.[1] It is recommended that in order to accomplish the right accuracy, the thermocouples should be calibrated against well-established standards such as the ones from the National Institute of Calibration Standards and Technology (NIST).[2]

There is a large variety of K thermocouples; as a general recommendation, it is suggested to use thin thermocouples (less than 1.5 mm) to eliminate any unwelcome effect. Yet, other thermocouples are suitable. In the event that custom-made thermocouples are used, it is mandatory to use ceramic covers to prevent any contact between the alloy and the thermocouple. The contact between the alloy and the thermocouple needs to be prevented as it introduces noise to the results and the thermal history of the sample will be invalid. It is recommended to use a low mass thermocouple/ceramic framework; however, for specific applications, other thermocouples are accurate enough. Additionally, the thermocouple shield (usually a ceramic) must not present phase transformation(s) in the investigated range of temperature. The K thermocouples have minimum thermal inertia and a low time constant making them ideal for thermal analysis. Both, the commercial and the custom-made, K thermocouples have a maximum working temperature of 1260 °C that is well above the melting and casting temperature of commercial Al-Si alloys. K thermocouples usually have an average error of less than ±0.6 °C in a temperature range of 300–750 °C [52]. It is recommended to record the thermal history at a rate of 5 Hz or higher.

2.3 Heat Exchange Conditions

In order to achieve high accuracy in thermal analysis, it is important to establish heat exchange conditions that comply with the model utilized for the baseline. The majority of the work presented in this section corresponds to cooling and solidification of samples as it is the portion of the process that can be associated to the properties of the final castings. In experimental set-ups samples can be solidified under natural heat exchange conditions, rapid or slow cooled to accomplish specific characteristics or properties. Here, different scenarios and ideas on how samples can be rapidly cooled, quenched, or cooled under natural heat exchange conditions are presented. The quenching process can be carried out by dropping the sample(s) into

[1] http://www.omega.com/temperature/.

[2] https://www.nist.gov/programs-projects/thermometry.

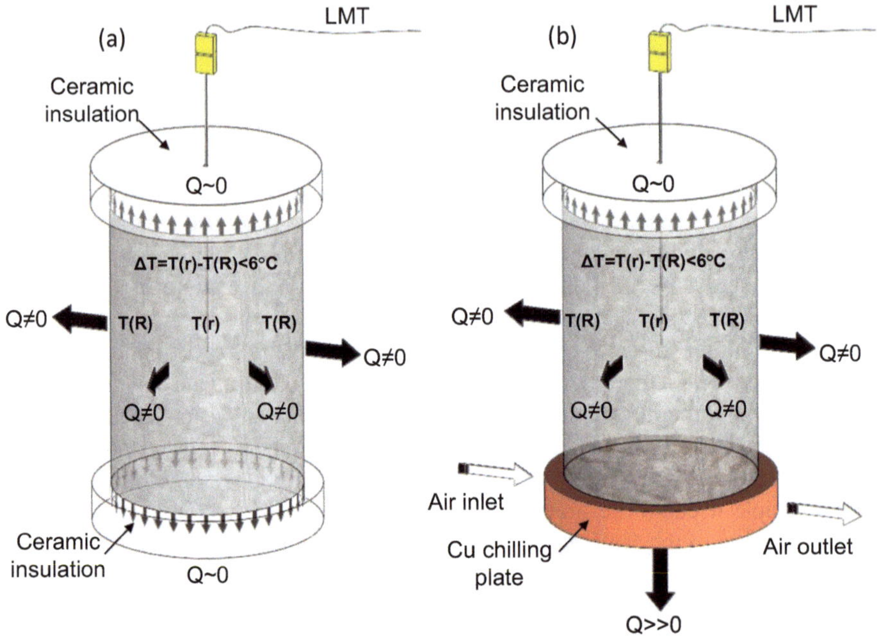

Fig. 2.1 Sketches of test samples solidified under (**a**) natural heat exchange conditions (heat released radially (1D)) and (**b**) directionally using the Cu chilling plate (heat released radially and vertically (2D))

water or brine (saturated water solution) or directionally chill using a high thermal conductor such as Cu. The effects of each of those environments are clearly explained and demonstrated in this chapter. Figure 2.1 shows a sketch of optimum heat exchange conditions for thermal analysis.

In an ideal case, during the heat exchange under natural heat exchange conditions, the heat should be released only in the radial direction (one dimension [1D]). The assessment of the heat released in one dimension is achieved by hindering the heat release from the top and the bottom of the samples by means of an insulator such as a ceramic (Fig. 2.1a). The heat released radially (1D) minimizes the gradient of temperature between the center and the wall of the sample. The gradient of temperatures can be measured using two thermocouples, one in the center of the sample and another one in the vicinity of the wall with at least 25 mm distance between them. As a general recommendation, the gradient of temperature ($\Delta T = T(r) - T(R)$) should be kept ≤6 °C, which is considered ideal for a sample of less than 10 cm in diameter. This gradient of temperature is used to comply with the respective Biot number for uniform temperature conditions [53, 54]. The Biot number determines the type of cooling behavior of the system under investigation, and this allows to select the proper thermal analysis algorithm.

2.4 Experimental Results and Thermal Analysis Algorithms

The cooling curves for a hypoeutectic (e.g. 319) and a hypereutectic (e.g. 390) alloys are provided in Fig. 2.2. In the graphs, the reactions taking place during solidification are identified. The stoichiometry of those reactions is provided in Table 2.1. Both alloys are commercial and have similar composition except for the Si and Al contents. In general terms, the main observable difference in those samples is the first solidifying phase. For the 319 alloy the phase is α-Al and for the 390 alloy, it is primary Si. Another marked difference is the longer solidification time for the 390 alloy, that is attributed to the larger amount of Si. The next solidifying phase is Fe-rich. Following this precipitation path, there is a series of eutectics that are rich on: Mg, Cu and Ni. All those reactions were detected in both alloys. It is important to mention that the precipitation time and temperature change with composition. In Table 2.2 are given the solidification reactions for each of phases as identified in the cooling curve and the numbers correspond to those in the respective cooling curve (Fig. 2.2). The phase identification presented here matches with that reported in the literature [3, 18, 55, 56].

Figure 2.3a, b shows the cooling, first derivative, and baseline for hypoeutectic (319) and hypereutectic (390) Al-Si alloys. Both alloys were solidified under natural heat exchange conditions using the 1D model. The cooling curves were recorded using only one thermocouple located in the center of the sample. Figures 2.2 and 2.3 show cooling curves where all the precipitation reactions are identified. The Al-Si hypereutectic alloys are more complex and less investigated, which is why in

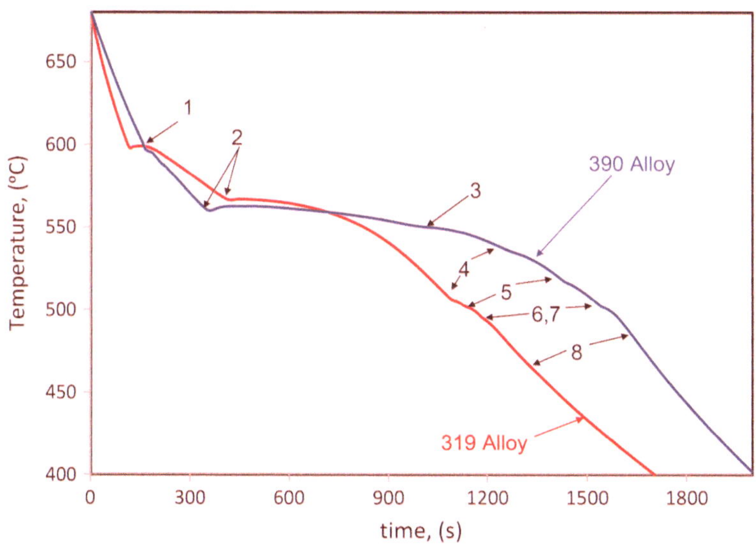

Fig. 2.2 Cooling curves for the 319 and 390 alloys solidified under natural heat exchange conditions. The numbers are the solidification reactions taking place at the respective temperature and times. The identification of each point is given in Table 2.2

Table 2.2 List of reactions identified in samples for the 319 and 390 Al-Si hypo and hypereutectic alloys solidified under natural heat exchange conditions.

ID. #	Alloy	
	319	390
1	α-Al, liquidus	Primary Si, liquidus
2	$Al_{15}(Fe, Cr, Mn)_3Si_2$	$Al_{15}(Fe, Cr, Mn)_3Si_2$
3	Al-Si eutectic	Al-Si eutectic
4	Mg_2Si eutectic	Mg_2Si eutectic
5	Al_3Ni Eutectic	Al_3Ni Eutectic
6	Al_3CuNi Eutectic	Al_3CuNi Eutectic
7	Al_2Cu Eutectic	Al_2Cu Eutectic
8	Solid	Solid

The Fe-rich phase is not clear in the 319 alloy, nevertheless its presence is usually identified. Potentially this is due to the faster cooling of this alloy when compared to the 390 alloy

this chapter a more descriptive analysis of them is presented. This statement is based on the fact that if we compare Fig. 2.3a, b the differences among them can be identified as the precipitation of the primary Si in the hypereutectic alloys. The rest is relatively identical. However, there is a characteristic that is rarely mentioned or investigated, which refers to the Si agglomerates that are covered in detail in this chapter. All the algorithms provided herein are valid for both alloys; yet, one must be careful in the analysis of the Si agglomerates and primary Si particles.

Figure 2.4 is an example of a sample solidified under 2D conditions. In this case, the sample is chilled from the bottom and the heat is also allowed to be released radially. The 2D environment develops a two-dimensional temperature gradient, one radial and the second vertical, as proposed in Fig. 2.1b. The curves in Fig. 2.4 clearly show that this practice compromises the accuracy of thermal analysis for Al-Si alloys. This effect results in the observed lack of accuracy to determine the nucleation of solidifying reactions that is attributed to a multi-axial temperature gradient. Further, the precipitation of the secondary phases (Mg, Cu, and Ni eutectics) can be identified, but it is not possible to resolve each phase. This is a good exercise to show the importance of the heat exchange environment. This effect is not related to a lack of accuracy or resolution of the thermal analysis system. The recorded curves demonstrated that the secondary phases are not allowed to precipitate or complete and therefore it is expected that the microstructure and physical characteristics of this sample are different to a sample solidified in 1D. Thermal analysis is still recommended for those cases.

The analysis presented in this report is based on the Newtonian model. To use this baseline calculation procedure, the characteristics of the thermal system (TS), sample, and cup, have to be considered. The TS has to comply with the Newtonian cooling model; therefore, the temperature gradient must be held as uniform as possible in both, radial (r) and vertical (Z), directions. It is recommended to use a system with low thermal resistance and low mass to guarantee a radial heat released. At the same time the bottom and the top of the sample should be insulated (see Fig. 2.1). In order to carry the analysis in 2D, it is necessary to use different algorithms. Here we will present examples of both cases to show the differences. The Newtonian

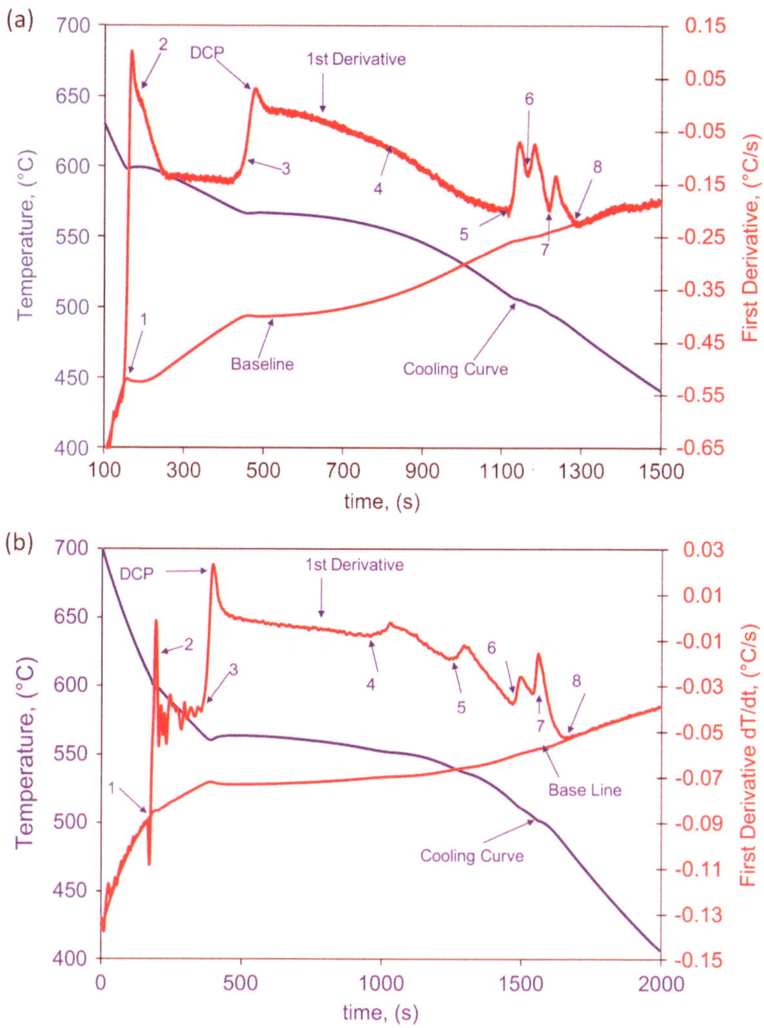

Fig. 2.3 Cooling, first derivative, and baseline curves for (**a**) a hypoeutectic alloy, e.g. 319 and (**b**) a hypereutectic alloy, e.g. 390. Both alloys were solidified under natural heat exchange conditions

model is only valid for cooling rates that are limited to a Biot number of Bi <0.1. Equation 2.1 is the model used to determine Bi. The Bi depends on the dimensions, overall heat transfer coefficient, and thermal conductivity of the material.

$$Bi = \frac{hL}{\lambda_b} \tag{2.1}$$

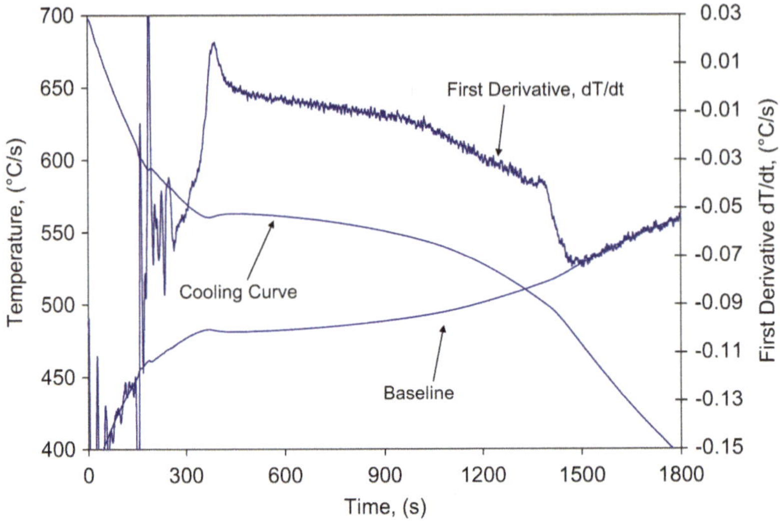

Fig. 2.4 Cooling, first derivative, and baseline curves of the 390 alloy solidified directionally on the Cu chilling plate (radial and vertical heat exchange (2D))

Where: Bi is the dimensionless Biot number,
h is the overall heat transfer coefficient,
L is the characteristic length and
λ_b is the thermal conductivity of the body

During the solidification the thermal capacitance of the system should be temperature independent. Under these assumptions the energy balance equation is given in Eq. 2.2:

$$\frac{dT_C}{dt} = \left(-\frac{UA}{\rho C_p V} \right) \left(T_C - T_\infty \right) + \frac{1}{\rho C_p V} \frac{dQ_L}{dt} \tag{2.2}$$

Where: T_c and T_∞ are temperatures at the center and wall of the test sample, respectively,
t is time,
ρ is the density of the molten alloy,
C_p is the heat capacity at a constant pressure,
U is the heat transmission coefficient, and
A and V are the area through which the heat is released and volume of the test samples, respectively.

The solution of this equation for $T_C(t)$ represents the cooling curve, in cases where no metallurgical reactions occur, when $Q_L = 0$ and when:

$$T_C \geq T_{C,liq} \; and \, T_C \leq T_{C,SOL} \tag{2.3}$$

Under those conditions Eq. 2.2 can be reduced into Eq. 2.4, as follows:

$$\frac{dT_C}{dt} = \left(-\frac{UA}{\rho C_p V} \right) \left(T_C - T_\infty \right) \tag{2.4}$$

Solving Eq. 2.4 will determine the time dependent, $T_{C,}$ which has to be determined in the portion of the curve before solidification starts and after it has been completed. This algorithm is stated in Eq. 2.3. It is logical that the heat transfer coefficient (U) is time and temperature dependent. This U obeys the same continuous function before and after the solidification. Another term that is time and temperature dependent is ρC_{P}. However, it fluctuates in a narrow range that for the most part, and for practicality, it is considered constant. For a sample that experiences no phase transformations, the $U(T_C)$ function can be calculated based on the T_C and the calculated dT_C/dt, if and only if the right side of Eq. 2.5 fits in polynomial terms of T_C by using a standard regression.

$$\frac{dT_C}{dt} = \sum_{n}^{i-0} a_i \left(T_C \right)^n \tag{2.5}$$

Using Eq. 2.6 one can determine the dT_C/dt values for all recorded values of T_C. This is again in the absence of reactions and the method is known as the baseline eq. (BL):

$$\left(\frac{dT_{C,zero}}{dt} \right) = BL\left(T_C \right) \tag{2.6}$$

This equation is clearly dependent on time because T_C is changing dynamically. Figure 2.5 is the same as the example shown in Fig. 2.3b, but here we present the cooling curve and its first derivative along with the baseline and fraction solid curves. Figure 2.5 is a typical curve for Al-Si hypereutectic alloys such as the 390, solidified under natural heat exchange conditions and under radial heat released or 1D.

The fraction solid is calculated based on the integral under the curve between the first derivative and the baseline. Eq. 2.7 represents the area within both curves that is associated with the latent heat of solidification.

$$\text{Total Area} = \int_{Liq}^{Sol} \frac{dT_C}{dt} - \int_{Liq}^{Sol} \frac{dT_{C,zero}}{dt} \tag{2.7}$$

The fraction integral is determined by calculating the portion of the integral from liquidus to the specific time (i) and temperature (j) characteristic point or reaction under analysis. This portion of the integral is then used to determine the corresponding fraction solid. The determination of the apparent fraction solid of the Mg-, Ni-, and Cu-enriched phases is carried out similarly, but in this case, the baseline needs

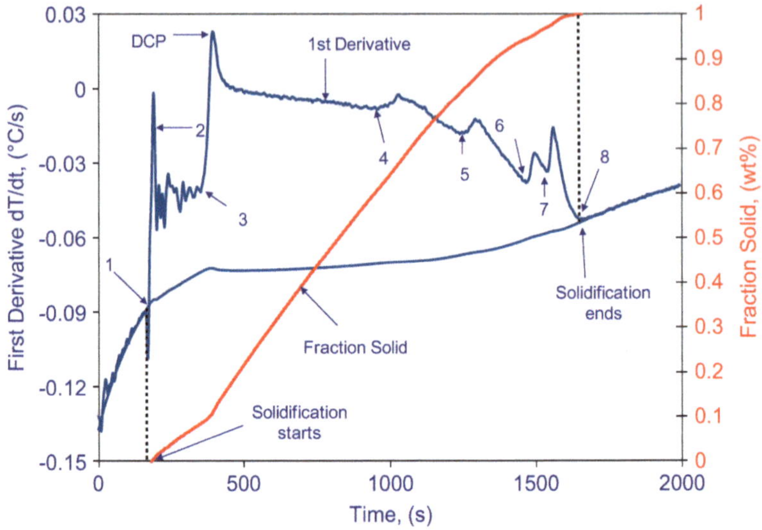

Fig. 2.5 Example of a cooling curve, first derivative, baseline, and fraction solid curves for the 390 Al-Si hypereutectic alloy. The sample was solidified under natural heat exchange conditions, with an approximate cooling rate of 0.08 °C/s

to be calculated using the Al-Si eutectic reaction as a reference curve ($BL^{Al\text{-}Si}{}_E$). The $BL^{Al\text{-}Si}{}_E$ is made by the regression of the Al-Si eutectic reaction and the portion of the curve following the solidus point, divided by the integral of the total area. Equation 8 shows the calculation for the area after liquidus for the time *i*.

$$\text{Area}\,|_{\text{tLiq}\langle t\langle \text{tSol}\,and\,\text{TLiq}\rangle T\rangle \text{TSol}} = \int_{Liq}^{t} \frac{dT_C}{dt} - \int_{Liq}^{t} \frac{dT_{C,zero}}{dt}\; or \sum_{t,T=i}^{t,T=Liq} \frac{dT_C}{dt} - \sum_{t,T=i}^{t,T=Liq} \frac{dT_{C,zero}}{dt} \quad (2.8)$$

Finally, the fraction solid calculated by the fraction of the area for the time and temperature "*i*" divided by the total area (integral) is expressed by Eq. 2.9.

$$fs = \frac{\text{Area}\,|_{\text{tLiq}\langle t\langle \text{tSol}\,and\,\text{TLiq}\rangle T\rangle \text{TSol}}}{\text{Total Area}} \quad (2.9)$$

Figure 2.6 shows two sections of the thermal analysis curves for the 390 alloy. The regions are selected to spot the "primary Si agglomerates" and the secondary reactions taking place during solidification of this alloy. The Si agglomerates are nuclei of Si atoms that form quasi-crystalline regions in the liquid state. In those regions there is a possibility to determine the so-called apparent fraction solids (afs). The afs is the amount of solidifying phase at a specific time and can be used to determine the amount of solid fraction for a specific phase at any given time. Here it is important to point out that the nucleation of the primary Si particles takes place at the liquidus line. From the liquidus temperature the phases start precipitating in

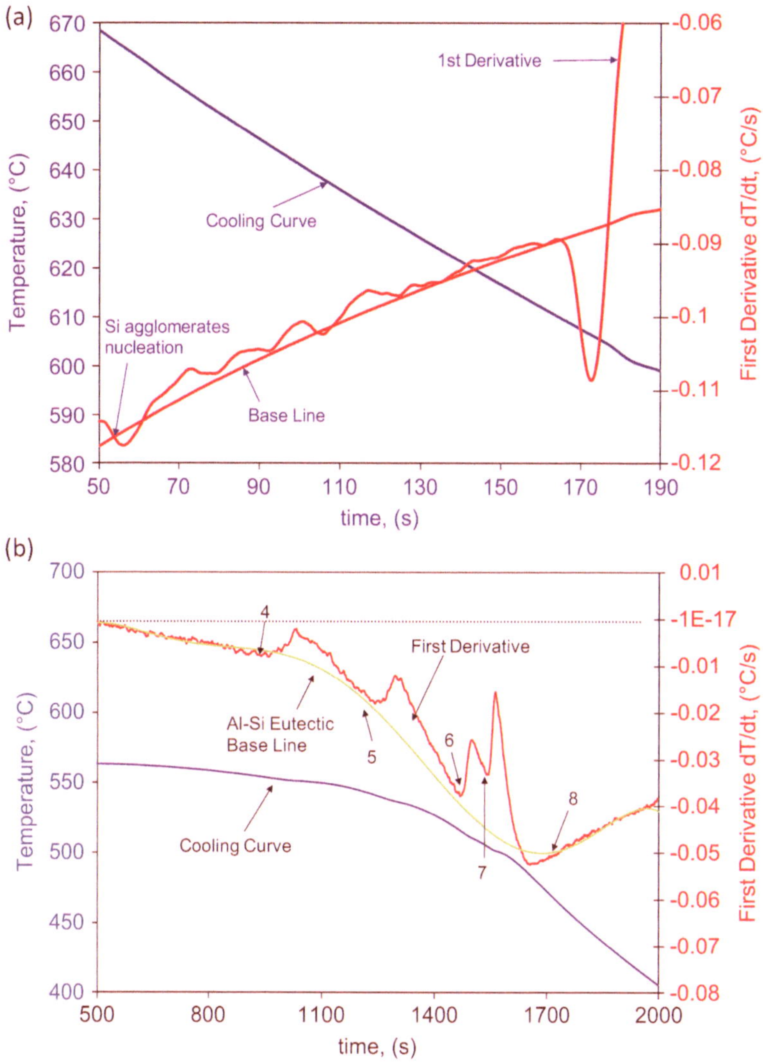

Fig. 2.6 Section of the cooling curves depicting apparent fractions that solidified during cooling (**a**) near liquidus (Si agglomerates) and (**b**) Cu-rich eutectics

solid form, but not before. Here we need to clarify that for Al-Si hypoeutectic alloys this model applies to an equilibrium or near equilibrium environment. In non-equilibrium conditions, it is possible to observe precipitates at temperatures above, but near, liquidus.

The Si agglomerates in the Al-Si hypereutectic alloys have been reported in the literature [13, 16, 18, 57–66]. The presence of the Si agglomerates has been observed at temperatures as high as 1075 °C that is approximately 500 °C above the equilibrium temperature for alloys such as the 390. This phenomenon has been reported in other alloys including silver [67], cast iron, etc. Usually the condition is that the precipitating phase has a highly covalent bonding. The exact nucleation temperature of the Si agglomerates nucleation is unclear. However, Robles et al. [52] reported the Si agglomerates in the 390 alloy at temperatures as high as approximately 66 °C above its respective liquidus temperature, or approximately 666 °C for the commercial 390 alloy. In this research, the identification was carried out by means of thermal analysis; therefore, other more sophisticated equipment is able to find the Si agglomerates at higher temperatures.

Depending on the alloy (e.g. 393) the silicon agglomerates are found with thermal analysis at temperatures as high as 130 °C above the liquidus. The Si agglomerates in this alloy is approximately 25 wt% Si. The equivalent or apparent fraction "solid" corresponding to the Si agglomerates in the 390 alloy was determined as $fs_{\mathrm{Si}}^{\mathrm{AGGL}} = 0.065$ wt%. Therefore, the presence of those agglomerates is almost negligible making their detection hard and complex or intricate demanding sophisticated systems, which here is where thermal analysis demonstrates its outstanding potential [57]. More sophisticated systems have been used such as neutron diffraction. Even though the nucleating fraction of the Si agglomerates is relatively small, they are key in the modification of the primary Si. The Si agglomerates have weaker bonding than primary Si, and therefore, they are susceptible to liquid treatment such as stirring and vibration [13]. This also explains why the modification of these alloys is effective when liquid and chemical (e.g. additions of AlP [52, 68, 69]) treatments are applied at temperatures above liquidus (usually >100 °C).

Figure 2.6b shows the precipitation of the secondary phases including the Mg_2Si and Cu-rich eutectics. The crystalline structure of the constituents of the Al-Si alloys is mainly FCC (e.g. Al, Cu, Ni, Si). The Al, Cu, and Ni structure is also known as "*A1*". Their respective lattice parameters are $a_{Al} = 4.0490$ Å, $a_{Cu} = 3.597$ Å, $a_{Ni} = 3.499$ Å [70–72]. Si, on the other hand, is "*A4*" (diamond-like) with a lattice constant of $a_{Si} = 5.4282$ Å [73]. This in a way explains the higher afinity among Al, Cu and Ni when compared to these elements and Si. The solidification process for Al, Cu, and Ni takes place in a normal fashion, which means it is 3D and relatively random. However, Si solidification path is layered (2D), that is, a significantly slower process taking longer solidification time. This process requires larger amounts of energy or enthalpy of transformation [74, 75]. In other words, during nucleation of Al, Cu, and Ni, the atoms are attached randomly without a preference location. For Si, the atoms are attached in an ordered manner and need to be allocated in specific positions to form lines of atoms followed by planes that pile up epitaxially. This type of solidification for primary Si and/or Si eutectic slows down the solidification process and this explains why an Al-Si hypereutectic alloy requires longer melting and solidification times. By looking at the integral between the baseline and the first derivative (or latent heat of solidification), the exothermic energy released by an Al-Si eutectic is significantly larger than any of the secondary reactions.

In the case of the solidification of Al, it is observed that it coarsens rapidly. In the case of Al-Si hypoeutectic alloys, the α-Al is the first to form. In the Al-Si hypereutectic alloys, the first phase to solidify is primary Si. During the solidification of Si, the surrounding region impoverishes in the presence of this element. As a result, the solidification of the α-Al phase is observed. This is a non-equilibrium phase or behavior. This aluminum is also known as the α-Al halo and it is identified in the surroundings of some primary Si particles. The next phase to precipitate, for both alloys, is the Al-Si eutectic. The fraction solid increases in approximately 10 wt% during the precipitation of the Al-Si eutectic in the first 4 °C. The next solidification step takes up to 40% of the solidifying fraction in approximately 10 °C.

Djurdjevic et al. [76, 77] and MacKay et al. [78] reported that using thermal analysis the af_s for the Mg-, Ni-, Cu-enriched phases can be determined. This procedure requires the determination of the integral among the baseline and the first derivative, but in this case, the baseline is calculated using the Al-Si eutectic reaction as a "zero" or reaction-free curve. Figure 2.5 shows the methodology for the assessment of the "apparent fraction solid (af_s)" for the Mg-, Ni-, Cu-enriched phases. The importance to have a complete analysis of these phases is the proper identification of their effects on strengthening before and after heat treatments such as T5, T6, and T7.

Figure 2.6b shows the Al-Si eutectic baseline which is a hypothetical model where only the secondary reactions take place and ends presumably at the solidus temperature. Therefore, any increase in the amount of heat will result in a change in the slope of the curve. In general those changes are exothermic (latent heat of solidification). The amount of exothermic heat during the nucleation of a phase is proportional to the amount of precipitating phase, latent heat, and enthalpy of transformation. The prediction of the strengthening effect of the particular phase(s) can be assessed by determining the af_s for all or the independent phases. At the same time, if there is a detrimental effect of a phase, the unfavorable effect can be quantified by the respective latent heat energy. For instance, a direct application of the af_s would be the effect in mechanical properties after natural or artificial aging.

2.5 Applications of the Thermal Analysis Algorithms

In the following text, a descriptive methodology for the interpretation of the thermal analysis results is provided. This procedure describes the steps needed to identify the characteristic temperatures for each of the reactions or identified phases in Al-Si alloys. The following exercise is an example of how the thermal analysis is applied in practice and its interpretation. Figures 2.7, 2.8, and 2.9 show the procedure recommended to identify the characteristic temperatures and critical points for each reaction. These figures are divided in sections to properly convey the interpretative procedure. Furthermore, the main difference among the 319 alloy and the 390 alloy will be the precipitation of the α-Al instead of the primary Si, respectively.

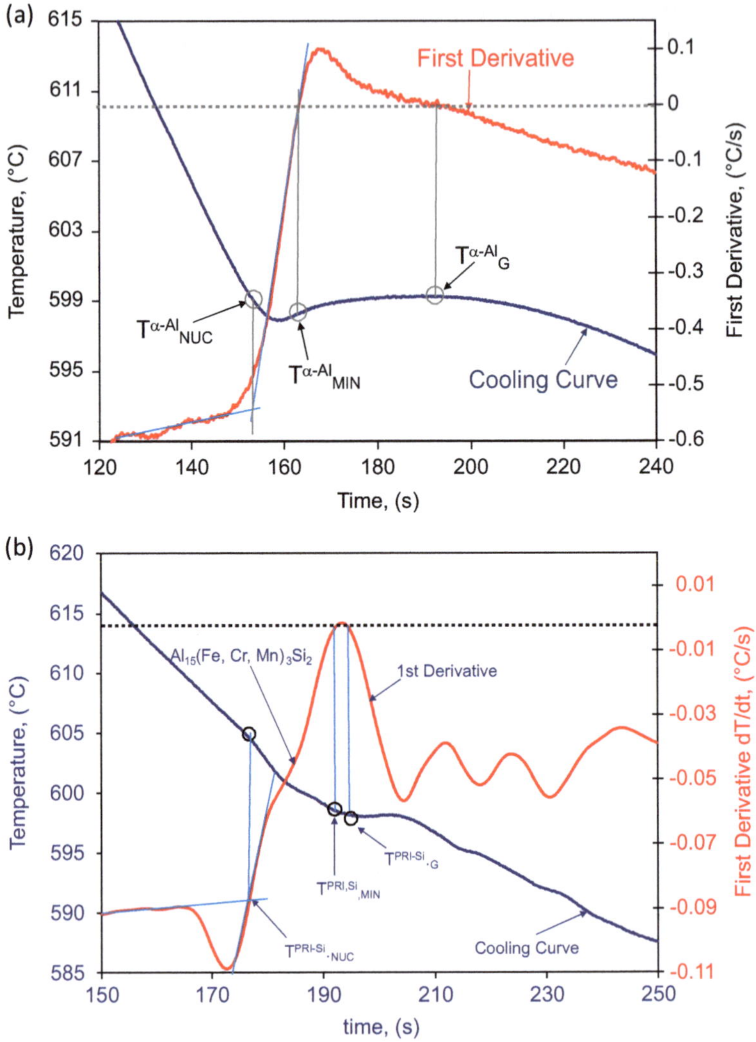

Fig. 2.7 Determination of the characteristic points, $T^{PRI\text{-}Si}_{NUC}$, $T^{PRI\text{-}Si}_{MIN}$, and $T^{PRI\text{-}Si}_{G}$, during the solidification of Al-Si (**a**) hypoeutectic and (**b**) hypereutectic alloys under natural heat exchange conditions

Following each figure, the procedure to identify the respective reaction(s) in the graph is described in Tables 2.3, 2.4, and 2.5, respectively. Each table contains all the details and specific characteristics of each reaction. The exact temperature and time for each reaction is determined in two different ways. In the first one, the coordinates are identified in the first derivative by the intersection of regression curves that are

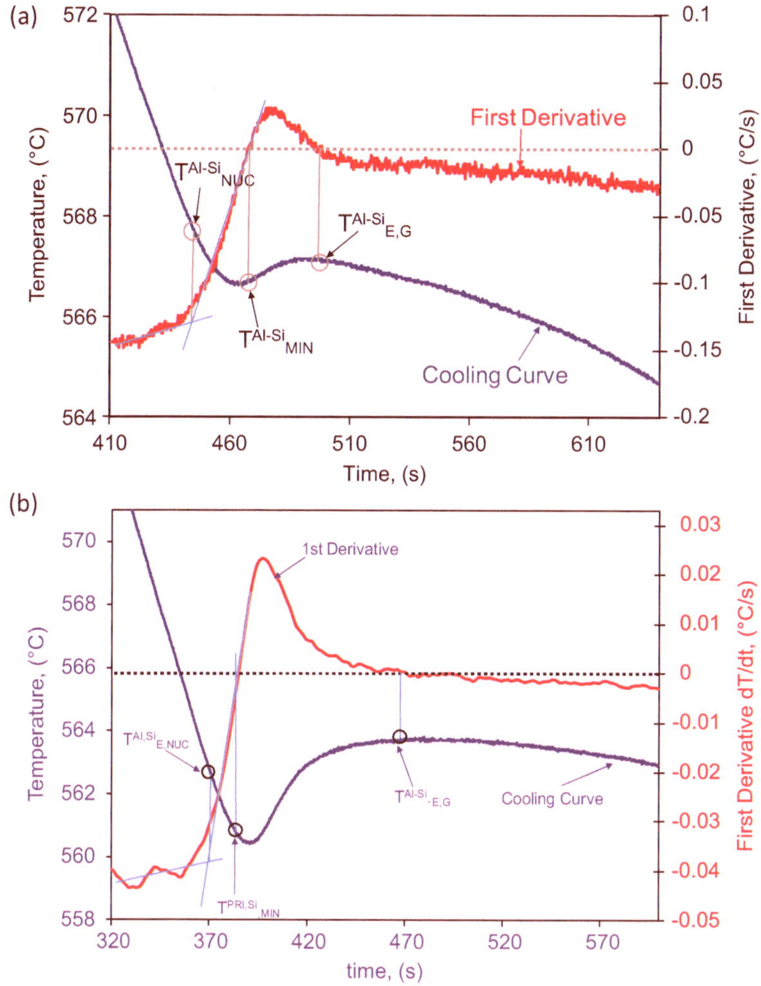

Fig. 2.8 Determination of the characteristic points, $T^{Al\text{-}Si}_{E,NUC}$, $T^{Al\text{-}Si}_{MIN,}$ and $T^{Al\text{-}Si}_{G}$, found during the solidification of Al-Si (**a**) hypoeutectic and (**b**) hypereutectic alloys under natural heat exchange conditions

calculated using the data on each side of the inflection point. In the second method, the point is identified by the intersection of the first derivative with zero. The first procedure is valid for the following reactions: $T^{PRI\text{-}Si}_{NUC}$, $T^{Al\text{-}Si}_{E,NUC}$, $T^{Mg2Si}_{E,NUC}$, $T^{Al3Ni}_{E,NUC}$, $T^{Al\text{-}Ni\text{-}Cu}_{E,NUC,}$ and $T^{Al2Cu + Al\text{-}Cu\ FE}_{E,NUC}$. The second procedure is valid where the first derivative intersects with zero: $T^{PRI\text{-}Si}_{MIN}$, $T^{PRI\text{-}Si}_{G}$, $T^{Al\text{-}Si}_{E,MIN,}$ and $T^{Al\text{-}Si}_{E,G}$.

The undercooling of the primary Si nucleation ($T^{PRI\text{-}Si}_{MIN}$) in Al-Si hypereutectic alloys is not as deep as the one presented for hypoeutectic Al-Si alloys systems such

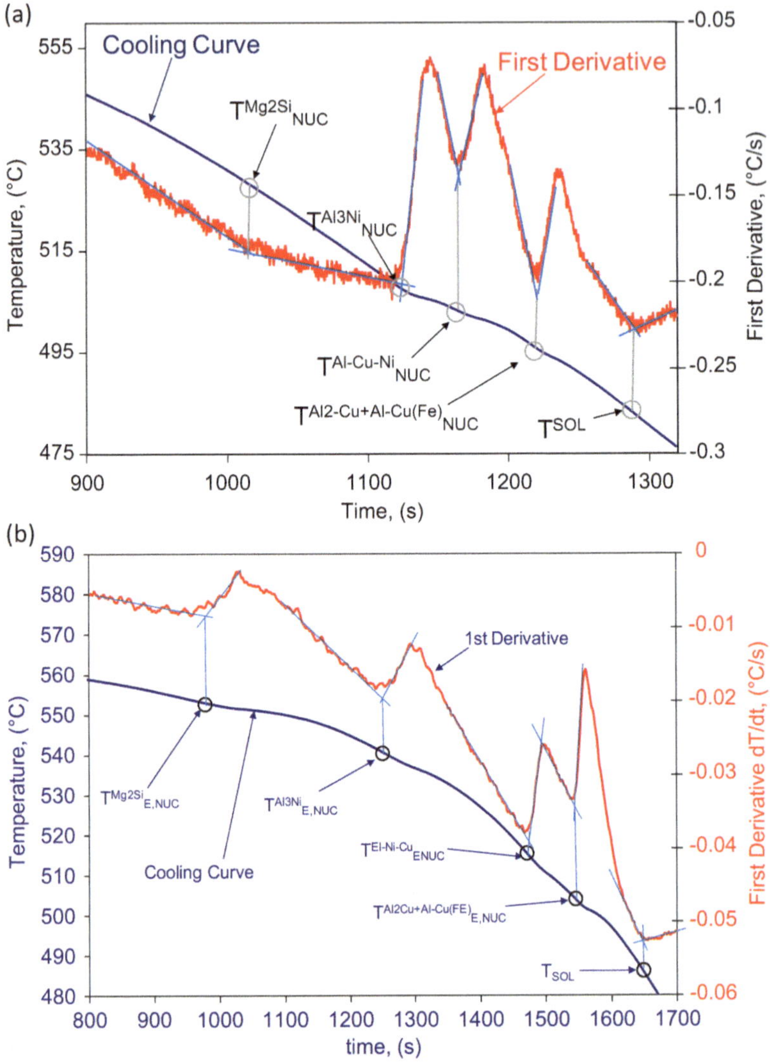

Fig. 2.9 Determination of the characteristic points, $T^{Mg2Si}_{E,NUC}$, $T^{Al3Ni}_{E,NUC}$ $T^{Al-Ni-Cu}_{E,NUC}$, and $T^{Al2Cu + Al-CuFE}_{E,NUC}$ found during the solidification of Al-Si (**a**) hypoeutectic and (**b**) hypereutectic alloys under natural heat exchange conditions. Note: FE in the $T^{Al2Cu + Al-CuFE}_{E,NUC}$ reaction means Fine Eutectic instead of the element iron or Fe

as the 319 alloys. Therefore, the characteristic points T^{PRI-Si}_{MIN} and T^{PRI-Si}_{G} are not significant enough to be considered as key point to identify any microstructural changes using thermal analysis. This may be the result of a "pre-nucleation or heterogeneous nucleation [52]" due to the coarsening of the Si agglomerates rather than a true precipitation including nucleation and growth.

Table 2.3 Definition of the characteristic points associated with the primary Si solidification of the 390 alloy

Symbol	Name	Detection algorithm	Definition
319 $T^{\alpha\text{-Al}}_{NUC}$ 390 $T^{PRI\text{-}Si}_{NUC}$	Liquidus α-aluminum nucleation Primary Si nucleation	This point corresponds to the intersection of the two local regression lines or slopes of the first derivative (Fig. 2.7).	At this point the α-Al or primary Si particles start the nucleation process.
$T^{PRI\text{-}Si}_{MIN}$	Primary Si undercooling	This point is determined by the intersection of the first derivative curve with the first intersection with the zero value of the first derivative (Fig. 2.7).	At this point, the nucleated primary Si particles have grown beyond a critical size and the latent heat release balances the heat extracted from the test sample.
$T^{PRI\text{-}Si}_{G}$	Primary Si growth	Second intersection of the first derivative with zero (Fig. 2.7).	Maximum point of recoalescence.

Table 2.4 Definition of the characteristic points associated with the Al-Si eutectic solidification of the 319 and 390 alloys

Symbol	Name	Detection algorithm	Definition
$T^{Al\text{-}Si}_{E,NUC}$	Al-Si eutectic nucleation temperature	This point corresponds to the intersection of two slopes when the first derivative changes (Fig. 2.8).	At this point the Al-Si eutectic phase starts to precipitate.
$T^{Al\text{-}Si}_{E,MIN}$	Al-Si eutectic minimum temperature	This point corresponds to the local intersection of the first derivative with the zero level (Fig. 2.8).	At this temperature the nucleated Al-Si particles try to reach a critical radius, releasing the enthalpy of transformation absorbed during the melting process, thus temperature increases.
$T^{Al\text{-}Si}_{E,G}$	Al-Si eutectic growth	Second intersection of the first derivative with zero (Fig. 2.8).	At this point the growth of the Al-Si eutectic occurs.

Table 2.5 Definition of the characteristic points associated with the precipitation of Mg-, Cu-, Ni-, Mn-, Fe-, and Si-rich phase for the 319 and 390 alloys

Symbol	Name	Detection algorithm	Definition
$T^{Mg2Si}_{E,NUC}$ $T^{A3Ni}_{E,NUC}$ $T^{Al\text{-}Ni\text{-}Cu}_{E,NUC}$ $T^{Al2Cu+Al\text{-}CuFE\ (THREE)}_{E,NUC}$	Mg_2Si, Al_3Ni, Al-Cu-Ni Al_2Cu and $Al\text{-}Cu_{Fine\ Eutectic}$ Eutectic phases	All reactions were identified using SEM/EDX analysis and correspond to the Mg_2Si and three Al-Mg-Fe-Si-Cu-rich phases. The above-mentioned reactions are identified by the intersection of their respective slopes (Fig. 2.9).	Temperature at which the precipitation of the Mg_2Si- and Cu-Ni-rich phase begins.
T_{SOL}	Solidus	Intersection of the two local slopes (Fig. 2.9).	End of solidification.

2.6 Thermal Analysis, Quench Testing, and Optical Microscopy

In this section, an exercise (case study) using the alloy 390 is presented as it is more interesting and less investigated than the Al-Si hypoeutectic alloys (e.g. 319). Samples of this alloy are quenched from various temperatures of approximately 100 °C above liquidus and at various temperatures in between liquidus and solidus. Quench temperatures are selected based on the nucleation temperature of each precipitation reaction. The following are the quench temperatures: 730 °C, 690 °C, 660 °C, 630 °C, 560 °C, 554 °C, 525 °C, 500 °C and at full solidification (<480 °C) (Fig. 2.10). All the samples are quenched in saturated brine (H_2O + NaCl) solution at near room temperature. The intention of presenting only the 390 alloy is to provide a whole spectrum of an alloy that contains a moderate amount of primary Si. The reason is that the primary Si in other Al-Si hypereutectic alloys may undertake the majority of the microstructure, which could be the case of the 393 alloy or other alloys with more than 25 wt% Si. Some recommended literature for Al-Si alloys includes references 7, 23, 52, 78–84.

The samples used for this test are between 3 and 10 g, which are relatively small to be able to release the latent heat as fast as possible and freeze at a specific phase and to prevent the potential of further phase precipitation. The experimental graph with all the cooling curves for all the investigated samples (one sample per temperature) are shown in Fig. 2.10. The rapid response observed in the cooling curves is due to the freezing in brine in few seconds. The main intention is to preserve the

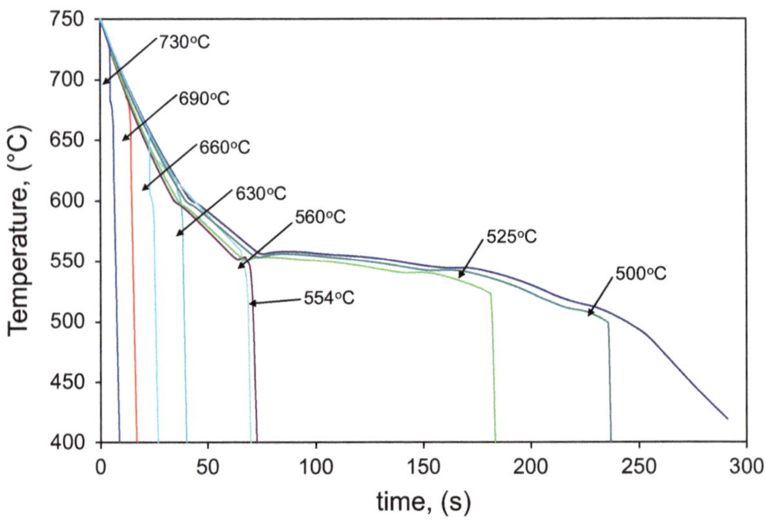

Fig. 2.10 The cooling curves of the 390 alloy with overlapping cooling curves for the quenching experiments

characteristic microstructure that exists at the temperature right before quenching. The quenching temperatures are selected a few degrees above the nucleation temperatures identified in Fig. 2.7b through Fig. 2.9b and the respective Tables 2.3, 2.4, and 2.5. The solidification temperature for each reaction along with its corresponding accumulative fraction solid is provided in Fig. 2.11.

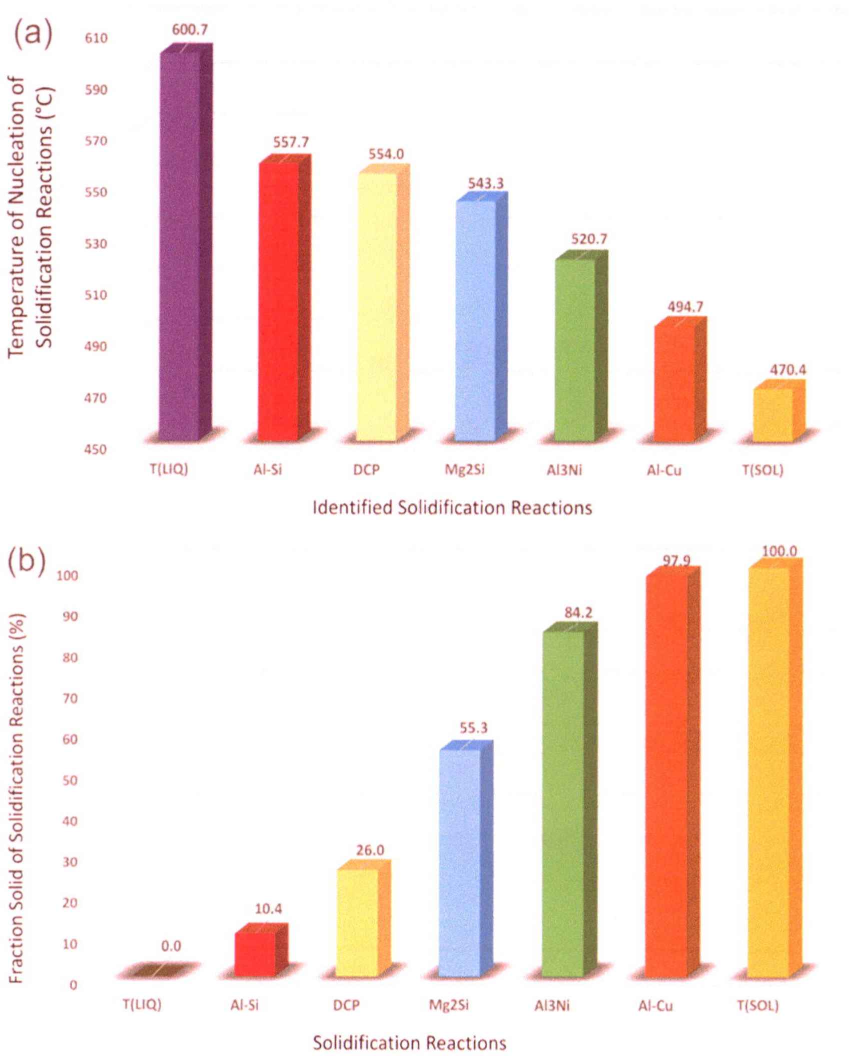

Fig. 2.11 Histograms showing the characteristic reactions: (**a**) temperatures and (**b**) fraction solids for the reference 390 alloy for the quenched samples at the temperatures presented in Fig. 2.10

2.7 Characterization: Optical and Scanning Electron Microscopy

The micrographs shown in Fig. 2.12 present the microstructure of the samples quenched above liquidus. The temperatures were selected to demonstrate the presence of the Si agglomerates in molten Al-Si hypereutectic alloys. This sequence of images shows the coarsening of the Si agglomerates and their transformation into primary Si particles. The Si agglomerates are observed at temperatures above liquidus and transform into primary Si particles at the liquidus temperature. At the Al-Si eutectic the primary Si particles' growth is hindered and the Al-Si eutectic nucleation and coarsening begins. Those particles grow quasi-linearly as temperature decreases. The analysis of the coarsening of the primary Si particles for the 390 and 393 alloys is presented in Figs. 2.13 and 2.14. This behavior is not characteristic of a specific composition, but it is rather common for the Al-Si hypereutectic family.

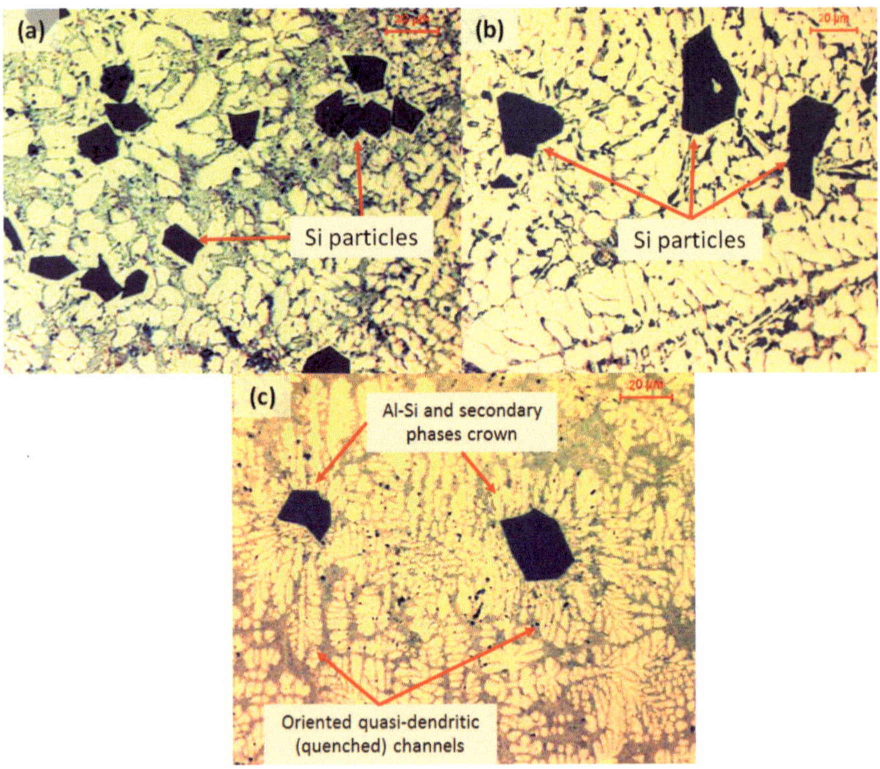

Fig. 2.12 Micrographs of the 390 quenched from different temperatures above liquidus (**a**) 730 °C, (**b**) 660 °C, and (**c**) 620 °C. All micrographs were made at 500× magnification

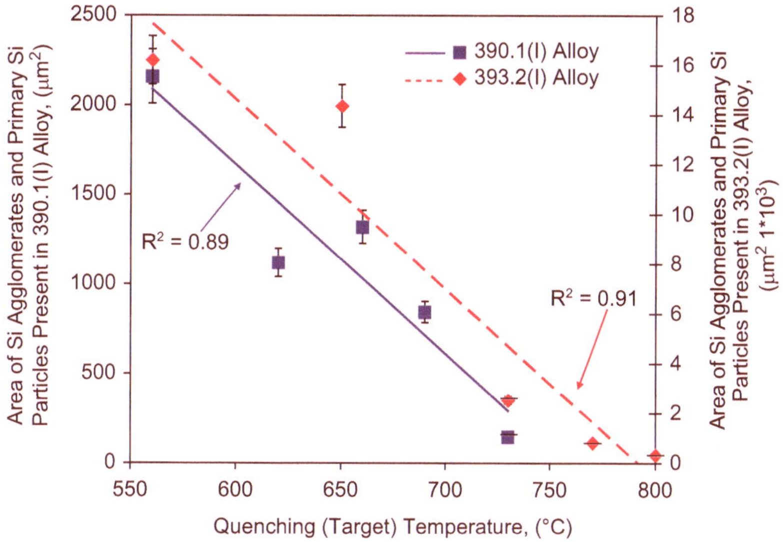

Fig. 2.13 Coarsening analysis of the Si agglomerates and primary Si particles for the 390 and 393 alloys. The 393 alloy contains up to 25 wt% Si

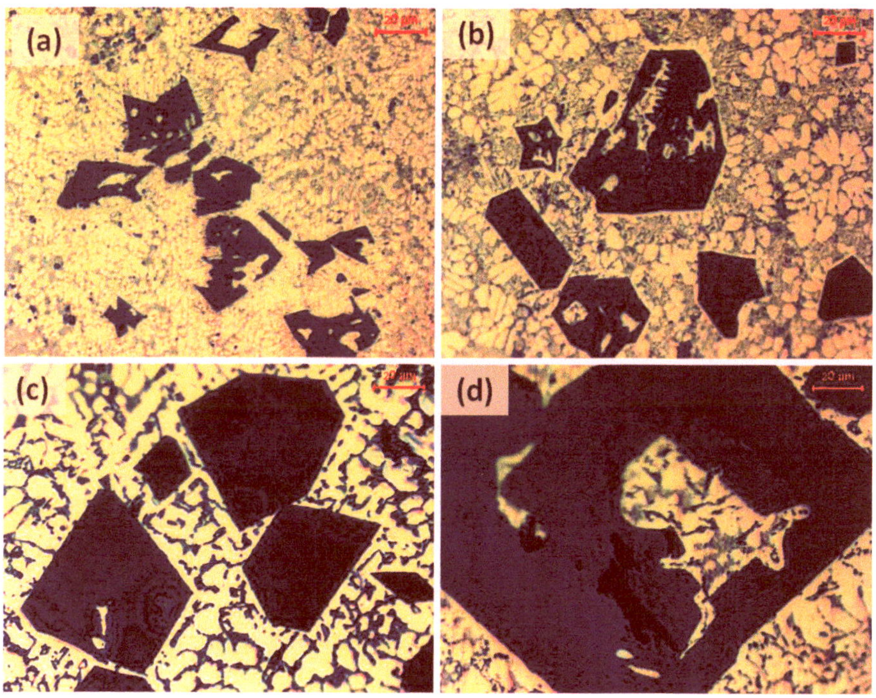

Fig. 2.14 Sequence of micrographs for the 393 alloy that contains approximately 25 wt% Si. The respective quenching temperatures are (**a**) 800 °C, (**b**) 770 °C, (**c**) 730 °C, and (**d**) below liquidus at 650 °C. Micrographs taken at 500× magnification

The above description applies to an alloy solidifying under close to equilibrium conditions. However, in many cases, the conditions may change. Further growth may occur, but it will be limited, at temperatures below the Al-Si eutectic (Fig. 2.15). This explains why the primary Si refinement is always carried out at temperatures above liquidus using a Si modifier such as P (usually in the form of AlP) that is effective at temperatures above liquidus [85]. The following references are examples of the primary Si refinement by means of mechanical, chemical, thermal, and electromagnetic and vibration treatments [3, 7, 13, 19, 56, 57]. The images in Fig. 2.15 are provided as evidence of the complete development of the coarsening of the primary Si particles and the nucleation of the Al-Si eutectic.

As the quenching temperature approaches liquidus, the Si agglomerates coarsen and Si-rich regions are formed in the vicinity of the Si particles along the solidification front. Those Si-rich regions are surrounded by α-Al. This combination of phases forms a "crown" along the Si agglomerates. This crown surrounds the solidified Si agglomerates and is orientated perpendicularly to the edges of the Si agglomerates forming a feeding pattern and the back diffusion of the α-Al. The secondary phases (Mg, Cu, Ni, Fe-rich, etc.) form oriented and dendritic-like "channels" that extends beyond the crown. Those channels are attributed to the natural convection of the alloy. The entire mechanism is responsible for transporting the liquid or semi-molten alloy carrying the imbalances in composition in the remaining molten alloy that is supersaturated.

The sections along the "crowns" are the locations where the α-Al halo develops. The mechanism is evident in Fig. 2.12c. The α-Al halo is a region starving on Si but rich on Mg, Fe, Mn, Cr, Ni, and Cu. This effect generates segregation in the vicinity to the crown. Therefore, the crown area takes care of transporting Al-rich molten metal in and the rest of the elements are transported out in an Al-Si suspension. The Al-rich matter becomes the α-Al haloes due to impoverish of Si and its segregation effects taking place in the surrounding area (Fig. 2.12c). The growth direction for the α-Al dendrites or haloes is the same. Though, the main differences are the crystalline structure between these two phases. This does not

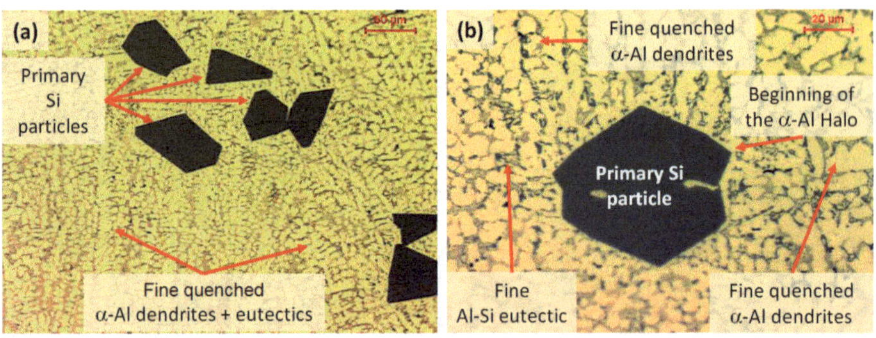

Fig. 2.15 LOM micrographs of the 390 samples quenched at 560 °C (~40 °C below liquidus) at (**a**) 200× and (**b**) 500× magnification

allow a coherent solidification between both elements. Based on the Hume-Rothery rules this is a system that cannot behave isomorphous or does not have the potential for a complete solubility [86, 87]. For alloys such as the 390, there is a rapid increase in fraction solid from the Al-Si eutectic to the dendrite coherency point from approximately 10 wt% to more than 25 wt% and this change is completed in only 4 °C (Table 2.6). The dendrite coherency point is the moment where the solidifying alloy forms a "skeleton" of solids that are interconnected. This skeleton builds up the strength of the solidifying alloy [88–90].

The sample quenched from a temperature of 560 °C (Fig. 2.15) does not present further coarsening of the primary Si. The Si particles have the typical polyhedral shape, and at this point, the above-mentioned crown becomes unclear (Fig. 2.15b). This is the beginning of the formation of the α-Al halo at the edge and the quasi-dendritic regions are fading and unclear. However, the Al-Si eutectic had not yet been fully developed. The quenching of the presently described test

Table 2.6 Solidification sequence of the 390 alloy based on thermal analysis in a series of quenching experiments and characterization carried out with optical and electron microscopy

Summary of reactions and corresponding characteristic points	Temperature range (°C)	Fraction solid (%)
(Reaction #1) L ⇒ Primary Si $T^{PRI\text{-}SI}_{NUC} = 600.5$ °C $f_S{}^{PRI\text{-}SI}{}_{NUC} = 0\%$	$(T^{PRI\text{-}SI}{}_{NUC} - T^{AL\text{-}SIE}{}_{,NUC}) = \Delta T^{PRI\text{-}SI}{}_{SRAN}$ 37.0	$f_S{}^{PRI\text{-}SI}{}_{NUC} - 0\% = af_S{}^{PRI\text{-}SI}$ 6.9
(Reaction #2) L ⇒ Al-Si (Eutectic) $T^{Al\text{-}Si}{}_{E,NUC} = 563.5$ °C $f_S{}^{Al\text{-}Si}{}_{E,NUC} = 18.1\%$	$(T^{Mg2Si}{}_{E,NUC} - T^{Al\text{-}Si}{}_{E,NUC}) = \Delta T^{Al\text{-}Si}{}_{SRAN}$ 11.8	$f_S{}^{Al\text{-}Fe\text{-}Mg\text{-}Si\text{-}Cu(one)}{}_{E,NUC} - f_S{}^{Al\text{-}Si}{}_{E,NUC} = af_S{}^{Al\text{-}Si}$ 56.9
(Reaction #3) L ⇒ Al + Si + $Al_{15}(Fe,Cr,Mn)_3Si_2$	Temperature for the $Al_{15}(Fe,Cr,Mn)_3Si_2$ somewhere between 563 and 525. Solidification ends before or at Solidus (T_{SOL})	Between 18.1 f_S and 89.4 f_S
(Reaction #4) L ⇒ Al + Si + Mg_2Si $f_S{}^{Mg2Si}{}_{E,NUC} = 97.4\%$	$T^{Al\text{-}Ni}{}_{E,NUC} - T^{Mg2Si}{}_{E,NUC} = \Delta T^{Mg2Si}{}_{SRAN}$ 11.9	$f_S{}^{Al\text{-}Ni}{}_{E,NUC} - f_S{}^{Mg2Si}{}_{E,NUC} = 21.9$
(Reaction #5) L ⇒ Al + Si + Al-Ni $f_S{}^{Al\text{-}Ni}{}_{E,NUC} = 89.8\%$	$T^{Al2Cu}{}_{E,NUC} - T^{Al\text{-}Ni}{}_{E,NUC} = \Delta T^{Al2Cu}{}_{SRAN}$ 29.2	$f_S{}^{Al2Cu}{}_{E,NUC} - f_S{}^{Al\text{-}Ni}{}_{E,NUC} = 8.0$
(Reaction #6) L ⇒ Al + Si + Al-Cu (Fine Eutectic (FE)) $f_S{}^{Al\text{-}Ni}{}_{E,NUC} = 97.5\%$	$T^{Al\text{-}Cu}{}_{FE,NUC} - T^{Al2Cu}{}_{E,NUC} = \Delta T^{Al\text{-}Cu}{}_{SRAN}$ 14.2	$f_S{}^{Al2Cu}{}_{E,NUC} - f_S{}^{Al\text{-}Ni}{}_{E,NUC} = 7.8$
(Reaction #7) L ⇒ Al + Si + $Al\text{-}Cu_{FE}$ + $Pb_{Particles}$ $f_S{}^{Al\text{-}CuFE + Pb}{}_{E,NUC} = 100\%$	$T^{SOL} - T^{Al\text{-}Cu}{}_{FE,NUC} = \Delta T^{Al\text{-}CuFE}{}_{SRAN}$ 22.8	$f_S{}^{SOL} - f_S{}^{Al\text{-}Cu}{}_{FE,NUC} = 2.6$

samples was at a temperature of 550 °C, which is ~5 °C above the Mg_2Si eutectic phase (545 ± 0.7 °C). Using SEM/BSE the Mg_2Si was successfully identified and is presented in Fig. 2.16. The Mg_2Si eutectic particles are surrounded by quenched Cu- and Ni-enriched regions. In this micrograph, it is observed that the Cu- and Ni-rich phases are already solidified, while the Mg_2Si has a fade appearance, implying that its solidification has not yet been completed.

The microstructure of a sample quenched from 550 °C (58% of fraction solid) has clearly defined primary Si particles, α-Al halo(s), Al-Si eutectic, and coarse α-Al dendrites. The respective microstructures are presented in Fig. 2.17. The temperature is near the dendrite coherency point. From this point on, no

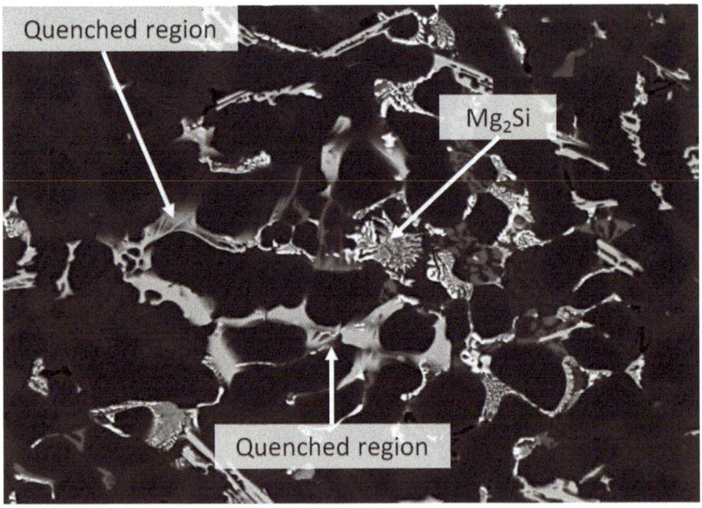

Fig. 2.16 SEM/BSE micrographs of the 390 sample quenched at 550 °C. In this micrograph, the Mg_2Si surrounded by non-equilibrium Cu- and Ni-rich phases is shown. 1000× magnification

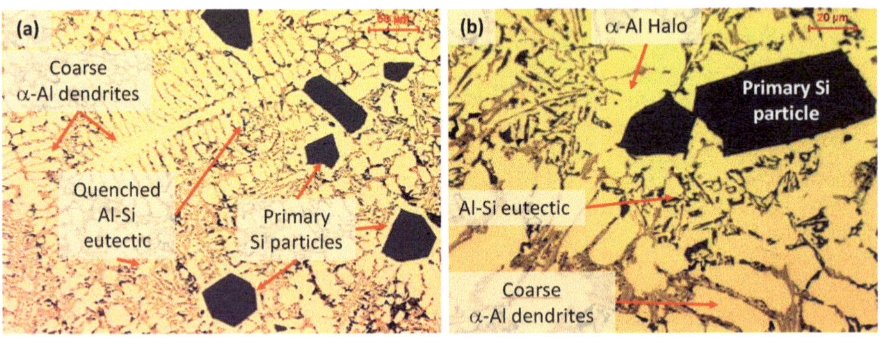

Fig. 2.17 LOM micrographs of the 390 reference sample quenched at 550 °C (~50 °C below liquidus) at (**a**) 200× and (**b**) 500× magnification

apparent coarsening of the dendrites is observed and the only potential growth is based due to the interdendritic feeding. The free flow of the remaining liquid is hindered except for the remaining liquid in the cavities within the dendrite. Furthermore, the Al-Si eutectic will continue coarsening during the remaining solidification and until the sample is completely solid.

The quenching temperature of 525 °C is selected because this is near the nucleation of the Ni- and Cu-rich phases (points 4–6 in Fig. 2.2). The corresponding fraction solid for the 390 alloy of approximately 80%. Some of the reactions are usually overlapping, and it is difficult to carry their full identification. For instance, in Fig. 2.9a, a change in slope between the $T^{Al\text{-}Ni\text{-}Cu}_{E\ NUC}$ and the $T^{Al2\text{-}Cu\ +\ Al\text{-}CuFE}_{E\ NUC}$ is observed, which makes a challenging identification of each phase. In this case, the exothermic reaction among the Al-Ni-Cu eutectic and the Al_2-Cu + Al-Cu(FE) phase overlap. FE stands for fine eutectic (see Table 1.3). The exact nucleation temperature of Al-Cu(FE) is unknown, but it is somewhere between 550 °C and 525 °C. The characteristic microstructures for those samples are provided in Fig. 2.18 and Fig. 2.19. These micrographs show that the Al-Ni-Cu eutectic,

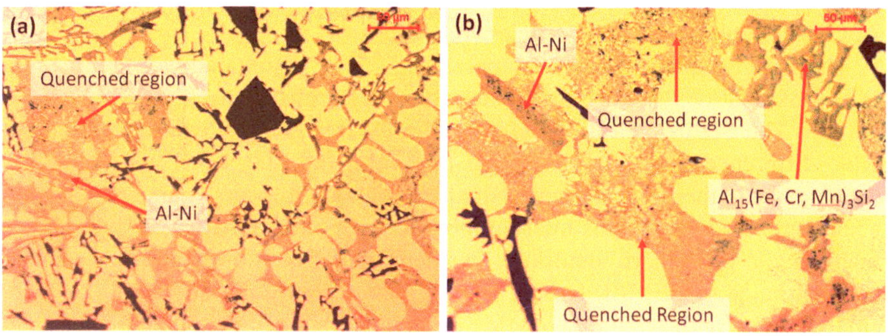

Fig. 2.18 Micrographs of the 390 quenched at 525 °C (**a**) 200× and (**b**) 500× magnification

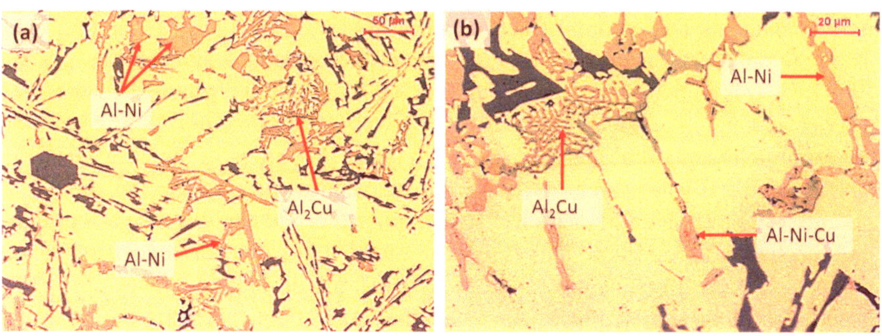

Fig. 2.19 Micrographs of the 390 quenched at 500 °C at (**a**) 200× and (**b**) 500× magnification

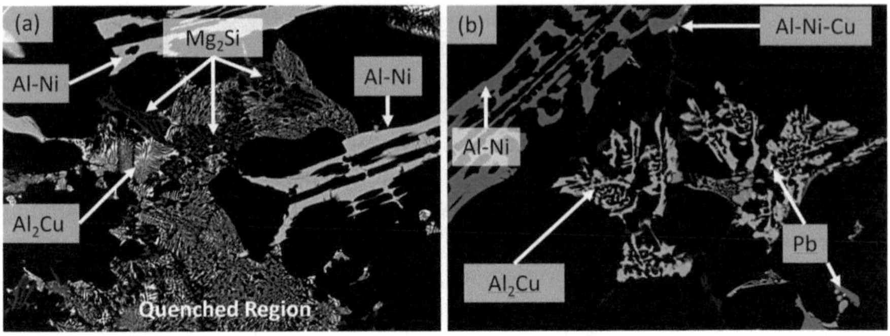

Fig. 2.20 SEM/BSE micrographs of the 390 samples quenched at 500 °C (1000× magnification)

the Al_2Cu, and the Al-Cu fine eutectic (Al-Cu_{FE}) are not well developed as they were frozen from the liquid. The Al_2Cu is identified in Fig. 2.19b at the center of the microstructure that was carefully made near the center of the cast sample. This phase solidified right after the Al-Ni-Cu eutectic. The last phases to solidify are the Al-Cu_{FE} followed by the Pb particles (Fig. 2.20b).

The samples quenched at a temperature of 500 °C contained a fraction solid of approximately 96% and a secondary dendrite arm spacing (SDAS) of 28.2 ± 1.8 μm. The SDAS measured before and after this point showed almost negligible differences (Fig. 2.19). Before this point the SDAS has not been fully developed and therefore it seems "smaller", but in reality, its coarsening has not been completed yet. In this micrograph the Al-Cu-Ni, the Al_2Cu, and the Al-Cu fine eutectic phases are identified. Further characterization is carried out with the assistance of SEM/BSE (Fig. 2.20). This not only provides images under composition contrast, but also helps in chemical and stoichiometry identification of each phase.

A summary of the solidification pathway for the 390 alloy is presented in Table 2.6. The findings in this summary table are based on thermal analysis and metallographic results. An analysis of the Mg-, Ni-, and Cu-enriched phases was performed using the apparent fraction solid (af_s^{Mg2Si}, af_s^{Al3Ni}, af_s^{AlNiCu}, and af_s^{Al2Cu}). The schematic demonstration of the apparent fraction solid is given in Fig. 2.5 and Fig. 2.6b. It is important to understand the behavior of these phases because they are rich in Cu, Ni, and Mg and are well known for their potential as precipitation hardening elements that increase the mechanical properties in the alloy [91–96]. The af_s has a direct correlation with the mechanical properties, and there is a potential to develop an algorithm that combines thermal analysis, metallography, and mechanical properties. This is a recommended exercise for empirical evaluations of castings. In many cases, the models that are developed between the characterization and testing are linear as in the case of the Si particle coarsening presented in Fig. 2.13.

References

1. Zamani, M., and S. Seifeddine. 2016. Determination of optimum Sr level for eutectic Si modification in Al-Si cast alloys using thermal analysis and tensile properties. *International Journal of Metalcasting* 10 (4): 457–465. https://doi.org/10.1007/s40962-016-0032-8.
2. Golbahar, B., E. Samuel, A.M. Samuel, H.W. Doty, and F.H. Samuel. 2014. On thermal analysis, macrostructure and microstructure of grain refined Al-Si-Mg cast alloys: Role of Sr addition. *International Journal of Cast Metals Research* 27 (5): 257–266. https://doi.org/10.1179/1743133614Y.0000000109.
3. Hernandez, F.C.R., and J.H. Sokolowski. 2006. Thermal analysis and microscopical characterization of Al-Si hypereutectic alloys. *Journal of Alloys and Compounds* 419 (1–2): 180–190. https://doi.org/10.1016/j.jallcom.2005.07.077.
4. Gloria, D., and J.E. Gruzleski. 1999. A study of the thermal analysis parameters applied to the grain refinement of Al-Si casting alloys. *Light Metals* 1999: 315–329.
5. Eguskiza, S., A. Niklas, A.I. Fernandez-Calvo, F. Santos, and M. Djurdjevic. 2015. Study of Strontium fading in Al-Si-Mg and Al-Si-Mg-Cu alloy by thermal analysis. *International Journal of Metalcasting* 9 (3): 43–50.
6. Aparicio, R., C. Gonzalez-Rivera, M. Ramirez-Argaez, G. Barrera, and G. Trapaga. 2013. Newton thermal analysis of unmodified and strontium modified Al-Si alloys. *Kovove Materialy-Metallic Materials* 51 (4): 211–220. https://doi.org/10.4149/km_2013_4_211.
7. Lu, S.Z., and A. Hellawell. 1995. Modification of Al Si alloys - Microstructure, thermal-analysis, and mechanisms. *Jom-Journal of the Minerals Metals & Materials Society* 47 (2): 38–40.
8. Nogita, K., and A.K. Dahle. 2003. Effects of boron on eutectic modification of hypoeutectic Al-Si alloys. *Scripta Materialia* 48 (3): 307–313. Pii S1359-6462(02)00381-0. https://doi.org/10.1016/S1359-6462(02)00381-0.
9. Nogita, K., S.D. McDonald, and A.K. Dahle. 2004. Eutectic modification of Al-Si alloys with rare earth metals. *Materials Transactions* 45 (2): 323–326. https://doi.org/10.2320/matertrans.45.323.
10. Nogita, K., S.D. McDonald, K. Tsujimoto, K. Yasuda, and A.K. Dahle. 2004. Aluminium phosphide as a eutectic grain nucleus in hypoeutectic Al-Si alloys. *Journal of Electron Microscopy* 53 (4): 361–369. https://doi.org/10.1093/jmicro/dfh048.
11. Malekan, M., S. Naghdali, S. Abrishami, and S.H. Mirghaderi. 2016. Effect of cooling rate on the solidification characteristics and dendrite coherency point of ADC12 aluminum die casting alloy using thermal analysis. *Journal of Thermal Analysis and Calorimetry* 124 (2): 601–609. https://doi.org/10.1007/s10973-015-5232-6.
12. Chavez-Zamarripa, R., J.A. Ramos-Salas, J. Talamantes-Silva, S. Valtierra, and R. Colas. 2007. Determination of the dendrite coherency point during solidification by means of thermal diffusivity analysis. *Metallurgical and Materials Transactions A-Physical Metallurgy and Materials Science* 38A (8): 1875–1879. https://doi.org/10.1007/s11661-007-9212-8.
13. Robles Hernandez, F.C., and J.H. Sokolowski. 2006a. Comparison among chemical and electromagnetic stirring and vibration melt treatments for Al-Si hypereutectic alloys. *Journal of Alloys and Compounds* 426 (1–2): 205–212. https://doi.org/10.1016/j.jallcom.2006.09.039.
14. Robles Hernandez, F.C., and J.H. Sokolowski. 2006b. Thermal analysis and microscopical characterization of Al-Si hypereutectic alloys. *Journal of Alloys and Compounds* 419 (1–2): 180–190. https://doi.org/10.1016/j.jallcom.2005.07.077.
15. Robles Hernandez, F.C., and J.H. Sokolowski. 2009. Effects and on-line prediction of electromagnetic stirring on microstructure refinement of the 319 Al-Si hypoeutectic alloy. *Journal of Alloys and Compounds* 480 (2): 416–421. https://doi.org/10.1016/j.jallcom.2009.02.109.
16. Robles Hernandez, F.C., J.H. Sokolowski, and J.J. De Cruz Rivera. 2007. Micro-Raman analysis of the Si particles present in Al-Si hypereutectic alloys in liquid and semi-solid states. *Advanced Engineering Materials* 9 (1–2): 46–51.

17. Robles-Hernández, F.C. 2004. *Improvement in functional characteristics of Al-Si cast components through the utilization of a novel electromagnetic treatment of liquid melts*. PhD disertation, Mechanicla Engineering. Windsor: University of Windsor.
18. Robles Hernández, F.C., and J.H. Sokolowski. 2005. Identification of silicon agglomerates in quenched Al-Si hypereutectic alloys from liquid state. *Advanced Engineering Materials* 7 (11): 1037–1043.
19. Al-Helal, K., I.C. Stone, and Z. Fan. 2012. Simultaneous primary Si refinement and eutectic modification in hypereutectic Al-Si alloys. *Transactions of the Indian Institute of Metals* 65 (6): 663–667. https://doi.org/10.1007/s12666-012-0171-4.
20. Choi, H., and X.C. Li. 2012. Refinement of primary Si and modification of eutectic Si for enhanced ductility of hypereutectic Al-20Si-4.5Cu alloy with addition of Al2O3 nanoparticles. *Journal of Materials Science* 47 (7): 3096–3102. https://doi.org/10.1007/s10853-011-6143-y.
21. Li, C., X.F. Liu, and Y.Y. Wu. 2008. Refinement and modification performance of Al-P master alloy on primary Mg2Si Al-Mg-Si alloys. *Journal of Alloys and Compounds* 465 (1–2): 145–150. https://doi.org/10.1016/j.jallcom.2007.10.111.
22. Emadi, D., and L.V. Whiting. 2002. Determination of solidification characteristics of Al-Si alloys by thermal analysis. *Transactions of the American Foundry Society* 110 (1): 285–296.
23. Bäckerud, L., G. Chai, and J. Tamminen. 1990. *Solidification characteristics of aluminum alloys: wrought alloys*. Vol. 1. 1st ed. Oslo: AFS/Skan Aluminium.
24. Bäckerud, L., E. Król, and J. Tamminen. 1986a. *Solidification characteristics of aluminium alloys*. Oslo: Skanaluminium.
25. Veldman, N.L.M., A.K. Dahle, D.H. St John, and L. Arnberg. 2001. Dendrite coherency of Al-Si-Cu alloys. *Metallurgical and Materials Transactions A-Physical Metallurgy and Materials Science* 32 (1): 147–155. https://doi.org/10.1007/s11661-001-0110-1.
26. Chai, G.C., L. Backerud, T. Rolland, and L. Arnberg. 1995. Dendrite coherency during equiaxed solidification in binary aluminum-alloys. *Metallurgical and Materials Transactions A-Physical Metallurgy and Materials Science* 26 (4): 965–970. https://doi.org/10.1007/Bf02649093.
27. Dahle, A.K., P.A. Tøndel, C.J. Paradies, and L. Arnberg. 1996. Effect of grain refinement on the fluidity of two commercial Al-Si foundry alloys. *Metallurgical and Materials Transactions A* 27 (8): 2305–2313. https://doi.org/10.1007/bf02651885.
28. Lambrakos, S.G. 2016. Inverse thermal analysis of steel welds using solidification-boundary constraints. *Journal of Materials Engineering and Performance* 25 (6): 2103–2115. https://doi.org/10.1007/s11665-016-2084-6.
29. Mihalache, V. 2016. Thermal analysis of ball-milled Fe-14Cr-3W-0.4Ti-0.25Y(2)O(3) ferritic steel powder evidence for contamination from the air. *Journal of Thermal Analysis and Calorimetry* 124 (3): 1179–1192. https://doi.org/10.1007/s10973-016-5304-2.
30. Ferraz, G., A. Santiago, J.P. Rodrigues, and P. Barata. 2016. Thermal analysis of hollow steel columns exposed to localised fires. *Fire Technology* 52 (3): 663–681. https://doi.org/10.1007/s10694-015-0481-2.
31. Cheon, J., D.V. Kiran, and S.J. Na. 2016. Thermal metallurgical analysis of GMA welded AH36 steel using CFD-FEM framework. *Materials & Design* 91: 230–241. https://doi.org/10.1016/j.matdes.2015.11.099.
32. Campos, M., L. Blanco, and J.M. Torralba. 2006. Thermal analysis of prealloyed Fe-3Cr-0.5Mo sintered steel. *Journal of Thermal Analysis and Calorimetry* 84 (2): 483–487. https://doi.org/10.1007/s10973-005-6991-2.
33. Kosny, J., E. Kossecka, A. Brzezinski, A. Tleoubaev, and D. Yarbrough. 2012. Dynamic thermal performance analysis of fiber insulations containing bio-based phase change materials (PCMs). *Energy and Buildings* 52: 122–131. https://doi.org/10.1016/j.enbuild.2012.05.021.
34. Tlijani, M., et al. 2010. A simultaneous characterization and uncertainty analysis of thermal conductivity and diffusivity of bio-insulate material "Palm date Wood" obtained from a periodic method. *IOP Conference Series: Materials Science and Engineering* 13 (1): 012015.
35. Yang, H.S., D.J. Gardner, and H.J. Kim. 2009. Viscoelastic and thermal analysis of lignocellulosic material filled polypropylene bio-composites. *Journal of Thermal Analysis and Calorimetry* 98 (2): 553–558. https://doi.org/10.1007/s10973-009-0324-9.

36. Double, J.S. 1966. Differential thermal analysis of polymers. *Transactions and Journal of the Plastics Institute* 34 (110): 73.
37. Haly, A.R., and M. Dole. 1964. Differential thermal analysis of polymers undergoing glass transition. *Journal of Polymer Science Part B-Polymer Letters* 2 (3pb): 285–290. https://doi.org/10.1002/pol.1964.110020308.
38. Kojima, Y., M. Takahara, T. Matsuoka, and H. Takahashi. 2001. High-pressure differential thermal analysis of polymers. *Journal of Applied Polymer Science* 80 (7): 1046–1051. https://doi.org/10.1002/App.1188.
39. Goff, H.D. 1995. The use of thermal-analysis in the development of a better understanding of frozen food stability. *Pure and Applied Chemistry* 67 (11): 1801–1808. https://doi.org/10.1351/pac199567111801.
40. Kerr, W.L., and D.S. Reid. 1994. The use of stepwise differential scanning calorimetry for thermal-analysis of foods. *Thermochimica Acta* 246 (2): 299–308. https://doi.org/10.1016/0040-6031(94)80097-9.
41. Urzendowski, I.R., and D.G. Pechak. 1992. Characterization of food-packaging materials by microscopic, spectrophotometric, thermal and dynamic mechanical analysis. *Food Structure* 11 (4): 301–314.
42. Wetton, R.E., and R.D.L. Marsh. 1990. Dynamic mechanical thermal-analysis (DMTA) of food materials. *Rheology of Food, Pharmaceutical and Biological Materials with General Rheology*: 231–247.
43. Gotz, A., and K.G. Heumann. 1987. Heavy-metal trace determination with a compact thermal ionization quadrupole mass-spectrometer. 2. Analysis of food samples. *Fresenius Zeitschrift Fur Analytische Chemie* 326 (2): 118–122. https://doi.org/10.1007/Bf00468493.
44. Takahashi, K., K. Shirai, K. Wada, and A. Kawamura. 1978. Differential thermal-analysis applied to examining thermal-behavior of starch – Thermal-behavior of high molecular substances in foods. 1. *Journal of the Agricultural Chemical Society of Japan* 52 (5): 201–206.
45. Eberth, J.F., J.A. Neal, and F.C. Robles Hernandez. 2012. Evaluation of heat propagation through poultry in a reduced computational-cost model of contact cooking. *International Journal of Food Science & Technology* 47: 1130–1137. https://doi.org/10.1111/j.1365-2621.2012.02951.x.
46. Robles Hernandez, F.C., J.A. Neal, and U.S. Aldea. 2013. Thermal characterization of poultry for the development of a comprehensive device to monitor safe and proper cooking. *Journal of Food Process Engineering* 36 (2): 160–172. https://doi.org/10.1111/j.1745-4530.2011.00665.x.
47. Nafisi, S., R. Ghomashchi, J. Hedjazi, and S.M.A. Boutorabi. 2004. *Optimizing the melt treatment of piston alloys using thermal analysis technique*. SAE Technical Paper.
48. Djurdjevic, M., H. Jiang, and J. Sokolowski. 2001. On-line prediction of aluminum-silicon eutectic modification level using thermal analysis. *Materials Characterization* 46 (1): 31–38. https://doi.org/10.1016/S1044-5803(00)00090-5.
49. Hegde, S., and K.N. Prabhu. 2008. Modification of eutectic silicon in Al-Si alloys. *Journal of Materials Science* 43 (9): 3009–3027. https://doi.org/10.1007/s10853-008-2505-5.
50. Heshmatpour, B. 1997. Modification of silicon in eutectic and hyper-eutectic Al-Si alloys. *Light Metals* 1997: 801–808.
51. Apelian, D., and G.K. Sigworth, and K.R. Whaler. 1984. Assessment of grain refinement and modification of Al–Si foundry alloys by thermal analysis. *AFS Transactions* 92 (2): 297–307.
52. Robles Hernandez, F.C. 2004. *Improvement in functional characteristics of aluminum-silicon cast components through the utilization of a novel electromagnetic treatment of liquid melts*. Ph.D Thesis. (University of Windsor): Windsor.
53. Incropera, F.P. 2007. *Fundamentals of heat and mass transfer*. 6th ed. Hoboken: Wiley.
54. Incropera, F.P., and D.P. DeWitt. 2002. *Software tools and user's guides to accompany fundamentals of heat and mass transfer, 5th edition & Introduction to heat transfer, 5th edition*. New York: Wiley.
55. Hernandez, F.C.R., M.B. Djurdjevic, W.T. Kierkus, and J.H. Sokolowski. 2005. Calculation of the liquidus temperature for hypo and hypereutectic aluminum silicon alloys. *Materials Science and Engineering A-Structural Materials Properties Microstructure and Processing* 396 (1–2): 271–276. https://doi.org/10.1016/j.msea.2005.01.024.

56. Hernandez, F.C.R., J.H. Sokolowski, and J.D.C. Rivera. 2007. Micro-Raman analysis of the Si particles present in Al-Si hypereutectic alloys in liquid and semi-solid states. *Advanced Engineering Materials* 9 (1–2): 46–51. https://doi.org/10.1002/adem.200600173.
57. Hernandez, F.C.R., and J.H. Sokolowski. 2005. Identification of silicon agglomerates in quenched Al-Si hypereutectic alloys from liquid state. *Advanced Engineering Materials* 7 (11): 1037–1043.
58. Lu, D.H., Y.H. Jiang, G.S. Guan, R.F. Zhou, Z.H. Li, and R. Zhou. 2007. Refinement of primary Si in hypereutectic Al-Si alloy by electromagnetic stirring. *Journal of Materials Processing Technology* 189 (1–3): 13–18. https://doi.org/10.1016/j.jmatprotec.2006.12.008.
59. Robles Hernández, F.C., and J.H. Sokolowski. 2006c. Thermal analysis and microscopical characterization of Al-Si hypereutectic alloys. *Journal of Alloys and Compounds* 419 (1–2): 180–190.
60. Xu, C.L., H.Y. Wang, Y.F. Yang, and Q.C. Jiang. 2007. Effect of Al–P–Ti–TiC–Nd2O3 modifier on the microstructure and mechanical properties of hypereutectic Al–20 wt.%Si alloy. *Materials Science and Engineering: A* 452–453 (0): 341–346. https://doi.org/10.1016/j.msea.2006.10.114.
61. Xu, C.L., H.Y. Wang, Y.F. Yang, H.-Y. Wang, and Q.C. Jiang. 2006. Effect of La2O3 in the Al–P–Ti–TiC–La2O3 modifier on primary silicon in hypereutectic Al–Si alloys. *Journal of Alloys and Compounds* 421 (1–2): 128–132. https://doi.org/10.1016/j.jallcom.2005.11.034.
62. Wang, W.M., X.F. Bian, J.Y. Qin, and S.I. Syliusarenko. 2000. The atomic-structure changes in Al-16 pct Si alloy above the liquidus. *Metallurgical and Materials Transactions A-Physical Metallurgy and Materials Science* 31 (9): 2163–2168. https://doi.org/10.1007/s11661-000-0134-y.
63. Sadigh, B., M. Dzugutov, and S.R. Elliott. 1999. Vacancy ordering and medium-range structure in a simple monatomic liquid. *Physical Review B* 59 (1): 1–4. https://doi.org/10.1103/Physrevb.59.1.
64. Xiufang, B., W. Weimin, and Q. Jingyu. 2001. Liquid structure of Al–12.5% Si alloy modified by antimony. *Materials Characterization* 46 (1): 25–29. https://doi.org/10.1016/S1044-5803(00)00089-9.
65. Wang, W., X. Bian, J. Qin, and T.G. Fan. 2000. Study on atomic density changes in the liquid Al-Si alloys by X-ray diffraction method. *Journal of Materials Science Letters* 19 (17): 1583–1585. https://doi.org/10.1023/a:1006741626728.
66. Bian, X., and W. Wang. 2000. Thermal-rate treatment and structure transformation of Al–13 wt.% Si alloy melt. *Materials Letters* 44 (1): 54–58. https://doi.org/10.1016/S0167-577X(00)00011-2.
67. Inui, M., S.I. Takeda, Y. Shirakawa, S. Tamaki, Y. Waseda, and Y. Yamaguchi. 1991. Structural study of molten silver halides by neutron diffraction. *Journal of the Physical Society of Japan* 60 (9): 3025–3031.
68. Ye, H. 2003. An overview of the development of Al-Si-alloy based material for engine applications. *Journal of Materials Engineering and Performance* 12 (3): 288–297. https://doi.org/10.1361/105994903770343132.
69. Vijeesh, V., and K. Narayan Prabhu. 2014. Review of microstructure evolution in hypereutectic Al–Si alloys and its effect on wear properties. *Transactions of the Indian Institute of Metals* 67 (1): 1–18. https://doi.org/10.1007/s12666-013-0327-x.
70. Kirin, A., A. Tonejc, and A. Bonefačić. 1969. Change in the lattice parameter of aluminum under the influence of rapid quenching from the liquid state. *Scripta Metallurgica* 3 (12): 943–946. https://doi.org/10.1016/0036-9748(69)90247-6.
71. Larson, B.C. 1974. High-precision measurements of lattice parameter changes in neutron-irradiated copper. *Journal of Applied Physics* 45 (2): 514–518. https://doi.org/10.1063/1.1663274.
72. Bandyopadhyay, J., and K.P. Gupta. 1977. Low temperature lattice parameter of nickel and some nickel-cobalt alloys and Grüneisen parameter of nickel. *Cryogenics* 17 (6): 345–347. https://doi.org/10.1016/0011-2275(77)90130-8.

73. Windisch, D., and P. Becker. 1990. Silicon lattice-parameters as an absolute scale of length for high-precision measurements of fundamental constants. *Physica Status Solidi A-Applied Research* 118 (2): 379–388. https://doi.org/10.1002/pssa.2211180205.
74. Glicksman, M.E., and SpringerLink (Online service). 2011. *Principles of solidification an introduction to modern casting and crystal growth concepts*. New York: Springer Science+Business Media/LLC. https://doi.org/10.1007/978-1-4419-7344-3.
75. Stefanescu D.M., and SpringerLink (Online service). 2009. *Science and engineering of casting solidification*. 2nd ed. Boston: Springer-Verlag US. https://doi.org/10.1007/b135947.
76. Djurdjevic, M.B., W. Kasprzak, C.A. Kierkus, W.T. Kierkus, and J.H. Sokolowski. 2001. Quantification of Cu enriched phases in synthetic 3XX aluminum alloys using the thermal analysis technique. *AFS Transactions* 16: 1–12.
77. Djurdjevic, M., J. Sokolowski, and T. Stockwell. 1998. Control of the aluminum-silicon alloy solidification process using thermal analysis. *Metallurgy* 4: 237–248.
78. MacKay, R.I., M.B. Djurdjevic, H. Jiang, J.H. Sokolowski, and W.J. Evans. 2000. Determination of eutectic Si particle modification via a new thermal analysis interpretive method in 319 alloy. In *Transactions of the American Foundry Society and the one hundred fourth annual castings congress*, 511–520.
79. Samuel, A.M., P. Ouellet, F.H. Samuel, and H.W. Doty. 1998. Microstructural interpretation of thermal analysis of commercial 319 Al alloy with Mg and Sr additions. *Transactions of the American Foundrymen's Society* 105 (105): 951–962.
80. Mahfoud, M., A.K.P. Rao, and D. Emadi. 2010. The role of thermal analysis in detecting impurity levels during aluminum recycling. *Journal of Thermal Analysis and Calorimetry* 100 (3): 847–851. https://doi.org/10.1007/s10973-010-0742-8.
81. Arnberg, L., L. Bäckerud, and G. Chai. 1996. In *Solidification characteristics of aluminum alloys: Dendrite coherency*, ed. S.P. Thomas. Oslo: American Foundrymen's Society.
82. Bäckerud, L., E. Król, and J. Tamminen. 1986b. *Solidification characteristics of aluminum alloys: Wrought alloys*. Oslo: Skanaluminium.
83. Backerud, L. 1983. How does a good grain refiner work. *Light Metal Age* 41 (9-10):6–10.
84. Backerud, L., P. Gustafson, and M. Johnsson. 1991. Grain refining mechanisms in aluminium as a result of additions of titanium and boron, Part I. *Aluminum* 67 (9): 910–915.
85. Chen, C., Z.-X. Liu, B. Ren, M.-X. Wang, Y.-G. Weng, and Z.-Y. Liu. 2007. Influences of complex modification of P and RE on microstructure and mechanical properties of hypereutectic Al-20Si alloy. *Transactions of Nonferrous Metals Society of China* 17 (2): 301–306. https://doi.org/10.1016/S1003-6326(07)60089-2.
86. Hume-Rothery, W. 1952. *Atomic theory for students of metallurgy*. Lincoln: Institute of Metals.
87. Huggins, M.L. 1936. The structure of metals and alloys (Hume-Rothery, William). *Journal of Chemical Education* 13 (7): 350. https://doi.org/10.1021/ed013p350.2.
88. Arnberg, L., G. Chai, and L. Backerud. 1993. Determination of dendritic coherency in solidifying melts by rheological measurements. *Materials Science and Engineering: A* 173 (1):101–103. https://doi.org/10.1016/0921-5093(93)90195-K.
89. Chai, G., L. BÄckerud, T. RØlland, and L. Arnberg. 1995. Dendrite coherency during equiaxed solidification in binary aluminum alloys. *Metallurgical and Materials Transactions A* 26 (4): 965–970. https://doi.org/10.1007/bf02649093.
90. Dahle, A.K., and L. Arnberg. 1996. The rheological properties of solidifying aluminum foundry alloys – Overview. *Jom-Journal of the Minerals Metals & Materials Society* 48 (3): 34–37.
91. Gupta, A.K., D.J. Lloyd, and S.A. Court. 2001. Precipitation hardening in Al-Mg-Si alloys with and without excess Si. *Materials Science and Engineering A-Structural Materials Properties Microstructure and Processing* 316 (1–2): 11–17. https://doi.org/10.1016/S0921-5093(01)01247-3.
92. Wang, X., S. Esmaeili, and D.J. Lloyd. 2006. The sequence of precipitation in the Al-Mg-Si-Cu alloy AA6111. *Metallurgical and Materials Transactions A-Physical Metallurgy and Materials Science* 37A (9): 2691–2699. https://doi.org/10.1007/Bf02586103.

93. Edwards, G.A., K. Stiller, G.L. Dunlop, and M.J. Couper. 1998. The precipitation sequence in Al-Mg-Si alloys. *Acta Materialia* 46 (11): 3893–3904. https://doi.org/10.1016/S1359-6454(98)00059-7.
94. Tavitas-Medrano, F.J., A.M.A. Mohamed, J.E. Gruzleski, F.H. Samuel, and H.W. Doty. 2010. Precipitation-hardening in cast AL-Si-Cu-Mg alloys. *Journal of Materials Science* 45 (3): 641–651. https://doi.org/10.1007/s10853-009-3978-6.
95. Ibrahim, M.F., A.M. Samuel, H.W. Doty, and F.H. Samuel. 2017. Effect of aging conditions on precipitation hardening in Al–Si–Mg and Al–Si–Cu–Mg alloys. *International Journal of Metalcasting* 11 (2): 274–286. https://doi.org/10.1007/s40962-016-0057-z.
96. Mohamed, A.M.A., and F.H. Samuel. 2012. Chapter 04: A review on the heat treatment of Al-Si-Cu/Mg casting alloys. In *Heat treatment – Conventional and novel applications*, ed. F. Czerwinski. Rijeka: InTech.

Chapter 3
Metal Casting Process

3.1 Casting Processes: Overview

There are many casting processes utilized in both the aerospace and automotive industries today, and cost is the primary driving factor. Those cost based factors (e.g. raw metal cost, metal processing, heat treatment, etc.) consider not just their durability or life cycle in their intended application. Automotive castings have high production cycle factors, which are needed in order to supply engine plants and automotive assemblers. This fits perfectly well when you consider just-in-time inventories that are a cost factor that can trim billions from the overall cost of delivering cars to dealership showrooms.

Aerospace applications for castings can be very different. Usually the complexity of the castings architecture is superior with more demanding mechanical strength, temperature cycles, more extreme environments, and higher dimensional stability than for automotive counterparts.

Because the capacity and needs are larger in the automobile industry than for the aerospace industry, the casting processes that support automotive applications tend to dominate in proportion of all castings made. The high pressure die casting (HPDC) process makes up about 80% of the entire foundry output that uses Al-Si alloys. HPDC process has a high cycle rate and uses more cost-effective secondary alloys. This is why a significant investment has been made by auto suppliers.

It would be prudent to briefly review some of the common aspects of the casting processes that will be prior to making a decision on which casting process is to be used for the intended application. Most have some degree of directional solidification mainly due to the fact that risers (also called feeders) have to be implemented. Risers are usually located on the top position of a casting during solidification and provide a thermal mass effect that induces directional solidification. Risers are a liquid aluminum reservoir to feed the last stages of solidification. The size of risers

© Springer International Publishing AG 2017

F.C. Robles Hernandez et al., *Al-Si Alloys*, DOI 10.1007/978-3-319-58380-8_3

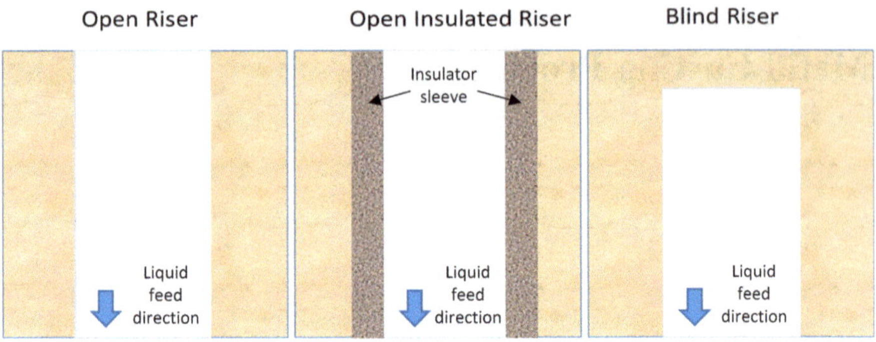

Fig. 3.1 Schematic of the three main types of risers that are used in aluminum casting process to provide direction solidification and liquid metal feed to the adjoining casting

as part of the overall total aluminum used to make one casting will depend on several factors such as: solidification rate, the alloy freezing range, effectiveness of the runner and gating design (rigging design), the size of the casting and the pouring temperature. The size of the risers ideally should be as small as possible to minimize the cost of the metal used and energy needed to melt the alloy. A general metric used widely in the metal casting industry is casting yield. The casting yield is the raw castings' weight divided by the poured metal (casting + riser + rigging weight).

Risers can have three basic formats, which of course will depend on the casting process used (Fig. 3.1). Open risers are used in many gravity-based pouring systems and they have the virtue of helping to determine the end of the pouring sequence. Because they are open, the top of the riser can lose heat, and when possible, this is mitigated by the incorporation of an insulator sleeve. Finally, there is what is called blind risers, which are closed at the top. The following are the two reasons to use a blind riser: (1) because the mold has a rotation process and liquid aluminum will drain out, and (2) liquid metal is delivered into the mold at high speed and/or pressure.

This is an important factor to consider with riser design or, more specifically, the chemistry of the alloy being used. As mentioned previously, Sr is an additive that helps refining Si structure from acicular to fibrous. Sr also, as a side effect, slightly modifies the solidification kinetics of the alloy. Sr forms a higher level of dispersed porosity throughout the castings and has the effect of reducing the feeding capacity of the riser. Conversely, a commercial grain refiner will have the effect of reducing disperse porosity and thereby enhances a risers feeding capacity. This method of quantifying the volume of the riser sink can be done by using a graduated cylinder filled with water. Pour the water into the sink until the water level is flush with the top of the riser. Then measure the volume by subtracting the water left in the graduated cylinder from the initial amount. An example of the riser volume determination method is shown in Fig. 3.2 [1].

Figure 3.3 shows the impact of the volume of the riser sink for the case of an automotive precision sand process. The cast structure which is directly below the riser will have higher porosity which can affect mechanical properties and potentially pressure tightness. One of the methods to rectify this is to either add grain refiner to the alloy melt having Sr additions, or if that cannot be done, increasing the size of the riser, which will add the needed metallostatic head and this will help drive down porosity.

Fig. 3.2 The method of using a graduated cylinder to quantify the volume sink size of the riser

3.2 High Pressure Die Casting Process

The earliest examples of die casting by pressure injection – as opposed to casting by gravity pressure – occurred in the mid-1800s. A patent was awarded to Sturges in 1849 for the first manually operated machine for casting printing type. The process was limited to printer's type for the next 20 years, but the development of other shapes began to increase toward the end of the century. By 1892, commercial applications included parts for phonographs and cash registers, and mass production of many types of parts began in the early 1900s.

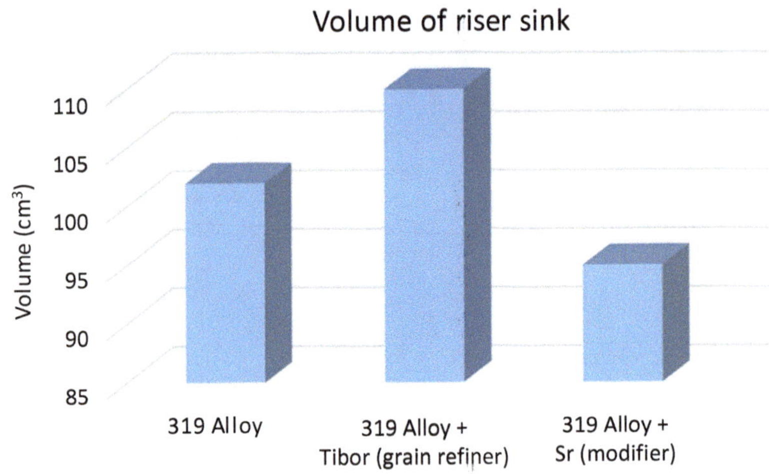

Fig. 3.3 The volume sink size in the same riser (seen in Fig. 3.2) in 319 alloy with grain refiner and Sr modifier [1]

The first die casting alloys were various compositions of tin and lead, but their use declined with the introduction of zinc and aluminum alloys in 1914. Magnesium and copper alloys quickly followed, and by the 1930s, many of the modern alloys still in use today became available. Today, 80% of all aluminum casting manufactured in North America are made using the HPDC process.

The HPDC process is very complex, requiring significant capital investment, complex dies with cooling circuits, controlled filling characteristics (gate velocities of close to 30 m/s), and pressures of 138 MPa in the die cavity. Figure 3.4 is a schematic of a typical die cast unit. The stages for a complete die cast cycle are

- Die preparation
  - die spray
  - die blow off
  - cast iron liner insertion
  - die closing
- Metal delivery
  - shot chamber filling
  - high pressure injection
- Phase transformation (liquid to solid)
- Casting extraction.

Probably the most critical stage of the die cast cycle is the injection phase, where there are three stages involved. The slow shot (first stage) is where the liquid metal is brought into the gating section of the die. This ensures that there is a minimal amount of trapped air in the shot sleeve and also ensures that liquid metal turbulence

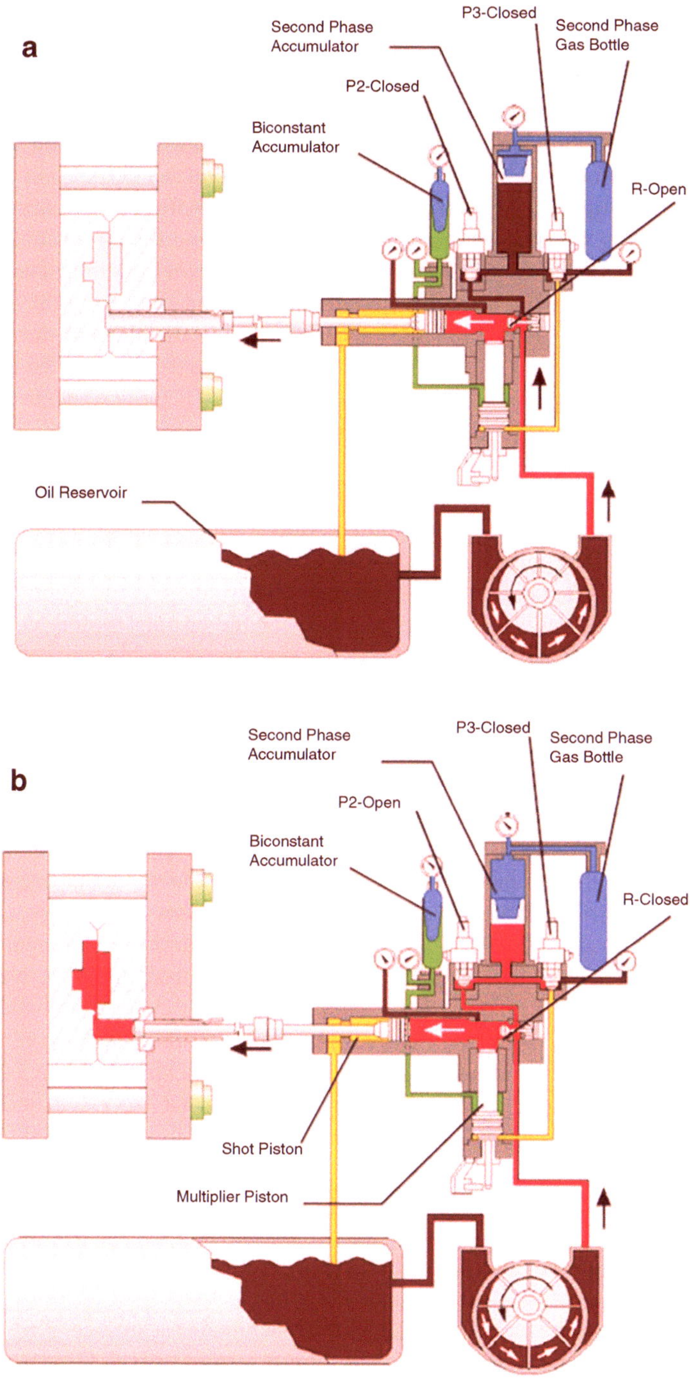

Fig. 3.4 Schematic of the HPDC process (**a**) before and (**b**) after shot entry into the mold

does not occur. This results in further air pockets being formed. The second stage is where the liquid metal in the shot sleeve is pushed into the die cavity by the plunger at high velocity. The filling rates are typically between 80 and 100 ms. As a reference, the blink of a human eye is between 300 and 400 ms, which is between 2.5 and 4 times faster. The third and final stage is intensification, where the semi-solid metal in the die is placed under incremental pressure to reduce internal porosity in the casting. This compensates for volumetric shrinkage. After solidification is complete, the casting is removed, mechanically degated, and then quenched in 90 °C water to lock in solute within the aluminum matrix and for handling. The casting is then thermally processing using a T5 treatment and then shipped to the customer's engine plant. The T5 heat treatment is also known as artificial aging that is applied at temperatures below 200 °C and the exact temperature depends on the alloy.

The typical advantages of die casting engine blocks is the lower unit cost when compared to other casting methods. High-quality die casting can achieve near net shape that lower casting yield and can help using thinner walls. Typical disadvantages are the limited internal cavity complexity possible. For example, open head deck design in engine blocks and the high velocity fill produces turbulence based internal defects. This greatly affects tensile and fatigue capabilities in terms of testing based on ASTM standards. An additional disadvantage is the significantly longer lead times and capital expenses required for die casting tooling compared to other casting processes.

One of the most critical and limiting aspects of the HPDC process, which deserves some attention, is the side effects of a rapid filling process just mentioned. The turbulence of the liquid fill will contribute to a distinctive oxide and porosity appearance, and this will look different from oxide damage that originates from prior to the casting cycle itself (e.g. shot chamber, ladle oxide lip, etc.). Figure 3.5 shows examples of analysis conducted using both X-ray radiography and metallography in a metallurgical lab. Typically the oxides and the voids that are enveloped can be asymmetrical in shape and the oxide extensions quite large and thick. Sometimes these defects can be seen in very close proximity of the ingate and can be the telltale observation that this oxide damage is from the shot chamber or upstream in the melting process. The sample seen in Fig. 3.5a is examined inside the cross-section from radiography. Figure 3.5b shows the true extension of the oxide and void defects directly below the defect seen on the machined surface. Figure 3.5c shows the higher magnification of the smaller voids that are associated with the larger voids seen by eye. These too are formed perhaps from the added turbulence of the oxides that originated from the shot chamber or upstream.

Figure 3.6 shows the other type of oxide and void damage that forms inside the mold itself, where coalescing metal fronts merge with each other at high rates resulting in oxide formation, with the air inside the mold itself being enveloped. Typically these defects will reside inside the cross-section of cast sections and rarely become exposed to the casting surface or the machined cast surface. This point will be important when discussing the role of fatigue testing of HPDC castings in both as the component itself and from using a protocol testing outlined by ASTM standards.

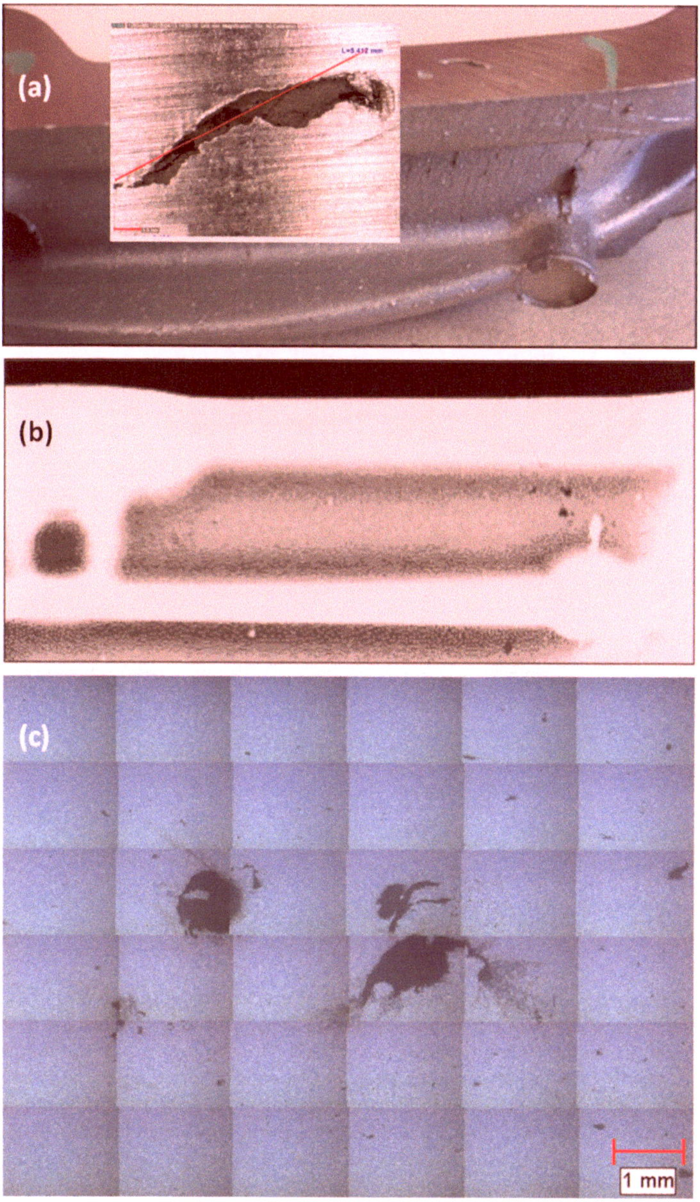

Fig. 3.5 (**a**) Top view of automotive HPDC with oxide defect below where an ingate is located. (**b**) Side view X-ray of casting seen in figure (**a**); more oxide damage below the ingate can be seen. (**c**) Light optical micrograph mosaic of the pores in more detail; in this case, the oxide skins associated with the pores can be seen

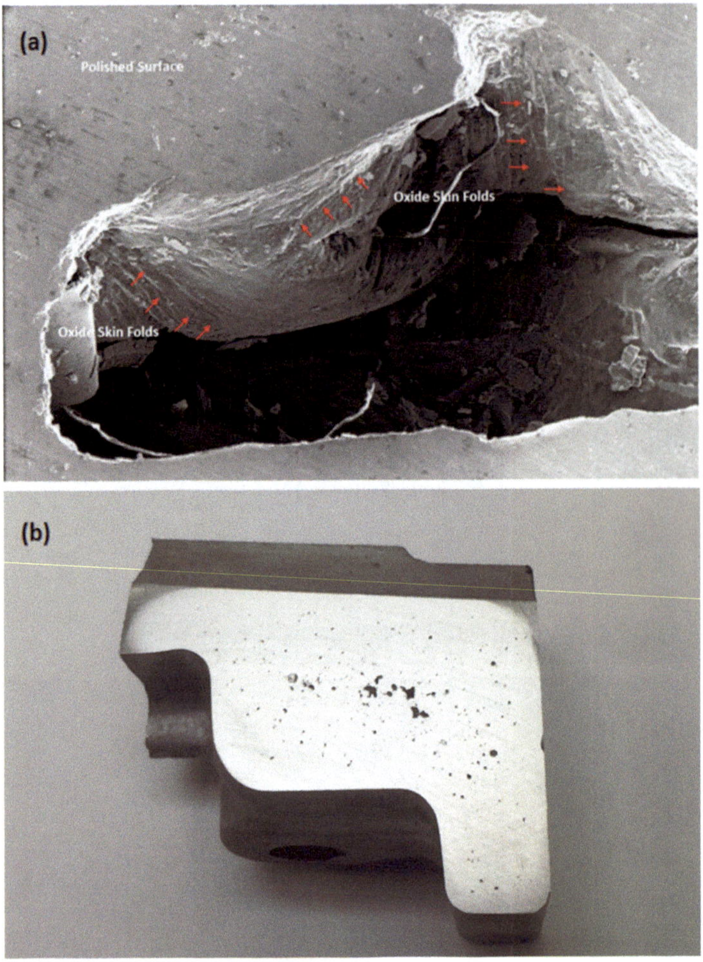

Fig. 3.6 (**a**) Scanning electron micrograph (secondary electron mode) of a typical pore found in a HPDC. Some oxide folds can be seen on the inner surface of the void. Typically at the fold points oxides may be presented extending away for the void. (**b**) The center cross-section of a HPDC engine block pan rail. The cluster of pores originates from turbulent filling of the mold cavity during casting. As oxide fronts violently fold, they encapsulate air pockets. The pore/oxide damage resides in the center because the liquid metal in contact with the die mold rapidly solidify to form a shell and push the resulting oxides and bubbles toward the center

One final point that needs to be made about the HPDC process is, because of the void formation just discussed, it is generally difficult to utilize at T6 or a T7 process (solution treat followed by aging/over aging). The solution stage is carried between 450 and 495 °C for 380 and 383 alloys and results in blistering on the cast surface. As a result, HPDC can only be processed using aging only (T5). This can be a limiting factor, as will be shown in subsequent section of the chapter; oxide damage can result in high scatter of mechanical test results, and a solution treatment followed by an aging cannot be used to improve the result. This process has lost favor to semi-permanent mold and precision sand processes because lower turbulence in mold filling occurs, and thus solution treatment can be applied. Unfortunately, it has a higher cost. To counter this limitation, some HPDC applications have attempted to use a partial vacuum draw inside the mold prior to metal entry. Once the process is optimized, this can improve casting soundness. However solution treating was not possible; current vacuum-assisted processes using high quality, low iron alloys are capable of producing structural components. Under those conditions, castings are resistant to be T6 or T7 without blistering. Another advantage is an extension of the high fatigue performance as well as improved mechanical properties.

3.3 Lost Foam Process

Lost foam casting (LFC) is a type of evaporative-pattern casting process that is similar to investment casting except that foam is used (Fig. 3.7). This process is similar to the lost wax process. This process takes advantage of the low boiling point of foam to simplify the investment casting process by removing the need to melt the wax out of the mold. Lost foam casting was invented in 1964 by M.C. Flemmings [2]. However, public recognition of the benefits of lost foam casting (LFC) was made by General Motors in the mid-1980s when it announced its new car line; Saturn would utilize LFC for production of all engine blocks, cylinder heads, differential carriers, and transmission cases [2]. Unfortunately, in 2015 General Motors retired their LFC operations as the casting did not meet the demands needed for current high performance engine blocks.

The final pattern is composed of approximately 97.5% air and 2.5% polystyrene. Pre-made pouring basins, runners, and risers can be hot glued to the pattern to finish it. Once completely assembled, it was referred to as the “foam cluster”.

Next, the foam cluster is coated with ceramic investment, also known as the refractory coating, via dipping. This coating creates a barrier between the smooth foam surface and the coarse sand surface. Secondly, it controls permeability, which allows the gas created by the vaporized foam pattern to escape through the coating and into the sand. Finally, it forms a barrier so that molten metal does not penetrate or cause sand erosion during pouring. After the coating dries, the cluster is placed into a flask with un-bonded sand. This sand is then compacted using a vibration table. Once compacted, the mold is ready to be poured.

This casting process is advantageous for very complex castings that would regularly require cores. It is also dimensionally accurate, maintains an excellent surface finish, requires no draft, and has no parting lines so no flash is formed. As compared to investment casting, it is cheaper because it is a simpler process and the foam is inexpensive. Risers are not usually required due to the nature of the process because the molten metal vaporizes the foam and the first metal into the mold cools more quickly than the rest; it results in natural directional solidification. Foam is easy to

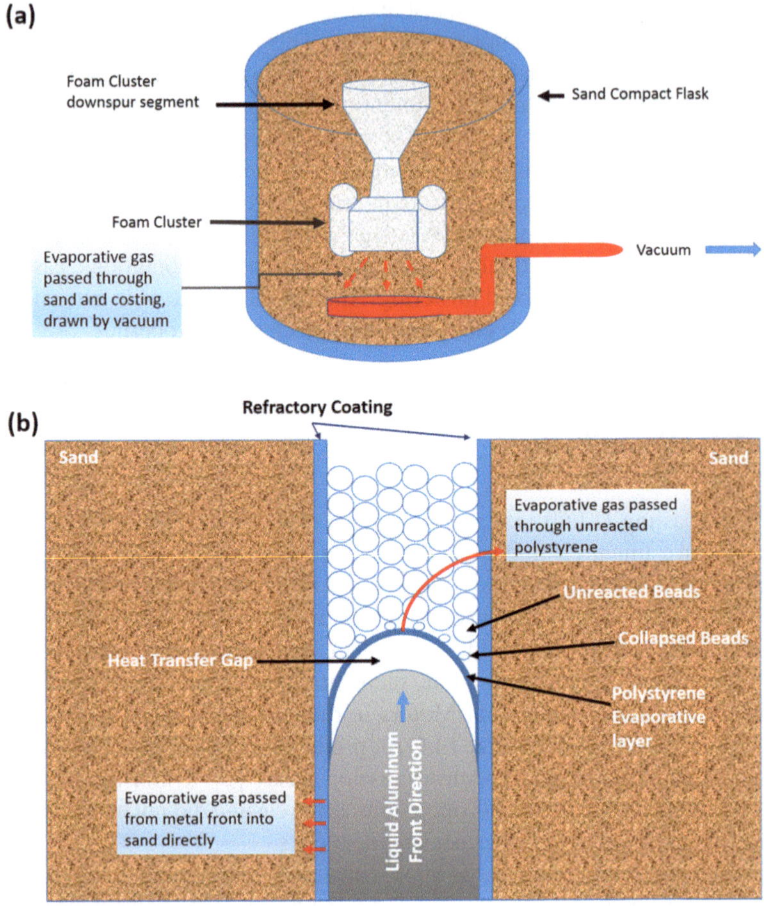

Fig. 3.7 (**a**) Schematic of the vacuum-assisted lost foam cell. (**b**) Schematic of the effect vacuum assisted method has on polystyrene decomposition

manipulate, carve, and glue. The flexibility of LFC often allows for consolidating the parts into one integral component; other forming processes would require the production of one or more parts to be assembled.

Typical lost foam aluminum castings experience a pressure gradient during filling due to the changing metal head pressure. For example, a 51 cm tall casting with a 61 cm tall sprue will experience a metal pressure of about 16.5 kPa at the casting bottom and about 2.8 kPa at the casting top. These pressures are the driving force that controls metal filling rate and removal of the pattern pyrolysis products from the casting cavity. Typical lost foam castings suffer from increased defect severity near the casting top where metal velocities and temperatures are lower.

Figure 3.7 shows the flask mold and sand with the polystyrene assembly ready for molten metal delivery. The bottom of Fig. 3.6 shows the process of the pattern pyrolysis as metal continually eats away during the filling stage. Evaporative

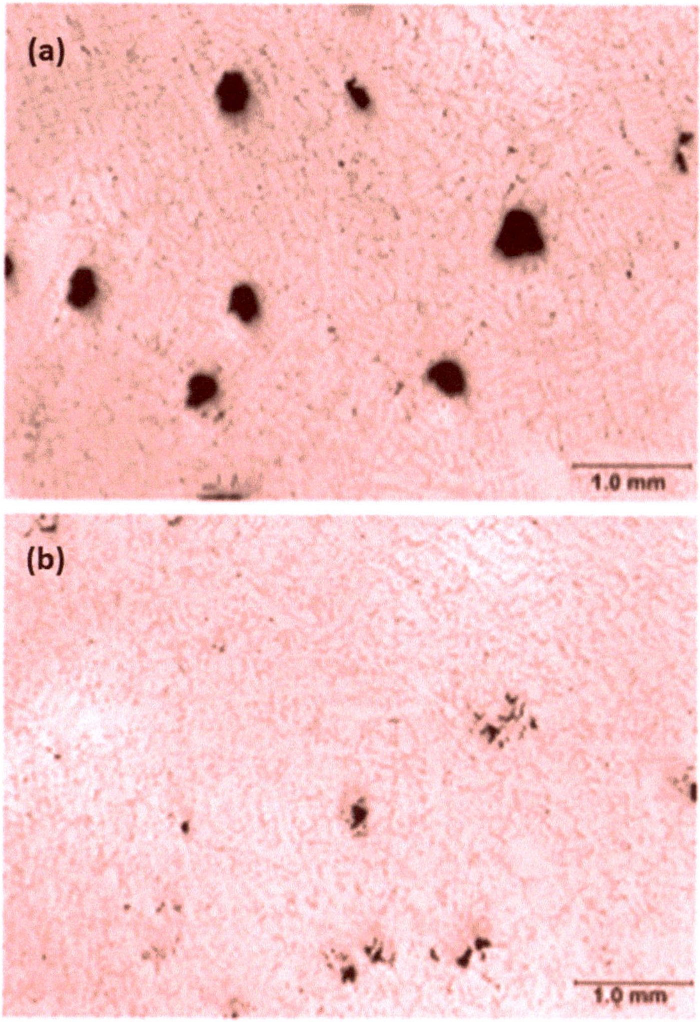

Fig. 3.8 Cast microstructures (**a**) without Vacuum Assist and (**b**) with Vacuum Assist – Lower porosity

by-products not only travel through the liquid stage of poured metal but travels through the coating into the sand. The pattern pyrolysis is a major contributor to cast structure discontinuities. Added to Fig. 3.6 is the integration of a vacuum system within the sand, which helps to further facilitate the removal of the evaporative by-products and reduces the deleterious effects on the solidified casting structure. This is a more expensive process added to the LFC process but allows this process to produce the higher cast component complexity while providing better casting structures than provided previously.

Figure 3.8 shows light optical micrograph images of cast microstructures utilizing the Vacuum Assist method for Lost Foam method. This indicated that the

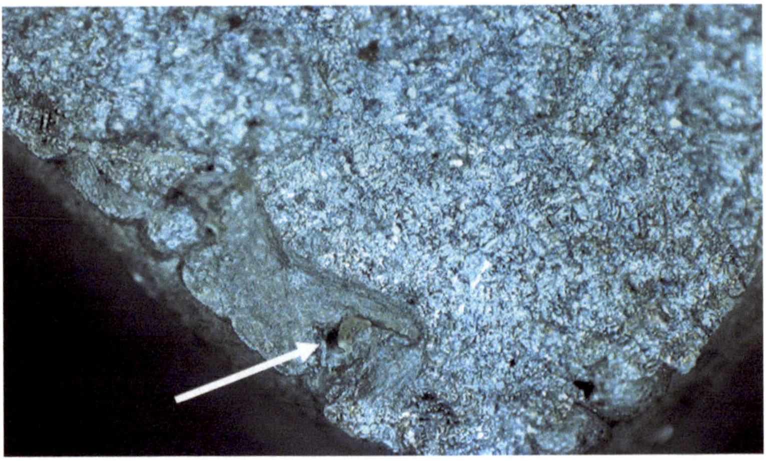

Fig. 3.9 Fracture surface of a head bolt column of an engine block, which failed in a dynamometer test due to the oxide fold defect indicated by an *arrow*

vacuum on the sand side of these castings has influenced the metal filling and pyrolysis product removal as illustrated in the schematic of Fig. 3.7b. This reduction in porosity has not only produced an improved casting microstructure, but also provides a higher tensile property result, and fatigue durability.

There are also several possible cost savings and cycle time improvements associated with the Vacuum Assist variant for Lost Foam (VALF). A lower melt temperature can be used with this process, which results in lower energy cost and increased furnace life. Finally, due to the presence of a vacuum the wall thickness can be decreased further from 4 to 4.5 mm to 3 mm. This makes the VALF process ideal for certain lighter weight components.

Vacuum assist coupled with the LFC process has several advantages over the resin-bonded sand with chills process, although the principles of vacuum assist could also be applied to this process. The major advantages of using the LFC process are the design freedom and part consolidation. The LFC process is an established process, and the implementation of vacuum is relatively simple. Production rates would not be affected. In addition, assuming the fold issues can be eliminated, pouring at lower temperatures offers significant energy and cost savings during melting and holding, extending the life of the melting and holding furnaces.

The type of defects that can be encountered using the LFC process relate to the foam cluster consistency in terms of the assembly accuracy, amount of coating used, and the glue used to lock the assembly components together. If excess glue is used, it may squeeze out of the side of the assembly and leave a negative imprint on the casting surface. This excess glue can also provide a retarding effect on the metal flow and produce a failure of metal fronts to coalesce fully as shown in Fig. 3.9. This particular defect provided a notch sensitivity in a head bolt column of an engine in a validation dynamometer test that lead to a tensile fracture that in turn lead to a leak path between the water jacket and outside of the block.

Fig. 3.10 Example of a glue blow defect in a bedplate

In Fig. 3.10, a bedplate casting made using LFC process can be seen; it had a blow type defect at a glue joint, which caused a leak path to occur in the crack chamber of a four cylinder engine dyno validation test.

3.4 Sand Casting and Precision Sand Casting Process

The sand casting process is a process where an assembly of cores are assembled together to form a one-time use mold where the aluminum in liquid state is introduced and then subsequently solidified. It has been one of the oldest metal casting processes used. Sand casting molding method was recorded by Vannoccio Biringuccio in his book published around 1540 [3], but aluminum was not used in this process until it became more cost-effective metal for use during the First World War.

The most common type of sand used is silica due to its cost-effectiveness. Zircon and chromite sand has been used in some limited applications. Sand used in the sand casting process needs to have tight controls on the grain size and distribution as this affects both compaction (and thus the amount of resin used to bind the grain together) and the permeability of the sand cores as metal fills the finished mold (air evacuates from the inside) and the burning of the resin as solidification progresses. Grain size can be quantified by an AFS GFN (American Foundry Society Grain Fineness Number). Most sand foundries will have a certified Sieve Tray system to check and monitor sand before using it in core make.

Core assembly process can be manual as with the case with aerospace aluminum castings where up to 100 cores may be involved in a complete mold, to a fully robotized assembly process such as can be found in the automotive foundries where between 10 and 25 cores may be used for a full core assembly. Many times sand cores are printed together like a jigsaw puzzle and may have glue points to ensure assembly remains in place when it is moved or manipulated.

During the pouring of the metal, it is delivered and filled into the mold via gravity pouring. Much work is involved in the design of the gating system to minimize turbulence and thus reduce oxide damage generation. Solidification rates can be generally slow and thus any oxide damage that results has even more deleterious impacts on mechanical properties. Shrinkage defects can occur more readily in the sand casting process as solidification fronts can lead to hot spots in thicker cast cross-sections. This can be suppressed with the implementation of integrated chills made from cast iron or copper.

The risers in gravity-based sand castings are usually open risers or open risers with insulating sleeves typically made from Fiberfrax. Blind risers are normally used in limited automotive applications, which have a roll process at the final stage of filling.

3.5 Aerospace Sand Casting Applications

Sand casting applications for aerospace can provide the unique ability to incorporate what would be many components into a single casting. This does provide a significant weight saving, reducing the need to welded joints and fittings. Some sand aerospace castings can be as large as 454 kg and are made using large molds that can have 100+ cores used in the final assembly. An example of a partially assembly sand mold for a Turbofan front engine frame is shown in Fig. 3.11.

Sand mold of this size may take up to a day for complete assembly and this then transported to the casting station where a batch melting process is used to either pour the mold with gravity or utilize a bottom filling process using low pressure. Figure 3.12 shows a schematic of the top view of an aerospace casting mold assembly with cope and drag sections prior to closing. Side views opened and closed are shown in Fig. 3.13. Most gravity-based systems can be augmented for low pressure or vacuum sand castings. Gravity-based variants will have lower casting yields (30–40%) due to the complexity of the runner, ingate, and risering architectures needed. In Fig. 3.12, a double runner system can be seen with its own ingate system and filters in place to

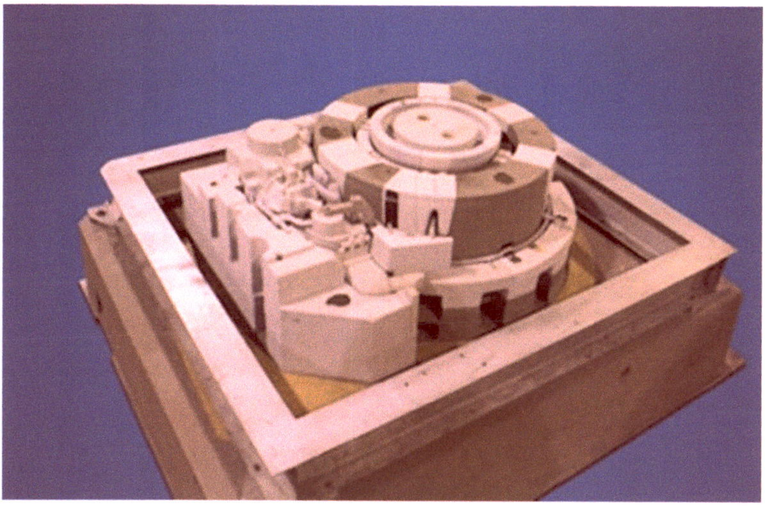

Fig. 3.11 Partial sand mold assembly (drag section) from Fansteel Dynamics. The casting is a Turbofan casting

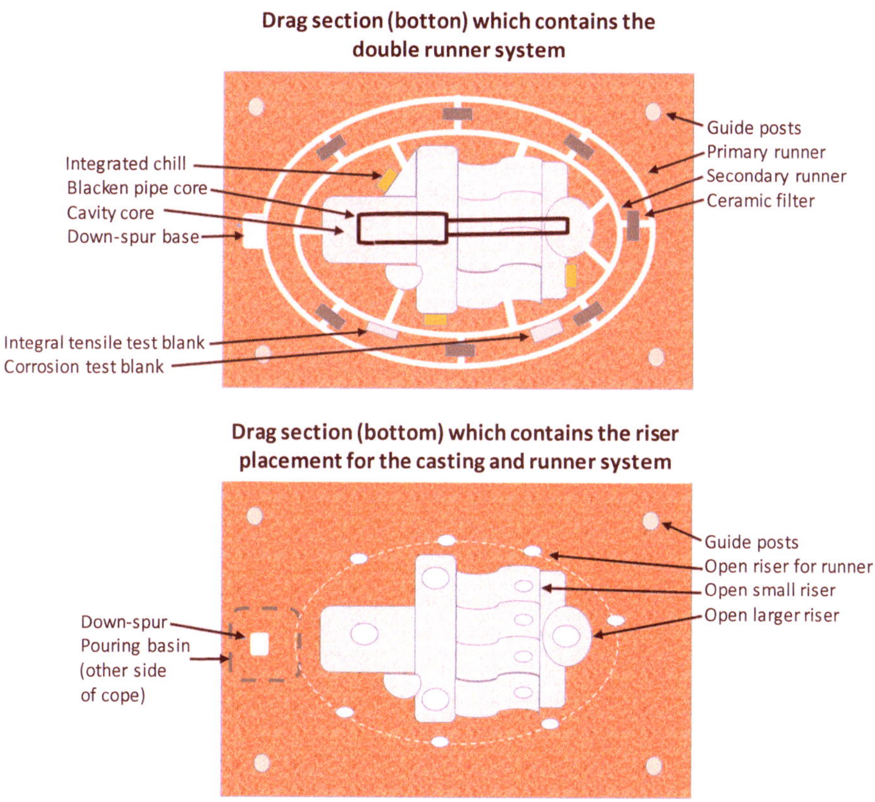

Fig. 3.12 Top view schematics of the drag and cope sections of a complex aerospace casting

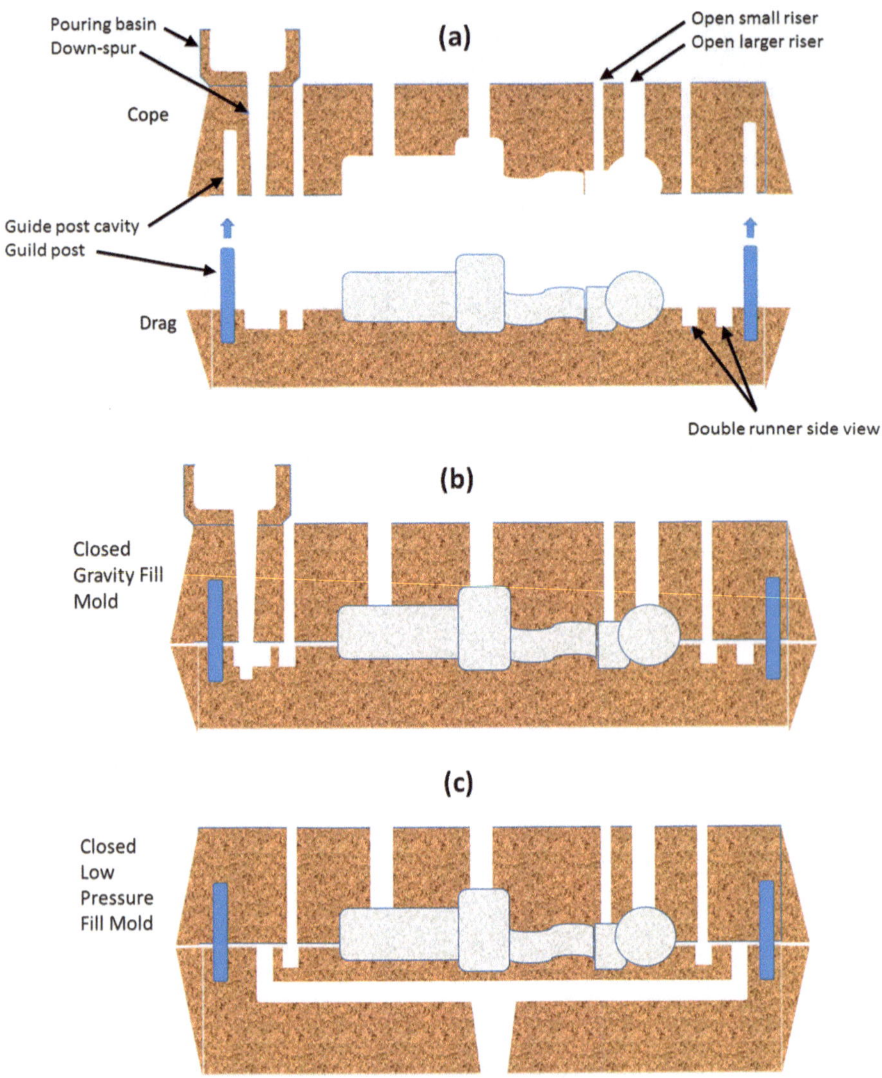

Fig. 3.13 Side view of the gravity-based system (**a**) opened and then (**b**) closed. (**c**) Low-pressure variant

collect oxides that may have initially formed from pouring basin and down-spur liquid metal entry. The secondary runner will then have more laminar liquid metal flow before entry into the mold itself. This is critically important as multiple metal fronts will coalesce inside the mold cavity, and it is important that the oxide film that envelopes the liquid metal front breaks away from the mold wall. This is preferred to oxides that fold into the cast cross-section and thus act as a new discontinuity. The multiple risers seen are meant to provide the needed feeding and directional solidification control to deter shrinkage formation, which in itself will form as another discontinuity and/or possible a leak path between cavities or to the outside of the casting. The risers are generally open as they allow pour operators to see that the mold has filled. In some cases, as these risers are open, insulating sleeves may be used for the riser to enhance feeding capacity.

Usually several integral chills are placed throughout the mold to rapidly chill sections, which cannot be connect to a riser. Solidification modeling such as Magma software may help development engineers to strategically place risers and chills. Nevertheless, most of the process is carried by trial and error testing on prototype castings. The cores may be hand modified to add more metal to existing risers, or small holes made to insert new chills. The prototype casting is made again and undergoes extensive quality checks before a final core make process and core design is finalized.

In Fig. 3.12, some of the cores are black colored which reflect the carbon blackening process used to help promote oxides from the metal front to break away and line the mold wall. The carbon blackening essentially makes the surface of the core smoother. Finally, not indicated in Fig. 3.12, a venting system is used for the final mold assembly. Most of the cores will have small holes drilled to better facilitate the escape of resin burning and deter core blow defects. Aluminum sand mold of the size shown in Fig. 3.11 can weigh thousands of pounds and will have an enormous amount of burning resin during mold filling and solidification. Complete solidification can take hours.

Figure 3.12 also shows the integration of tensile and corrosion test blanks in the rigging setup (attached to secondary runner system). This is usually a requirement by the aerospace customer, as mechanical testing and corrosion testing from an expensive casting has a significant cost impact. The blanks are removed as part of the rigging removal process and then placed in a container or bag so that the blanks and casting can be heat treated together. If welding repair is conducted and reheat treatment is required for the repaired casting, the blanks samples may follow with the casting to gauge the mechanical properties, or they may be routed to a metallurgical lab after the first heat treat cycle. Part of the preliminary development is to ensure that the secondary dendrite arm spacing (λ_2) of the tensile test blank is identical to critical sections of the casting itself.

In recent years, precision sand process for aerospace applications has improved in terms of further reducing oxide generation with the use of low-pressure metal delivery to the mold. This is shown in Fig. 3.13 with the modified runner system and with blind risers. The blind risers allow for the detection of mold being full when the pressure required to fill the liquid metal in the stock tube experiences an abrupt increase. In this case, instead of a pouring basin and down-spur system, metal in a low-pressure furnace is delivered from the bottom. A vertical spur is added into the runner and then molded at a controlled rate and allowing a more quiescent filling. In the meantime, the melt is sealed with inert gas, from which it is pumped to a stock tube that connects vertically to the mold.

The added feature of the low-pressure fill process is that pressure can be varied during the filling stage to reduce liquid metal turbulence even further. In some cases, if an existing casting previously cast using gravity is converted to low pressure, the secondary runner system may be removed and thus have the added factor of improved casting yield.

An important note should be made concerning the melting practices of the aluminum alloys for precision sand aerospace. Most of the alloys are the D357 alloy and the Fe is capped at 0.2 wt%, which tends to be sold at a premium as it is always made from primary aluminum. It was just mentioned that precision sand for aerospace can have a casting yield of 30–40%. This means that the remaining 60–70% will be remelted (sometimes called back scrap) as this will be the most cost-effective approach. As an example it is expected that 133 kg of the alloy is recycled for every 190 kg of casting plus riser and ingates. If the cost of the D357 alloy is 3.08 USD/kg (2015 cost snapshot); this means that there is 392 USD (minus melt loss) of metal value left over from riser and ingate removal. However, if melting practices such as ladle coatings, melting furnace coatings, low-pressure stock tub coatings, are not maintained to the highest standards possible, remelting may be difficult and costly. Fe pickup from improperly coated items used in melt handling can result in a slow incremental increase in Fe concentration past 0.2 wt%. If this occurs then rigging that is cut off may have to be sold at a discount to another foundry, a secondary smelter or a metal scrap dealer. Most aluminum castings made do not run that low in Fe concentrations and thus are able to more easily manage their remelting streams.

3.6 Precision Sand for Automotive Applications

Generally cast iron liners in the cylinder bores are integrated in the core make and assembly process. These liners will generally need to be heated just prior to liquid metal entry so that cold shut and misrun defects do not form. This is done using an inductive heating system which is conducted just prior to casting. The versatility of this process, like the lost foam process just mentioned, is the ability to cast aluminum components with complex component architectures with internal cavities.

In the example shown in Fig. 3.14, an engine block casting is illustrated. Metal entry in this case is conducted by the delivery of liquid aluminum from a pouring furnace via electromagnetic or mechanical pump. The fill of the mold is done slowly and quiescently in order to minimize turbulence and formation of undesirable oxide damage and pores in the casting section. When the mold has completely filled, a roll-over process is done which allows the last metal to enter the mold (which is also the hottest) to provide an effective riser for liquid metal feeding and promote directional solidification. Once the rollover has been completed, the last metal to enter the sand package (the hottest metal) is now in the risers. The chill on the other side of the casting then promotes a directional solidification from the bulkhead section toward the cylinder bore walls and then to the head deck where risers and ingates are attached. The main feature about sand castings with quiescent filling processes is that it generally reduces the potential for oxide formation, discontinuities, and pore

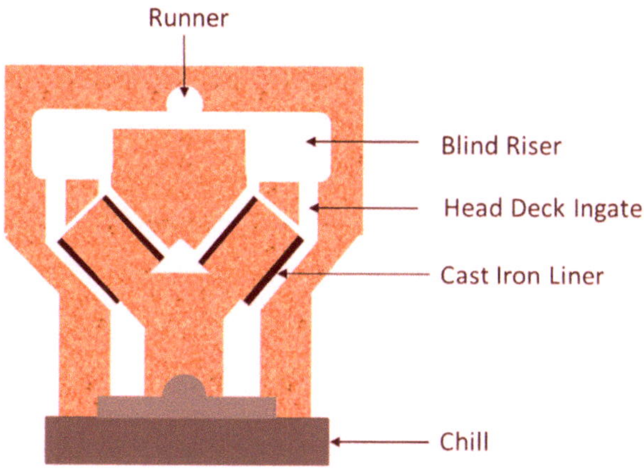

Fig. 3.14 Precision sand process – example for engine block casting with head deck gating and crank based chill

nucleation. Expectedly, this helps in reaching the highest mechanical properties and fatigue durability.

The process outlined cannot be made without a chill (lower capital cost) if a non-copper-based alloy system such as 319 alloy is used. Examples can be either the 356 alloy (used for the manufacture of alloy wheels) or the 357 alloy (used mainly in the aerospace castings). Figures 3.15 and 3.16 show the comparison of the chill and non-chill strategies, respectively. In the non-chill configuration, the risers can be placed on the bulkheads directly and thus provides a slightly higher casting yield.

After sand removal, ingates and riser are cut off and then a shot blasting operation is conducted to smooth out the rough sand cast surface. Following this, the sand castings are processed for heat treatment (T7) and then machined and quality checked for the customer's engine plant.

Casting defect associated with this process can be related to the sand mold assembly itself. Core chips (a dislodged fragment from a core) and core shifts will generally result in a scarp condition. Shrinkage can also form if a large enough oxide is present in a location near a hot spot region. These are illustrated in Figs. 3.17 and 3.18 for the core fragment cases and Fig. 3.19 for the shrinkage defect case.

The other factor that needs to be considered is the casting temperature used to make the casting. In automotive foundries, a continuous melt process is used and sometimes if the temperature control deviates from target or becomes too high, then the impact on the sand mold core sections is to evolve burnt resin much earlier and more intensely near the earlier stages of solidification. The pressure of the resin converting to gas will lead to a subsurface porosity formation which is not intrinsic to the metal quality (e.g. not hydrogen based). Upon review of casting cross-sections or X-ray radiographic audit checks, this could be misdiagnosed as melt quality issues. An example of this can be shown in Fig. 3.20.

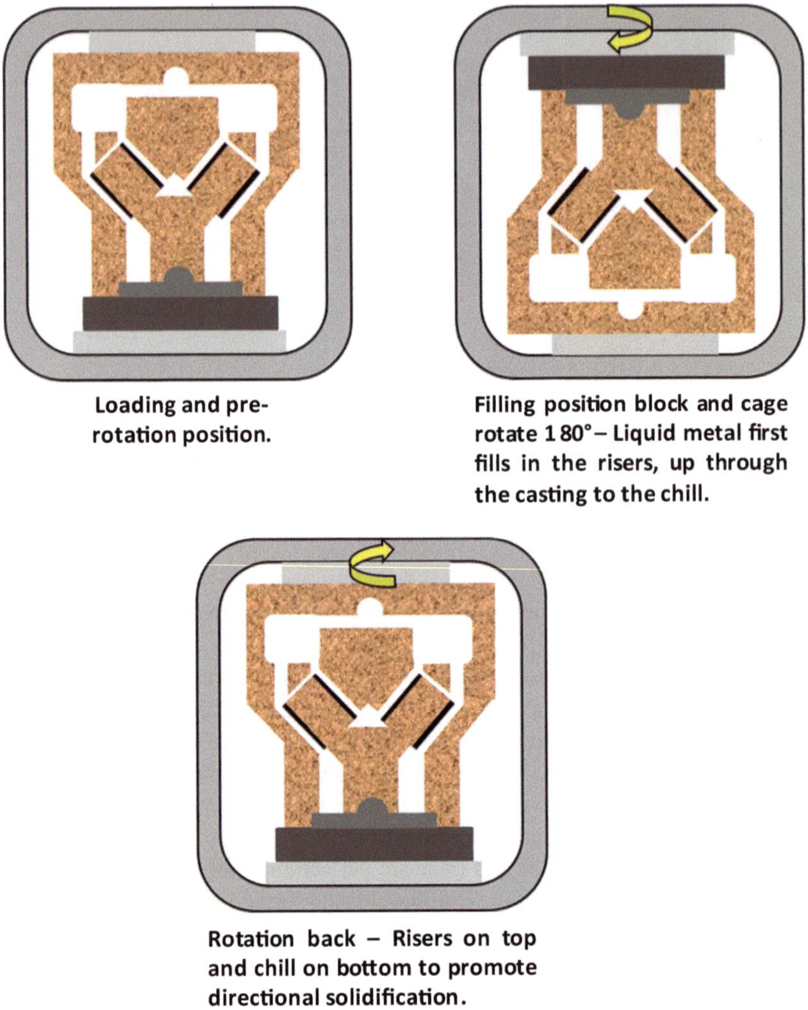

Fig. 3.15 Precision sand process with chill

3.7 Semi-Permanent Molding Process

The semi-permanent mold process (SPM) is also known as the gravity semi-permanent molding process (GSPM). This is typically used in the automotive industry for cylinder heads and engine blocks. In this process the mold is made out of steel, which provides a chill to the casting allowing critical cast sections to boost their mechanical properties. Other portions of the casting require inter cavities that could be made using a sand core. Sand cores can be inserted into the mold prior to

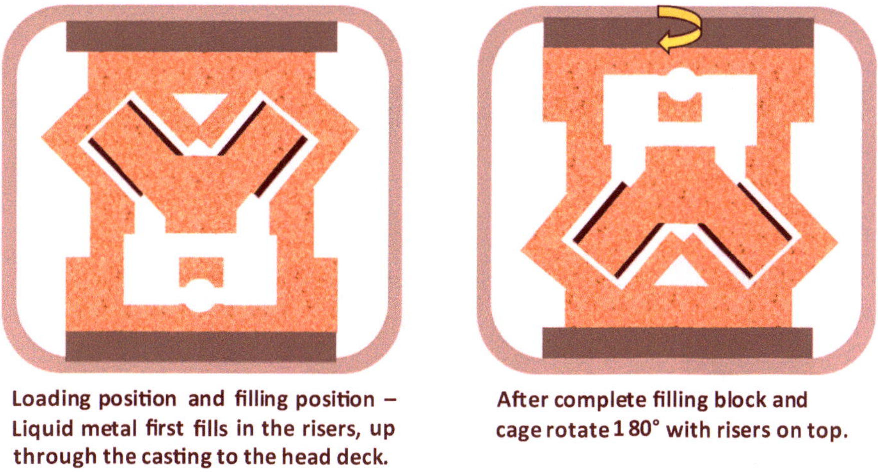

Fig. 3.16 Precision sand process without chill

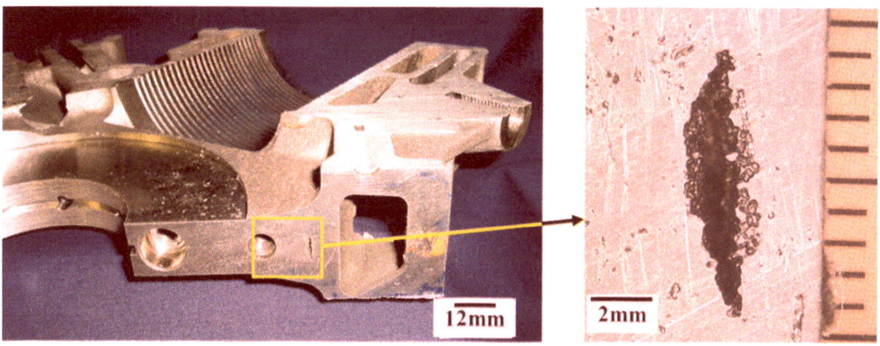

Fig. 3.17 Core fragment defect (core chip) with sand grains forming the outline of the void left behind. Most of the core fragment has fallen out after it was exposed after machining

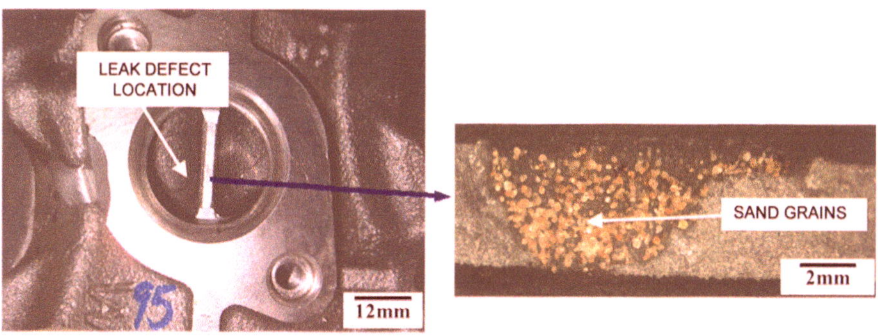

Fig. 3.18 Core fragment which became lodged between to cores and cased a leak path. The sand grains become evident embedded on the walls of the void left behind

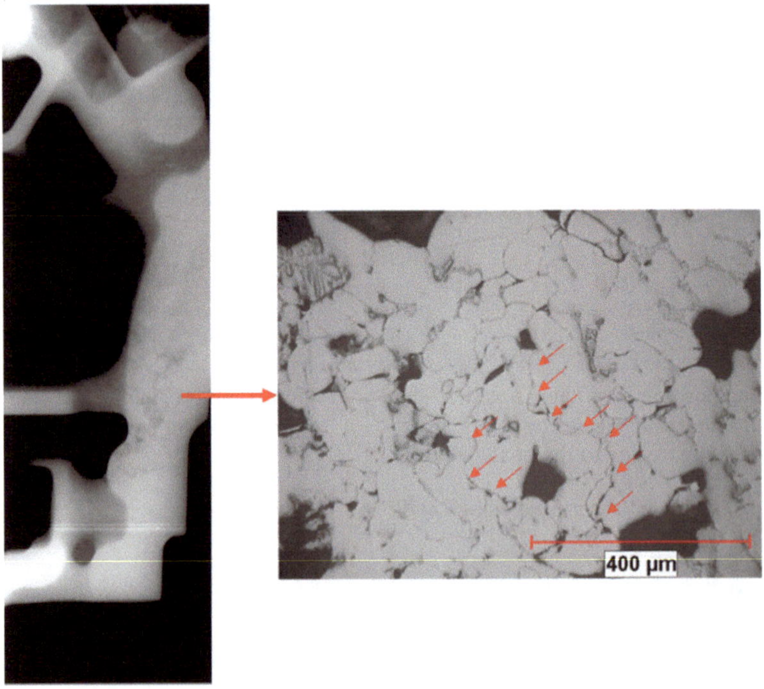

Fig. 3.19 Shrinkage defect caused by a young oxide defect, as seen in the light optical micrograph (*red arrows*)

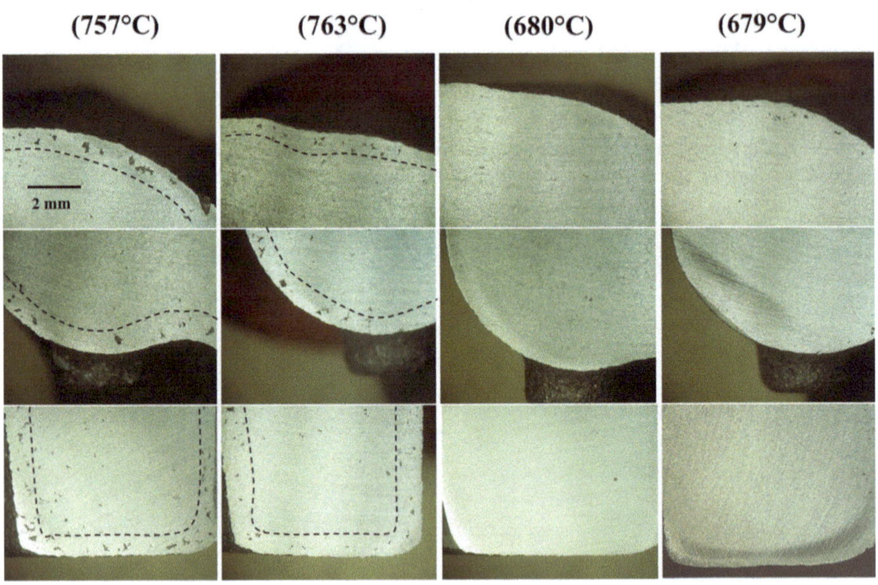

Fig. 3.20 Effect of liquid aluminum temperature of the evolution of core gas, which results in a subsurface rim of porosity

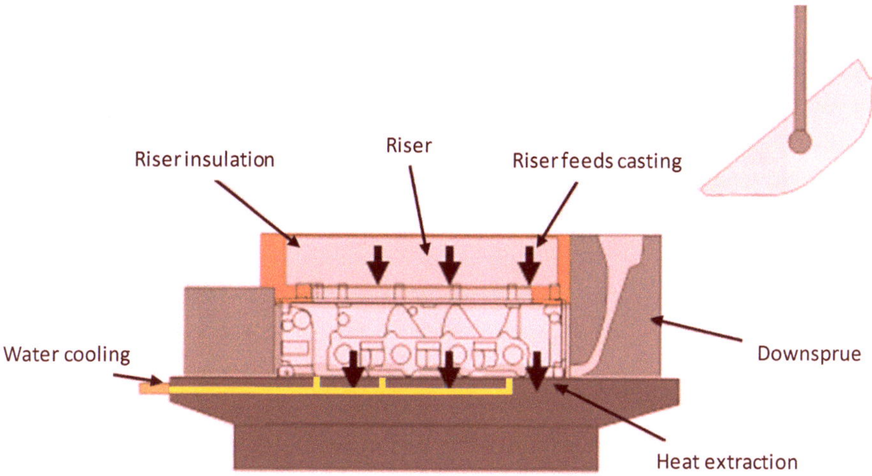

Fig. 3.21 Semi-permanent mold process

casting. An illustration of the SPM process used for a cylinder head application is shown in Fig. 3.21.

The most common alloys used for SPM castings are the 319 and the 356 alloys. SPM processing is typically a batch process (holding furnace) with gantry ladle delivery of liquid metal to the downspure and gating system. Solidification occurs fastest from the bottom of the casting, which is adjacent to the water cooled die mold, and then progressively solidifies up to the ingates and riser system. The water cooling typically needs to control the drag temperature within a target range between 230 and 370 °C. This ensures that no casting defects arise from unfavorable solidification conditions. To promote the directional solidification sequence the risers, formed by the sand core can be insulated. One example of insulation is Fiberfrax®.

Once solidification has been completed the sand removal can be either mechanical or thermal baked off in a thermal sand removal furnace. Following this, riser and gating removal is conducted. In some cases the semi-permanent mold make castings are heat treated with a solution treatment followed by an over-aged heat treat stage (T7). The desire to heat treat those castings is to improve the mechanical strength of a particular portion of the casting. An example for cylinder heads would be the combustion chamber face, which typically facing the bottom of the die mold.

A disadvantage that occurs over time is that inside the water lines of the drag will become coated with calcium, which comes out of solution when the water is rapidly turned to steam during the initial portion of the cooling cycle. To counteract this effect, each mold must have its water lines acid-cleaned on a preset frequency.

The defects usually seen in semi-permanent mold castings are similar to some seen in precision sand casting. Examples of defects include sand core chips, shrinkage defects, and oxides, among others. The shrinkage defects are caused by inadequate control of the mold temperature. If the die drag portion is on the low end of

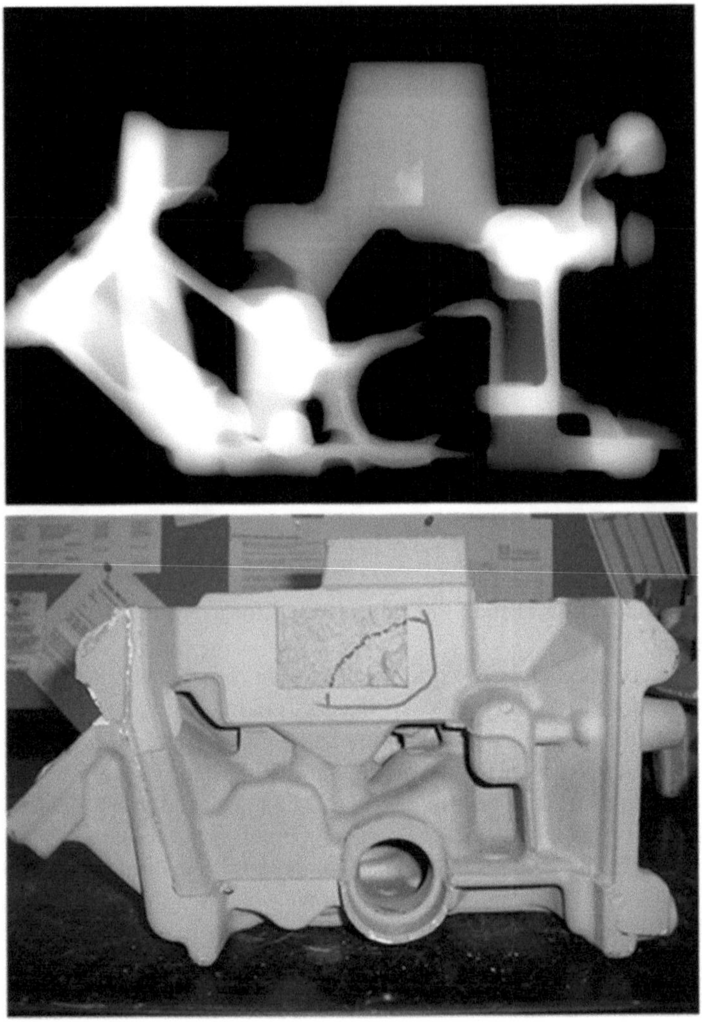

Fig. 3.22 Ceramic insert may have helped deter the formation of shrinkage

the temperature range, the steep temperature gradients set up can cause interdendritic feeding impairments. This intensifies the magnitude of a hot spot, leading to shrinkage. The temperature gradients have to be carefully controlled between the riser and the lower portion of the casting that faces the drag die.

Figures 3.22 and 3.23 show how the integration of an insulating insert in the die drag mitigates the effect of low die temperature toward the formation of a shrinkage defect. The low temperature condition may occur as part of a startup or restart of a mold used. The insulating insert reduces the number of castings that may need to be rejected at startup and improve the jobs-per-hour (JPH) performance.

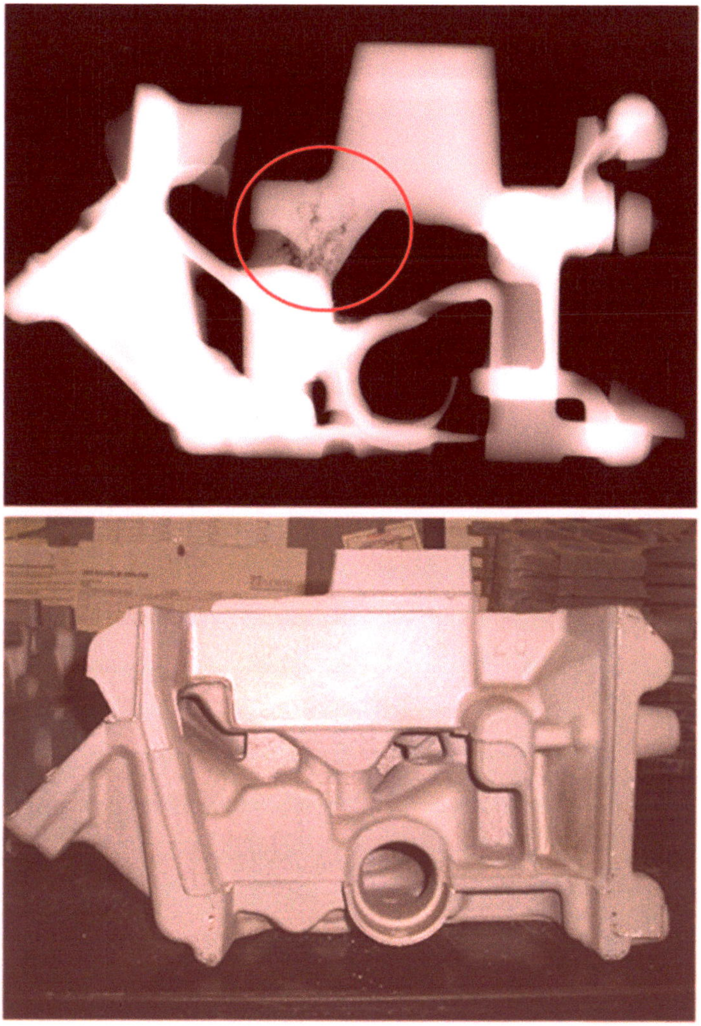

Fig. 3.23 The lack of a ceramic insert may have led to the formation of a S8 rated (as referenced from the ASTM E155 X-ray radiographic standard) shrinkage in the bolt boss/web area

An additional discussion should be made concerning the hot spot just mentioned as this is a more significant issue in the semi-permanent mold. This is found in most aluminum metal casting processes. Figure 3.24 shows in more detail the directionality of the solidification profile along the cam tower section of the same casting in which the shrinkage defect was resolved with the insulating insert. Secondary dendrite arm spacing (λ_2) profile is mapped out on a polished (2-D sides at 90° orientation) cross-section. Then, from the λ_2 array, an approximate solidification time could be established using the well-referenced mathematical relationship $\lambda_2 = 10T_f^{\ 0.33}$ [1]. What can be seen is that a natural hot spot forms at the base

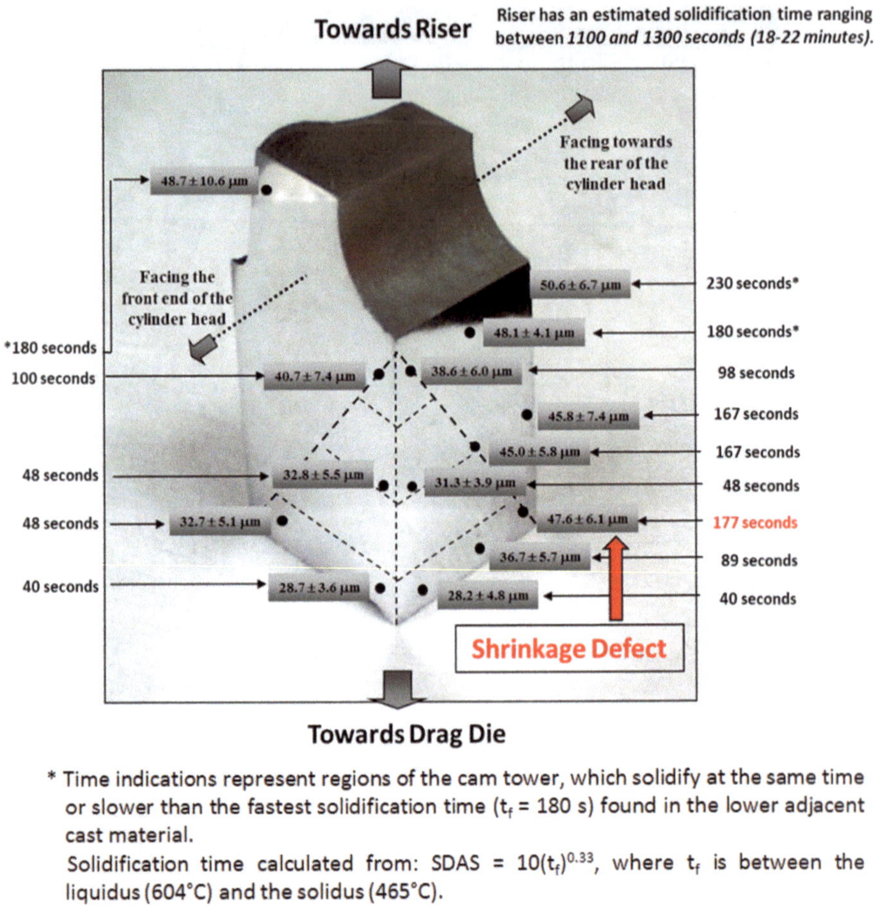

Fig. 3.24 Effect of solidification gradients as measured by λ_2 and the possible cause of shrinkage formation

of the cam tower (where 177 s solidification time was determined and is indicated in red) and above that location the solidification rate drops down to 166 s. The shrinkage indication is essentially formed below the cross-section and is again seen in Fig. 3.24.

3.8 Permanent Mold Process

The permanent mold process utilizes a two-mold system that closes before mold filling in its simplest form. Mold temperature is generally controlled by water cooling channels. The most common liquid metal delivery is gravity; however, low-pressure versions do exist [4]. The die mold is generally made using cast iron or tool steels due to their thermal fatigue resistance. Water cooling maintains target temperatures between 150 and 200 °C. A preheat may be required at the beginning and several castings must be performed before the first salable casting is made. This is generally done by either making “start-up castings”, which are made to generate the target temperature and isothermal conditions of the mold. The start-up castings are usually sent to remelt. A Belch Torch can also be used to heat the mold temperature, instead of making start-up castings, as long as the target temperature range is large enough.

This process may not help in accomplishing the isothermal conditions in the mold. A mold wash is generally used to deter castings from sticking to the mold and needs to be re-applied after making several castings. In some cases ejector pins may also be used to aid in casting removal after the mold has opened.

Venting is important as the air contained inside the mold after die closing has to be evacuated to deter porosity formation. Sometimes the small spacing between the die mating surfaces can be sufficient, but if this is not the case, then vents (small holes drilled in the die) must be employed.

The main advantage of this process is the surface finish that is possible. For this reason, a significant portion of the aluminum cast wheels made for the automotive industry are manufactured using the permanent-mold-type processes. The surface finish allows for anodizing and painting of the wheel surface, which is critical toward improving to look of cars in the showrooms.

As indicated, the gravity-based version for permanent mold process is the most common; however, depending on how the riser and gating are designed, a certain amount of liquid metal turbulence can be expected. This turbulence will lead to bi-film generation and pore development. The low-pressure variant of this process, commonly referred to as the low-pressure permanent mold (LPPM) process, provides a form of liquid metal delivery. In this process, the melt flows in a much more quiescent fashion and thereby produces more sound castings. Cycle times are lowered when compared to the permanent mold process. In addition, the feed stock tube, which directs the aluminum into the mold, has to be made from ceramic materials to deter Fe pickup. This inconvenience has been seen previously with cast-iron-based stock tubes. Ceramic stock tubes are more costly and can be subject to cracks which necessitate more frequency replacements.

Figure 3.25 shows examples of the permanent mold and low-pressure permanent mold casting processes respectively for aluminum wheels. Venting and water cooling channels are not illustrated in the figures but are important components of the process and cycle time controls. Riser strategy is generally different for the two forms of permanent castings, where in low-pressure permanent mold the risers are blind and the fill cycle is controlled by a computer-aided fill program. In permanent mold the risers are open and are used in some cases to signal visually to the fill operator that the mold is full.

The LPPM process provides a better casting in terms of X-ray radiography and mechanical properties. However, a balance must be made with the added capital investment required for LPPM and the unit price a customer will pay for the casting. The cost advantage offered by LPPM is the higher casting yield than for the permanent mold process.

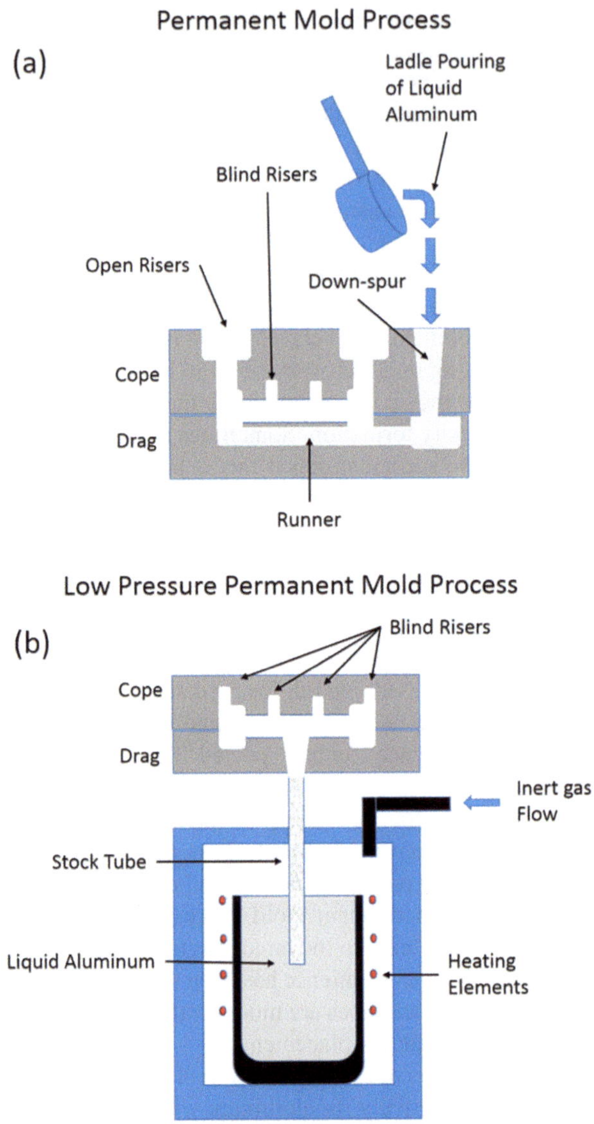

Fig. 3.25 (**a**) Gravity permanent mold process. (**b**) Low-pressure permanent mold process

3.9 Investment Casting Process

The investment casting process, also known as the lost wax process, has been around for 5000 years [5]. The investment casting process is used for a great variety of alloys, including Al and Ti [6]. Beeswax has been used in this process to form the intended casting. Today, more sophisticated reclaimable waxes are used [5, 7].

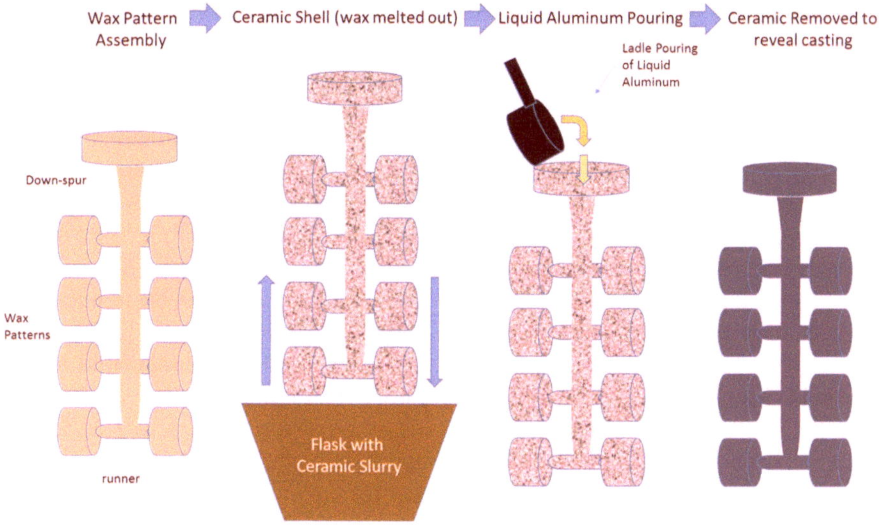

Fig. 3.26 Investment casting process

In simplest terms, a mold is used where wax is poured, allowed to thicken, and then is removed. The wax replica must also include the required gating and risering. The wax is then removed and undergoes a coating process with a wet ceramic slurry. The coating process is done by a dipping process. Removal of the wax replica with its initial ceramic coat and excess wet ceramic is allowed to drip off resulting in a uniform coating. Re-coats may be necessary in order to provide the target thickness of the ceramic mold. The types of ceramic used include silica, zircon, aluminum silicates, and alumina. The binder used for these ceramics can be ethyl silicate, colloidal silica, sodium silicate, and a hybrid of these controlled for pH and viscosity.

In the next step, the coated wax unit is stuccoed with coarse ceramic particles, by dipping it into a fluidized bed. Then the ceramic coating is allowed to harden, which can take between 16 and 48 h. Finally, the wax is melted out (dewaxing) leaving the ceramic mold (called the investment) empty, but with the shape of the cast piece. In order to minimize crack damage to the ceramic (due to differences in thermal expansion of the wax and ceramic mold), the heat up is conducted as fast as possible, driving surface melting of the wax replica to happen first. This allows the inner solid wax to expand without applying stress on the ceramic mold.

In order to drive out any moisture that may be left in the investment, and to remove residual wax, it undergoes a burnout and preheat stage (~1000 °C). After cool down, any investments that are discovered to have any minor cracks can be repaired. Figure 3.26 shows an illustration of the investment casting process.

Now that the investment is ready for casting, another preheat is applied in order to provide adequate fluidity into thin walled casting sections. Once the metal has been poured, the casting is allowed to solidify and then undergoes the divesting process. This is where the ceramic mold is removed using a controlled hammering process, media blasted or vibration stage.

In the aerospace industry, it is now common to apply a rapid cool-down stage by using liquid nitrogen at short intervals after the casting has been poured. This innovative process was first by Cercast (now under Howmet) in Montreal and had the trademark name the Sofia Process®. Not only relatively strong casting structure is achieved, but the ceramic mold removal (thermal shock) occurs. This process has driven thinner and thinner wall casting structures to be achieved, producing a microstructure that has vertically no porosity and fine grains. At the same time, with this process the highest mechanical properties seen in the industry can be attained. The typical alloy used in this process is D357. The D357 alloy contains higher target levels of Mg, which promotes a shorter freezing range and lower interdendritic feed demand. The rapid solidification also promotes a finer grain size.

The added benefit of the more recent investment casting process such as Sofia is that inroads are being made to replace components previously made from machining from stock aluminum or welding of wrought sections together to make a substructure door. The superior characteristics of this process allow for weight and unit cost savings. Both bode well for the cost-intensive aircraft industry. Investment casting process, due to its labor costs, and long cycle times, has not been a process that is used extensively in the automotive manufacturing of Al-Si castings.

3.10 Casting Process Selection

A casting selection process is critical to establish early or before a component program begins. Many cost factors such as tooling, required consumables, alloy costs, labor costs, thermal processing, and post-production operations are considered. Some post-production operations include weld repair, hot isostatic processing, and impregnation, among others. All of them and other factors contribute to the overall unit cost of an aluminum casting. In addition to this, a predicted scrap rate also becomes incorporated in the overall unit cost. Getting predictive scrap rates incorrect (e.g. higher than expected) can have profound impacts on a loss per unit casting.

Many factors can go into a casting process selection when there are options; however, there are many applications where there is only one valid process to work with. Around 80% of all aluminum castings today are made using the high pressure die casting process for the reasons outlined earlier. It is worth mentioning that the dominating factor of small size and volume components will steer clearly toward high pressure die casting process.

Precision sand, lost foam, and investment casting processes are clearly low-volume applications and are capable for complex internal geometries. Cost ranges for these types of castings can run between hundreds of dollars and up to 20 K dollars. Table 3.1 is from a publication by Mackay et al. [8], which compares basic comparative analysis of three castings processes used for the casting of aluminum engine blocks.

Table 3.1 Summary of casting processes for aluminum V6 engine blocks

Process element	Die casting	PSCP	PSCP – crank chill	Rationale
Microstructure	10–15 πm	Head deck: 40–50 πm Bulkhead: 55–65 πm	Head deck: 60–70 πm Bulkhead: 20–30 πm	Inherent process
Soundness	Low	Medium	High	Property increase
Typical scrap	10–15% In-house 3–5% Customer	3–5% In-house 0.10–0.30% Customer	3–5% In-house 0.10–0.30% Customer	Inherent process
Casting yield	70–75%	73%	55%	Design necessity
Heat treatment	T5	T5	T5 and T7	Property increase and dimensional stability
Alloy	AlSi10Cu2.5(1.0 Fe max)	AlSi7Cu3.5 (0.4 Fe max)	AlSi8Cu3 (0.6 Fe max)	Customer requirement
Impregnation	10–100% of product	0% of product	0% of product	Customer requirement
Durability	Low	Medium	High	Customer requirement
Metal temp	680–700 °C	730–760 °C	730 °C	Inherent process
Modifier/grain refiner	None	None	Sr/TiB	Improved castability

3.11 Heat Treatment Process for Aluminum Castings

The thermal processes of a casting is a critical stage that provides the required mechanical properties and dimensional stability in service conditions. Most castings will be designed based on a specific parameter: the yield point. Yielding reflects the point in which deformation or potential fracture will occur in a casting component. Dimensional stability is also important as all aluminum castings will be part of a larger assembly with tight tolerances and sealing particularly for the automotive and aerospace industries (e.g. engines). Excessive growth of a casting after it is in the field can result in poor functionality or potentially a warranty replacement.

There is the added factor that machining parameters used are set based on the castings' yield strength and hardness. One of the issues in machining is the formation of burrs which must break off in small portions so that they can be removed with ease. Longer burrs that form may remain on the casting after machining and run the risk of fracturing off later on and becoming logged inside of an internal cavity. The burrs can become dislodged due to circulating oil or coolant and impairing the field functionality of an engine, pump system, etc. Thus it does happen that while target mechanical properties are achieved, further adjustments may be made to support machining characteristics of the casting.

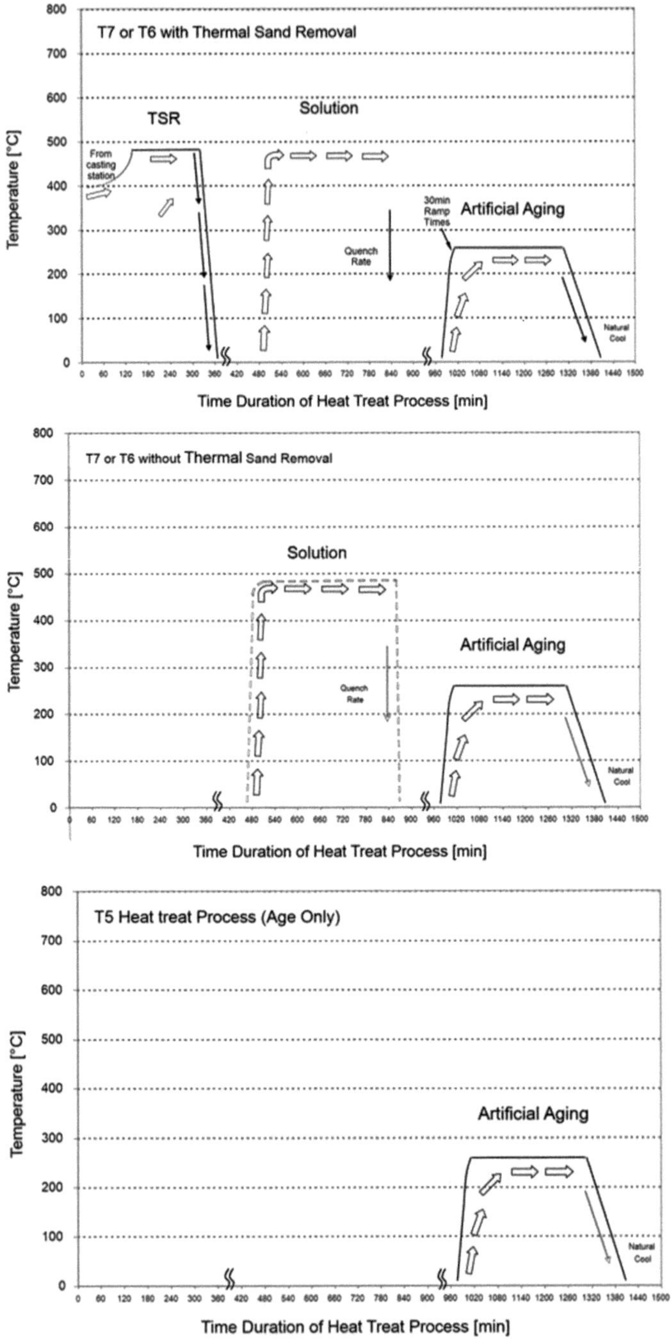

Fig. 3.27 Examples of thermal processing – T6/T7 and T5 only

Two basic type of heat treatment processes are used, one that has a high temperature stage (solution treatment) followed by a quenching, and then a lower temperature stage (aging). The Aluminum Association (AA) has provided temper designations to simply and concisely designate the heat treatments to manufacture aluminum castings. The three most commonly used heat treatment processes are illustrated in Fig. 3.27.

The solution stage is the main elevated temperature portion of the T6 and T7 processes, and it drives the dissolution of Al_2Cu and Mg_2Si phases. In this process, the intention is to have atomically free Cu and Mg by diffusion into the lattice and vacancies in the α-Al FFC matrix. After a set time at the solution temperature, a quench operation is implemented to lock both Cu and Mg in super-saturated solution. The aging process initiates the precipitation process of the α-Al matrix, resulting in pinning of dislocations which elevates the yield point.

In the HPDC process the subsurface porosity can result in blistering if a solution treatment is used, thus only an aging process is part of the thermal processing. The rapid solidification retains a significant amount of the Cu and Mg in α-Al matrix in solution. Therefore, a proper design and meaningful aging treatment can bring an improvement to the casting strength. Cost considerations associated with the heat treatment of Al-Si castings are important, as a typical per casting cost can range from 4 to 8 USD/pcs. Here, it is important to mention that modern technologies allow the manufacturing of high-performance casting with low iron. Those castings can be subjected to T6 or T7 heat treatments allowing further mechanical improvements, including high cycle fatigue performance.

References

1. Mackay, R.I. 2003. *Development of a new durable Al-Si alloy for the next generation of engine block casting*. Ph.D. Thesis, University of Windsor.
2. Flemings, M. 1964. Fluidity of metals-techniques for producing ultra-thin section castings. *The British Foundryman* 57: 312–325.
3. Williams, R.S. 1943. The pirotechnia of Vannoccio Biringuccio (1540) (Translated from the Italian by Cyril Stanley Smith and Margaret Teach Gnudi). *Journal of Chemical Education* 20 (3): 156.
4. Kréziak, G., C. Rigaut, and M. Santarini. 1993. Low pressure permanent mould process simulation of a thin wall aluminium casting. *Materials Science and Engineering A* 173 (1): 255–259.
5. Pattnaik, S., D.B. Karunakar, and P.K. Jha. 2012. Developments in investment casting process—A review. *Journal of Materials Processing Technology* 212 (11): 2332–2348.
6. Lee, T.J., et al. 2016. Titanium-zirconium binary alloy as dental implant material: Analysis of the influence of compositional change on mechanical properties and in vitro biologic response. *International Journal of Oral and Maxillofacial Implants* 31 (3): 547–554.
7. Taşcıoğlu, S., and N. Akar. 2003. A novel alternative to the additives in investment casting pattern wax compositions. *Materials and Design* 24 (8): 693–698.
8. MacKay, R., and D. Szablewski. 2010. The use of Weibull statistical method to assess the reliability of a development engineered automotive casting component. *International Journal of Metalcasting* 4 (4): 31–45.

Chapter 4
Powder Metallurgy

4.1 Introduction to Mechanical Milling

This chapter deals with the production of Al-Si alloys by powder metallurgy (PM) as well as the physical and mechanical properties developed by this processing method as compared to others. The main focus is Al-Si hypereutectic alloys, using the 393 alloy as a benchmark. Powder metallurgy (PM) is defined as the manufacturing process to consolidate fine powders, usually powders below 40 μm [1], to produce monolithic or solid components. This technique is attractive due to the superior properties that emerge from the refined and uniform microstructure that evolves during processing [1, 2]. Additional attributes to the method include its diversity in terms of manufacturability and the potential to obtain high-quality, complex, neat net-shaped parts that may require tight tolerances [3]. The technique is usually cost-effective when compared to the benefits that can be obtained. Mechanical milling and alloying have been around for decades; Benjamin et al. pioneered in this field in the decade of 1960s [4] to produce high-temperature super alloys followed by ceramics [5].

Powder metallurgy is an ancient method that is a predecessor of the widely used ingot metallurgy or general castings. The motivation is due to human kind's limitations to achieve high temperatures capable of melting materials as well as to build crucibles to hold the liquid melts [2]. It is known that in 3000 BC the Egyptians fabricated iron tools and weapons by hammering iron-rich particles that were heated to temperatures below their melting point [2]. We can use this example as the pioneering effort in pressing and sintering. Other outstanding examples include the Iron Pillar of Delhi, which was made in India about 1600 years ago.[1] The Iron Pillar of Delhi is approximately 6.5 tons of reduced iron powder (Fig. 4.1). Among other characteristics, this pillar has a high resistance to corrosion due to its composition

[1] https://en.wikipedia.org/wiki/Iron_pillar_of_Delhi.

© Springer International Publishing AG 2017

F.C. Robles Hernandez et al., *Al-Si Alloys*, DOI 10.1007/978-3-319-58380-8_4

Fig. 4.1 The Iron Pillar at Qutb Minar, New Delhi, India (Image courtesy: Sunil Malhotra)

[3, 5, 6]. Metal-pressing technology was also used by the Inca civilization in the manufacturing of gold, silver, copper, bronze, and tumbaga [7]. Tumbaga was identified in the pre-Colombian times in the Americas for decorative art. More recently, this alloy has been explored for its catalytic characteristics [8].

In the 1800s with the development of more advanced furnaces, powder metallurgy has been pushed forward particularly when platinum and tungsten parts were manufactured successfully for the first time [2, 3]. During the 1920s and 1930s it was possible to mass-produce tungsten carbides and porous bronze bushes for bearings and topper-graphite electrical contacts [3, 5]. In the 1940s, and especially during the Second World War, it was possible to develop new tungsten alloys, refractory metals, and a great variety of ferrous and nonferrous products. Here is when composites initiated a steady growth period that stayed constant through the postwar and until the early 1960s. Since then, growth of PM has expanded more rapidly, mainly because of three potential reasons: economical processing, unique properties, and high-volume production method [5]. Today the applications of PM are extensive, including structural, electrical, energy, automotive, aerospace, and defense, among other [3].

Aluminum powders began to be used in the production of wrought components in the 1940s and 1950s [9]. Aluminum products were reinforced with Al_2O_3 powders dispersed to improve the high-temperature strength. During the same time, the atomization of aluminum powders with high solute contents was investigated. These alloys utilized the benefits derived from rapid solidification that included alloying flexibility, increased solid solubility limits, fine microstructures, and homogeneity. From this research two types of alloy systems emerged: (i) dispersion-strengthened alloys and (ii) heat-treatable alloys (e.g., 2xxx and 7xxx series). The former family of alloys is known for its finely dispersed content of insoluble intermetallics. The latter family has fine grains and a refined microstructure. This resulted in remarkable improvements in strength, fracture toughness, and corrosion resistance.

In 1979, Voss [10] compared the advantages in microstructure and mechanical properties for commercial aluminum alloys (e.g., 2024 and 7075) produced by conventional methods and powder metallurgy. Among other advantages, it was demonstrated that powder metallurgy sponsors fatigue resistance benefits. The findings clearly show that reducing the grain size by an order of magnitude was the key for unprecedented improvements. Additionally, the manufactured alloys developed resistance to recrystallization having a clear increased strength isotropy and corrosion resistance. These improvements have generated controversy because some authors have reported similar strength results to those identified in samples produced by conventional casting methods [2]. In addition, the improvement in strength is attained at the expense of ductility.

4.2 Al-Si Alloys

Vijeesh and Narayan Prabhu [11] published a review concerning the Al-Si hypereutectic alloys. Aluminum alloys are the most abundant after steel and the Al-Si family is the most common comprising approximately 90% of the total aluminum castings [12–14]. This dominance is attributed to its well-accepted use by the automotive and aerospace industries. Some of the outstanding advantages of these alloys include their excellent castability, crack resistance, abundance, cost-effectiveness, and good technological impact. The properties of Al-Si alloys are strongly dependent on the casting process, composition, and melt treatment.

In Al-Si alloys, the content of silicon varies from 4.5 to 25 wt% [15]. According to the binary phase diagram shown in Fig. 4.2, the alloys can be further classified as hypoeutectic (<12 wt% Si), eutectic (12–13 wt% Si), and hypereutectic (14–25 wt% Si) depending on the silicon concentration [11]. Here, it is important to mention that ideally any composition above 14 wt% Si is hypereutectic. However, alloys containing more than 25 wt% Si are unusual or they are master alloys. Al-Si hypereutectic alloys are attractive, among other things, due to their low density, stiffness, relatively high-temperature resistance, wear resistance, and low coefficient of thermal expansion. These characteristics make them ideal for the transportation industry, particularly automotive, due to their ability to replace cast iron in engines and some of their parts. Typical applications of Al-Si hypereutectic alloys include high-performance connecting rods, rocker arms, cylinder, pistons, and valve retainers [11].

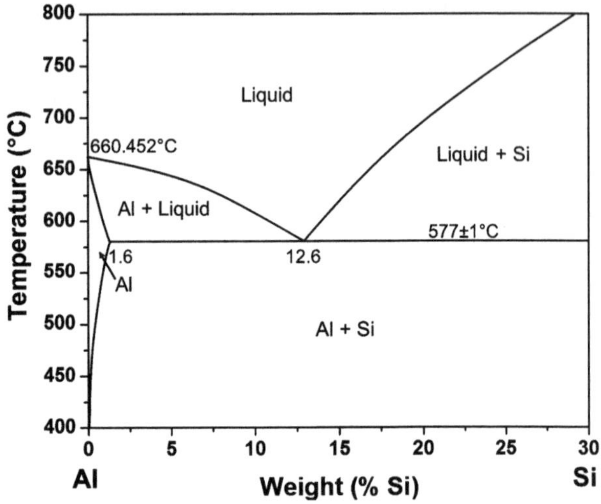

Fig. 4.2 Al-Si phase diagram

The boom of Al-Si hypereutectic alloys was initiated in the 1920s when these alloys were used for pistons and other automotive engine components [11]. Unfortunately, at that time the parts were brittle and hard to work. In 1933, phosphorous was added to refine primary silicon [16] that helped in the improvement of mechanical properties, sliding, and wear resistance. The result was successful enough to use the alloy for a wide range of applications such as pistons, bearings, engine cylinders, mold tools, and ram guides for briquette presses and cylinder liners [8]. In the following years, the A390 aluminum alloy was used as a cylinder block material by some automotive companies. The A390 alloy presented some casting challenges due to its lack of homogeneity, largely the primary silicon particles. Since then, extensive research work has been done in this field to develop it as an engine material. Nowadays, almost all automobile manufacturers use Al-Si alloys under different trade names [11, 17].

The reason for selecting Al-Si hypereutectic alloys as a cylinder block material was the presence of hard primary silicon particles in the microstructure of the alloy [11]. The alloy is chemically etched to expose individual primary silicon particles above the eutectic matrix and thereby enhance the wear resistance of the alloy. Studies reveal that the wear properties of the alloy mainly depend on the morphology and distribution of primary silicon in the matrix. GM pioneered the implementation of Al-Si alloys for engine blocks in commercial cars. There is controversy in the success of this effort and the Vega 2300 engine block [18–23], but no one can deny that nowadays most cars use Al-Si engine blocks.

4.3 Powder Production Methods

There are several methods to produce metallic powders [3, 5, 17, 24]; a classification of them is given in Fig. 4.3. The fabrication method affects the size, shape, microstructure, chemistry, surface nature (metallic or oxide), and cost [3].

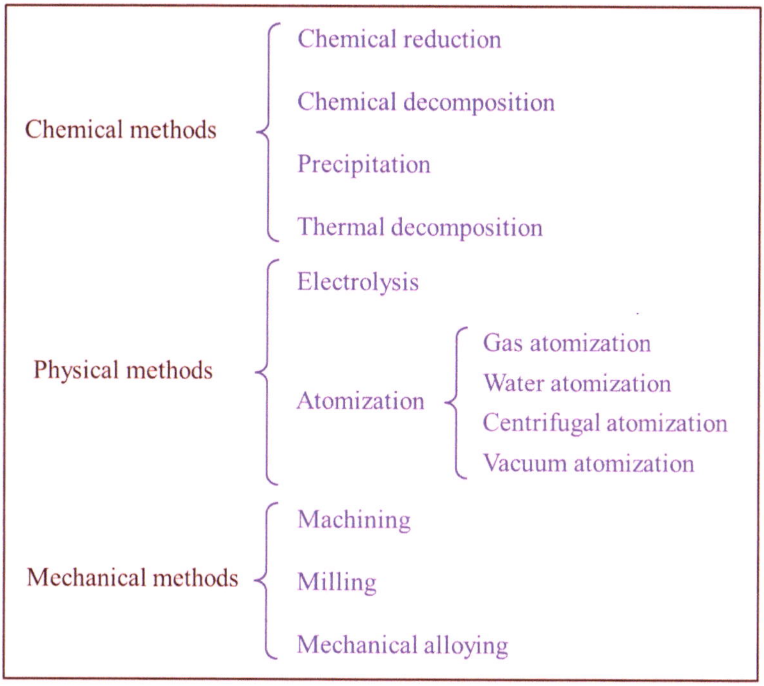

Fig. 4.3 Classification of the metallic powder production methods

The flexibility of powder metallurgy allows the use of the powders from virtually any manufacturer. Potentially what is important is the composition, oxidation (surface), and their morphology, including the particle size. One of the most common and well-accepted techniques is gas atomization. The following list indicates the characteristics of each of the alloying elements used in this exercise:

- Al (99.5% purity, −325 mesh),
- Cu (99.5% purity, −100 mesh),
- Mg (99.8% purity, −325 mesh),
- Ni (99.8% purity, −325 mesh),
- Fe (99+% purity, −200 mesh),
- Si, it was used in the form of lumps (98.4% purity, 1.2–2.5 cm).

The mixed composition used in this chapter is similar to the composition of the 393 alloy. The actual composition is included in Table 4.1. After Si, Cu is the element having the largest average particle size (approximately 150 μm). However, the average particle size for the other elements is between 74 (−200 mesh) and 44 μm (−325 mesh).

Table 4.1 Composition of the 393 alloy

Alloy	Al	Cu	Mg	Fe	Ni	Si
393	Bal.	0.9	1.0	1.3	2.3	22.0
AA [15]	Bal.	0.7–1.1	0.7–1.3	1.3	2.0–2.5	21.0–23.0

4.4 Mechanical Alloying

It is a general rule of metallurgy that the finer the particles, the better the benefits in mechanical properties such as strength. This theory was first proposed by the Hall-Petch model and has been demonstrated by numerous applications [25–27]. Usually the strengthening mechanism occurs by the presence of obstacles to dislocation motion, which in the case of a single-phase alloy these obstacles are the grain boundaries. In a more generic case, the obstacles are due to the presence of solute atoms, GP-zones (Guinier-Preston zones), precipitates, dispersed oxides, other dislocations, etc. [28–31]. As a general rule, one can say that the smaller the particles, the higher the benefits in strengthening and, in general, the mechanical properties. In addition, it is important to contemplate that there is a critical limit in the size of precipitates and dispersed particles within the matrix. Large particles can act as active sites for fracture, flaws stress concentrators, or crack initiators.

Mechanical alloying is a widely used method in powder metallurgy [4, 32–36]. In this method a "milling medium," usually steel balls, is added to a vial or container that is vigorously shaken to generate chaotic collisions among the media, vial, and powder particles. This chaotic process generates repeated impacts that cold-weld the powders and at the same time fracture the powders in a series of events between welding and fracturing. The result is an intimate mix of the raw powders with a high degree of refinement and homogeneity. The technique was initially developed to synthesize nickel-based superalloys strengthened by oxide dispersion [33, 37]. Nowadays, it is used as a solid-state powder processing method to make an extensive number of alloys and composites with unique microstructures. It is usually the case that the solubility limit of the alloying elements widens considerably, which is the example of Si in Al. This allows to produce hypereutectic alloys (e.g., 393) microstructurally comparable to the eutectic in terms of the high level of dispersion.

Figure 4.4 describes a typical processing method for powder metallurgy. There are several milling methods that differ from the other by the energy transferred to the powders during the process. The following are some of the well-known milling methods: ball mill, attritor, planetary, Simoloyer, and Spex. The horizontal ball milling is a low-energy system, followed by attritor, planetary, and Simoloyer that are mid-energy systems, and the Spex is well known for its high-energy capabilities. The order of the techniques mentioned above is not necessarily based on the energy transmitted during milling. There are several modifications to these systems to make them work under controlled atmospheres or under cryogenic conditions. The milling media is another parameter of great importance, and in order to keep the contamination levels to a minimum, one can test several methods of milling media to balance the level of impurities, time, and results [1, 4, 38, 39]. Figure 4.5 shows a commer-

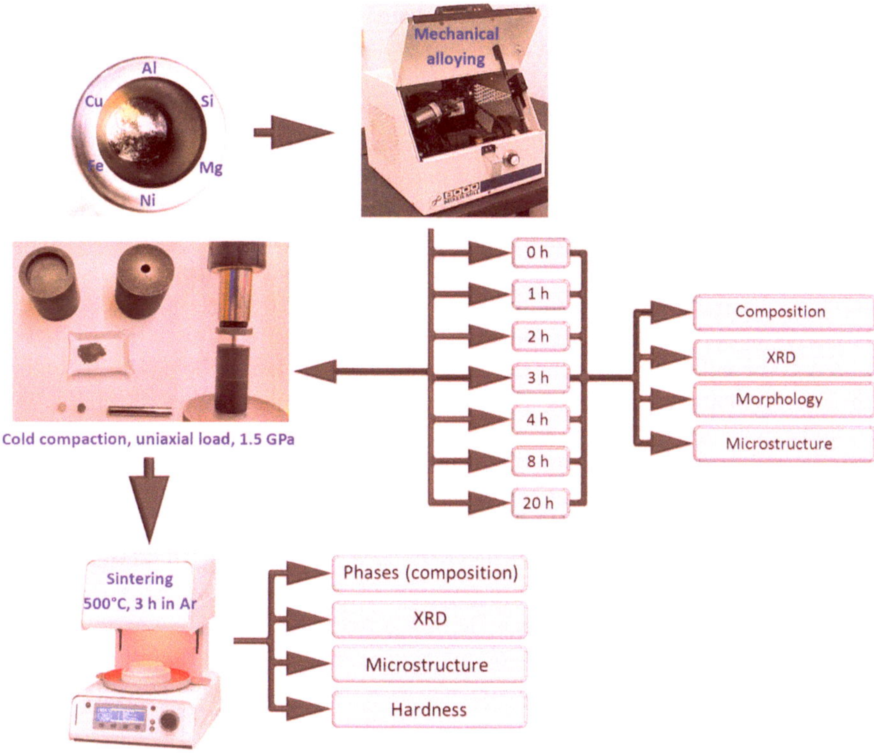

Fig. 4.4 Scheme of the general steps used to obtain and characterize Al-Si alloys

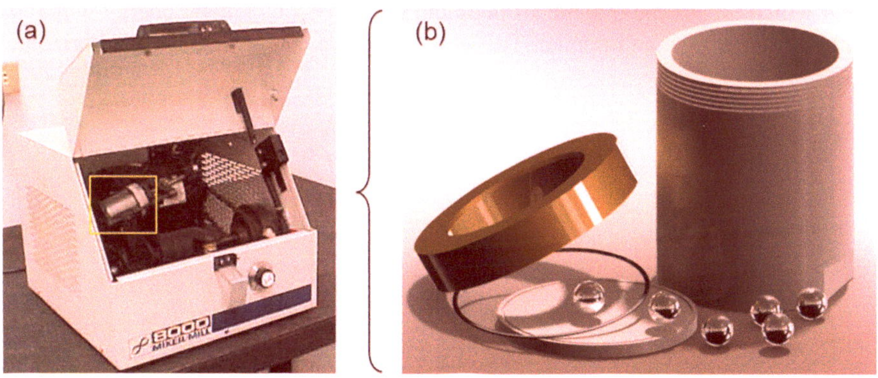

Fig. 4.5 (**a**) Spex-8000M Mixer/Mill® and (**b**) virtual model of the vial with milling balls (Image courtesy: Dr. R. Perez-Bustamante)

cial Spex-8000M mill, which is widely used for research. The typical capacity of this system is approximately 10 g. However, this changes depending on the density of the sample.

The most common milling media are hardened steel balls and vials. Another parameter that is key in the successful processing of powders is the ball-to-powder ratio that usually varies from 1:1 to 10:1 [1, 40, 41]. Ductile powders have the tendency to agglomerate and form lumps that can be prevented by adding a so-called "process control agent" [42]. The control agent is usually an organic lubricant that is added in small amounts (fraction of a percent) to prevent excessive cold-welding among the powders from keeping them loose [42–44]. Methanol is a common control agent. Other commercial additives exist and are used based on preference. When the material is subjected to milling, its surface area increases along with its affinity for oxygen and other gases identified in the atmosphere. If purity is critical, it is suggested to carry out the entire process under controlled conditions [45–48]. For instance, the milling process can be carried out using Ar atmosphere. The procedure involves a complete handling of the powders and milling media in a glove box.

Other recommendation to prevent contamination is to run a preliminary test using the vials and balls to mill a preliminary sample of the same composition or similar composition. This guarantees that all the milling medium is impregnated with the alloy. This preliminary trial can be carried out for a set period (2–10 h) to make sure that the media is well coated with the alloy. Milling time is critical and it has to be set based on the expected outcome of the final product(s). For metallic systems, the milling time can vary up to 50 h in Spex, but other low-energy systems (e.g., horizontal ball milling) can be used for several days [1, 49, 50]. Figure 4.6 shows the aspect of the powders before and after the mechanical alloying processed in a Spex-8000M mill.

Other methods are recommended for batches larger than a few grams. For instance, the Simoloyer® can be used to produce from 0.1 kg to several kilograms of alloy (Fig. 4.7). The principle of this mill is based on a horizontal, robust design rotor which allows the transfer of high and homogeneous kinetic energy when the ball impacts against an impeller. This equipment is a combination of the horizontal

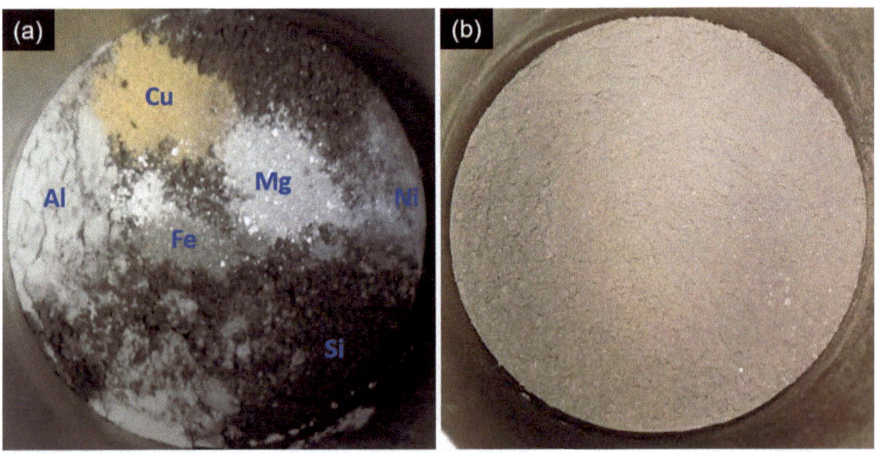

Fig. 4.6 393 alloy: (**a**) elemental, raw powders and (**b**) mechanically alloyed powders for 1 h

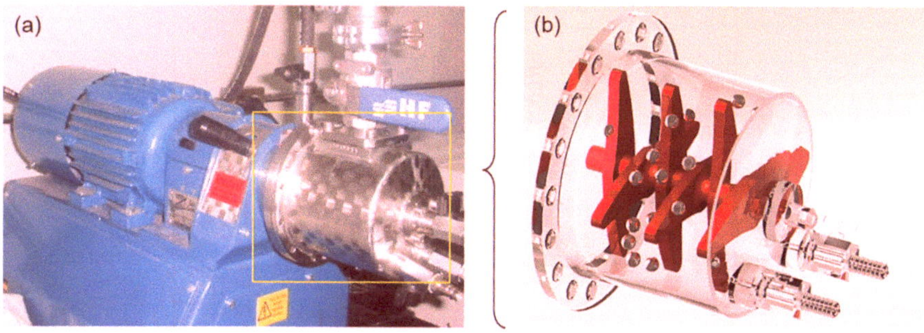

Fig. 4.7 (**a**) Simoloyer® mill and (**b**) virtual model of the vial assembly (Image courtesy: Dr. R. Perez-Bustamante)

ball mill and the attritor. In addition, the Simoloyer® does not have the dead zones typical of the horizontal ball mill and it can operate under controlled temperature and atmosphere. This device can be considered as a highly kinetic horizontal rotational ball mill. Typically, its base material is stainless steel and the balls are hardened steel. In this case, the milling ball-to-powder weight ratio can be higher (14:1).

Conventionally, the characterization of alloys involves a variety of complementary methods such as X-ray diffraction (XRD), scanning electron microscopy (SEM), energy dispersive spectroscopy (EDS), and optical microscopy. This combination is key to having a good understanding of the microstructure and mechanical properties of the alloyed products. Mechanical alloying is used to produce composites, but that does not limit or exclude the potential to produce alloys [42, 51–66].

Figure 4.8 presents the XRD results of the 393 alloy before and after milling. The XRD method is very sensitive to composition, but additions below 3 vol% are difficult to identify. For this reason, in the XRD patterns we can only see the Al and Si reflections. The Williamson-Hall method is used to identify the crystallite size in mechanically alloyed products. As a general finding we can say that the raw product has a grain size of 63.56 nm and the milled product 28.7 nm. This is typical of mechanical alloying and this level of refinement is expected in this process.

Figure 4.9 presents the morphology evolution of the as-milled powders at several milling times. Under those milling conditions, the powder comminution is evident: as the milling time is increased, a lower powder size is observed. The inset in Fig. 4.9d reveals that particles are composed of homogeneous submicron phases. The exercise presented in Fig. 4.9 is a good example of the capabilities provided by optical microscopy. It is obvious that for a better understanding of the microstructure characteristics, other techniques are recommended, e.g., SEM.

Figures 4.10 and 4.11 show the benefits in homogeneity and elemental distribution among the raw and milled samples. The spatial distribution of the elements in the powder samples was examined through SEM/EDS mappings. A mapping of the blended elemental powder mixture is shown in Fig. 4.10, where a heterogeneous distribution of the constituents is observed. Figure 4.11, on the other hand, shows a more refined and more homogeneous microstructure that is a typical benefit of mechanical milling.

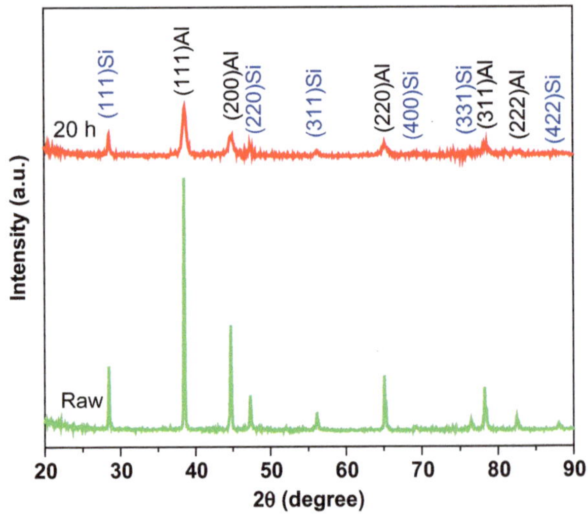

Fig. 4.8 X-ray diffraction patterns of the 393 alloy: raw and after milling for 20 h

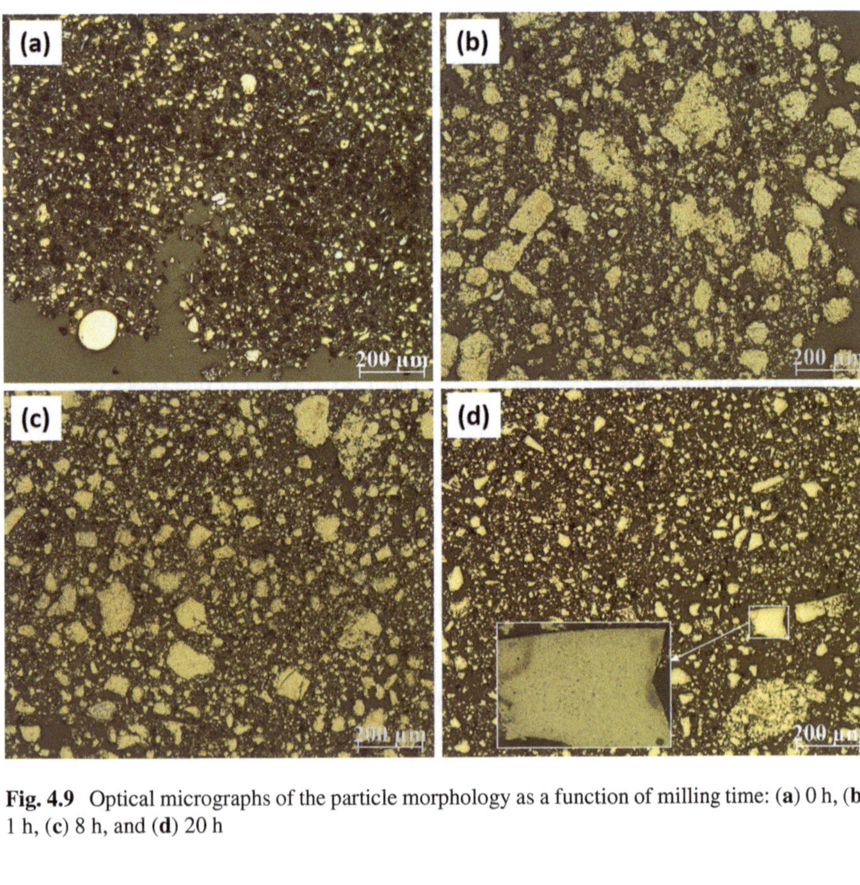

Fig. 4.9 Optical micrographs of the particle morphology as a function of milling time: (**a**) 0 h, (**b**) 1 h, (**c**) 8 h, and (**d**) 20 h

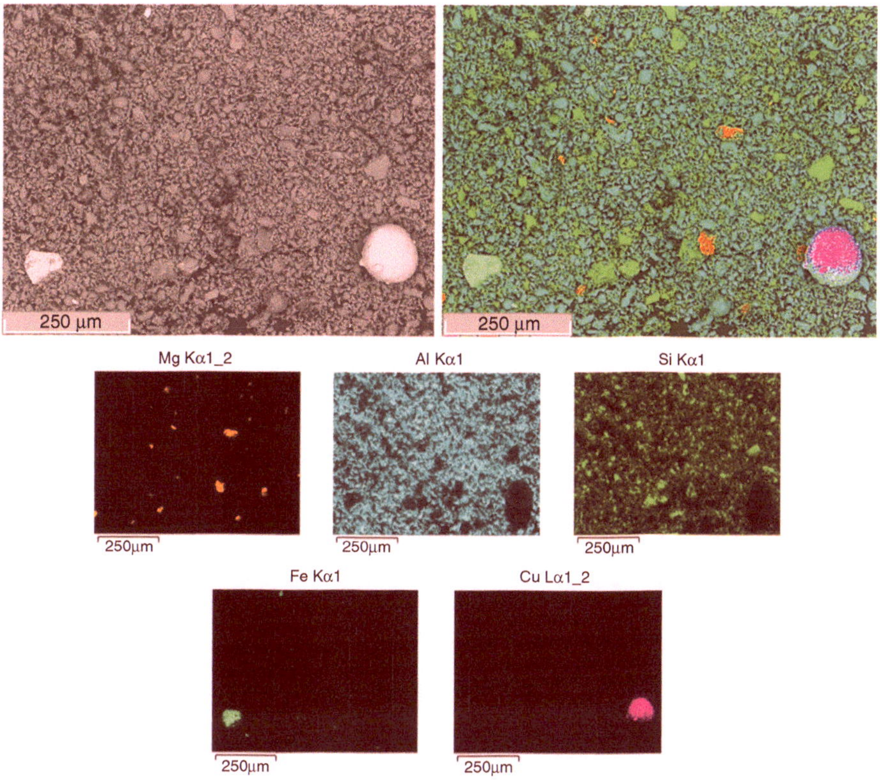

Fig. 4.10 SEM/EDS mapping analysis of the blended elemental powder mixture (0 h of milling)

4.5 Consolidation Methods

There are several consolidation methods: die compaction, conventional sintering, cold isostatic pressing (CIP), dynamic powder compaction, powder roll compaction, hot pressing, hot isostatic pressing (HIP), spark plasma sintering (SPS), powder extrusion, powder forging, powder injection molding, spray techniques, arc-melting process, and rapid induction sintering, among others [2, 3, 5, 17, 67, 68]. In the present chapter, only the cold compaction and conventional sintering methods were used to give shape to the powders synthesized by mechanical alloying. This process is outlined in Fig. 4.4. The cold compaction must be appropriately applied to ensure adequate green strength and structural integrity in the as-pressed product. Solid organic lubricants (stearic acid) are often applied on the compaction tooling or mixed with powders to facilitate handling of the sample and extend the die's life. For hard powders, it is suggested to use a binder during compaction to improve green strength [1]. Both the lubricants and binders are evaporated out of the compacts during heating while sintering, and in some cases it is suggested to pre-cook the powders to eliminate the binder.

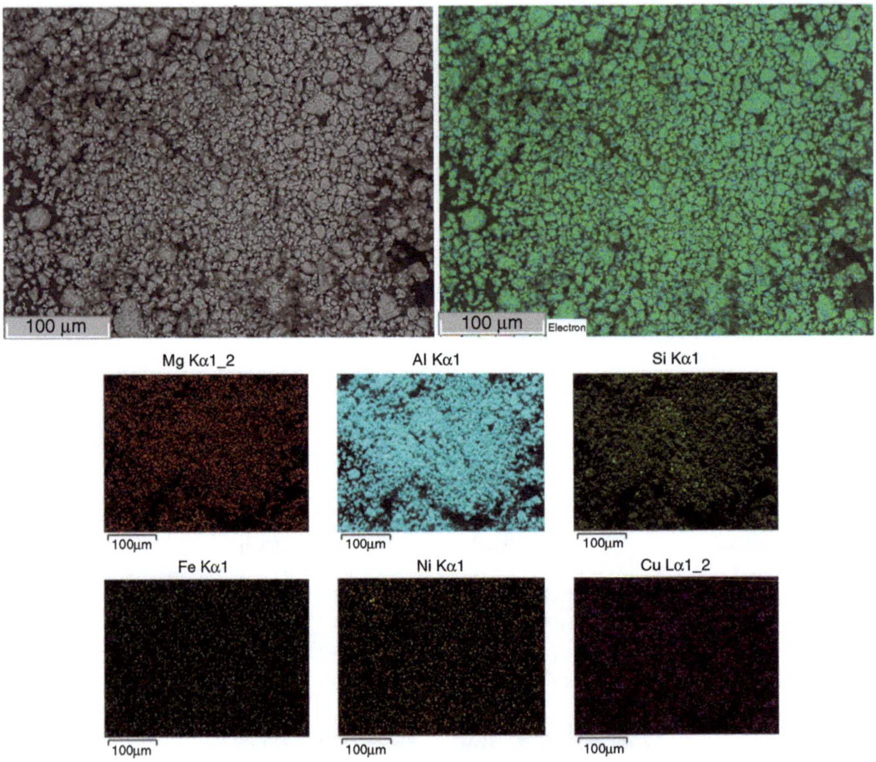

Fig. 4.11 SEM/EDS mapping analysis of the mechanically alloyed powders for 20 h

Figure 4.12 outlines the single cold compaction process. The powders are first loose with no strength and a large number of voids. As pressure is applied, the particles rearrange and the contact among them increases. In the case of ceramics, there is a need to add a binder, but ductile metals (e.g., Al-Si alloys) do not require it. The pressure increment is responsible for a better packing with the subsequent decrease in porosity, as well as improved intimacy among particles that increases the powder density. The compaction process also serves to induce elastic and in some cases plastic deformation.

The plastic deformation induces strain hardening of the compact and promotes recrystallization that is critical for sintering. Typical compaction stresses are usually above the elastic limit of the material; in the case of Al-Si hypereutectic alloys, it is between 300 and 350 MPa [69, 70]. Micrographs at the bottom of Fig. 4.12 correspond to the 393 alloy powders cold pressed at 1.5 GPa for 1 min. Figure 4.13 presents the appearance of a sample after being cold compacted. The applied pressure to form the cold compacts should be above the yield strength of the powders. However, here is where some of the Al-Si hypereutectic alloys lack in properties since they have limited elongation, particularly when the amount of Si is above the eutectic, limiting the elongation and in most cast alloys do not reach a deformation

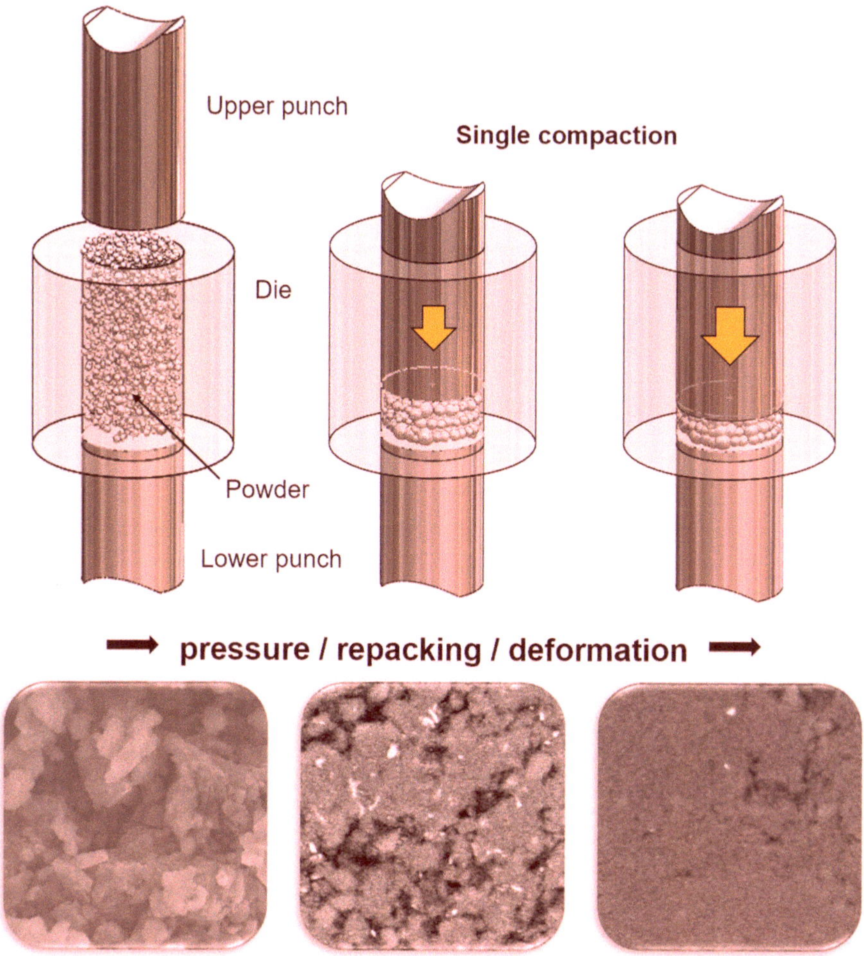

Fig. 4.12 Representation of the single compaction process

to have yield strength. This consolidation step is conducted to allow proper handling of the samples prior to the sintering process.

Figure 4.14 presents micrographs of the green compacts. Mechanical milling has as a side function the potential to sponsor better consolidation. This is important since it can easily help to determine the exact milling and consolidation conditions. For instance, all samples present good consolidation, except the sample milled for 20 h that exposes a greater number of pores. This is because the samples milled for longer times have limited plastic elongation and excessive plastic deformation. The severe deformation can be the root cause of the crack initiation sites and inhibits compaction optimization. Figures 4.15 and 4.16 present the SEM/EDS results for the consolidated samples in the raw and milled conditions, respectively.

Fig. 4.13 Mechanically alloyed powder (*left*) and green sample (*right*) of the 393 alloy

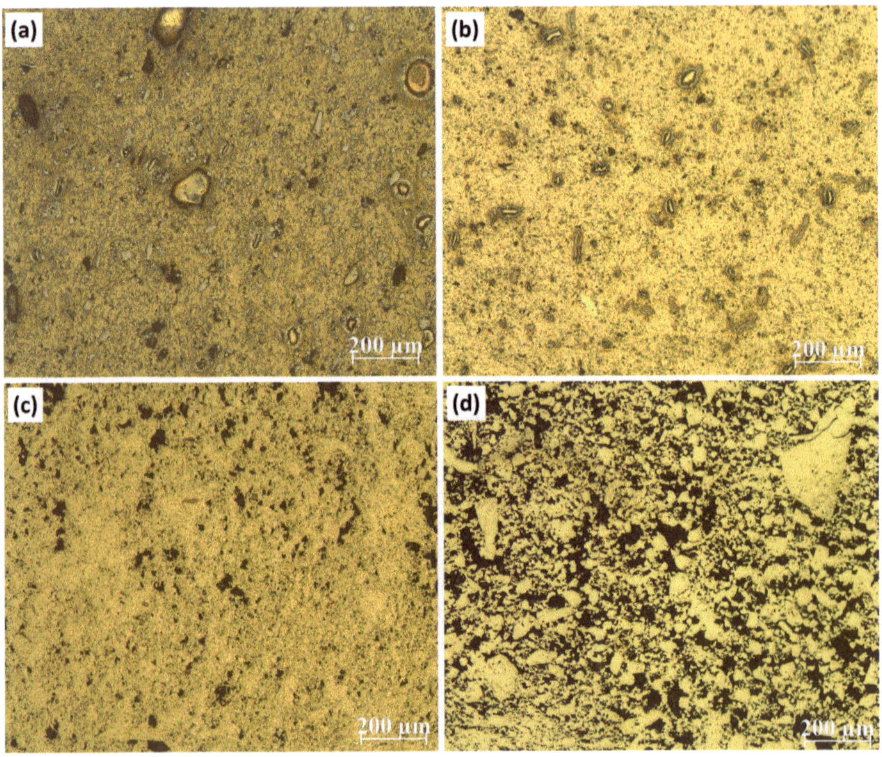

Fig. 4.14 Optical micrographs of the green compacts as a function of milling time: (**a**) 0 h, (**b**) 1 h, (**c**) 8 h, and (**d**) 20 h

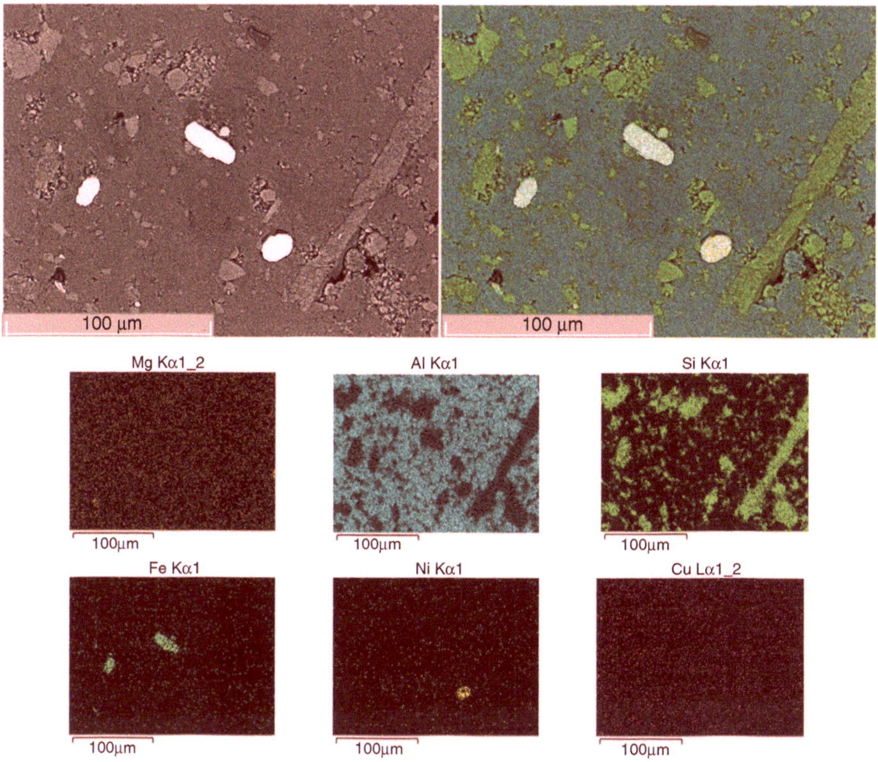

Fig. 4.15 SEM/EDS mapping analysis of the cold compacted elemental powder mixture (0 h of milling)

4.6 Sintering

Sintering is a process that can be carried out under different conditions; but for metallic systems with high affinity to oxygen the general recommendation is to use a protective atmosphere. Yet, there are some intrinsic characteristics (e.g., size) in the samples that may prevent this type of procedure. Sintering is considered the process by which an assembly of loose particles are compacted under pressure or simply confined in a container at a homologous temperature of approximately $T_{sintering} = 0.8T_M$ (where T_M is the melting temperature in absolute scale) [5]. In some cases, chemical reactions occur during sintering and these are designed to enhance bonding in order to form a coherent body. This process is accelerated with temperature. Sintering or densification is used to transform the compacted powder into a consolidated sample that occurs by diffusion and assists in reducing porosity and coarsening the powders [71]. The resultant material has improved properties (e.g., strength) and in most cases they are comparable or superior to the materials produced by casting [72, 73].

Figure 4.17 presents the sequence of the sintering process. In a general way, the first stage corresponds to a bonding among the particles (Fig. 4.17a), which starts

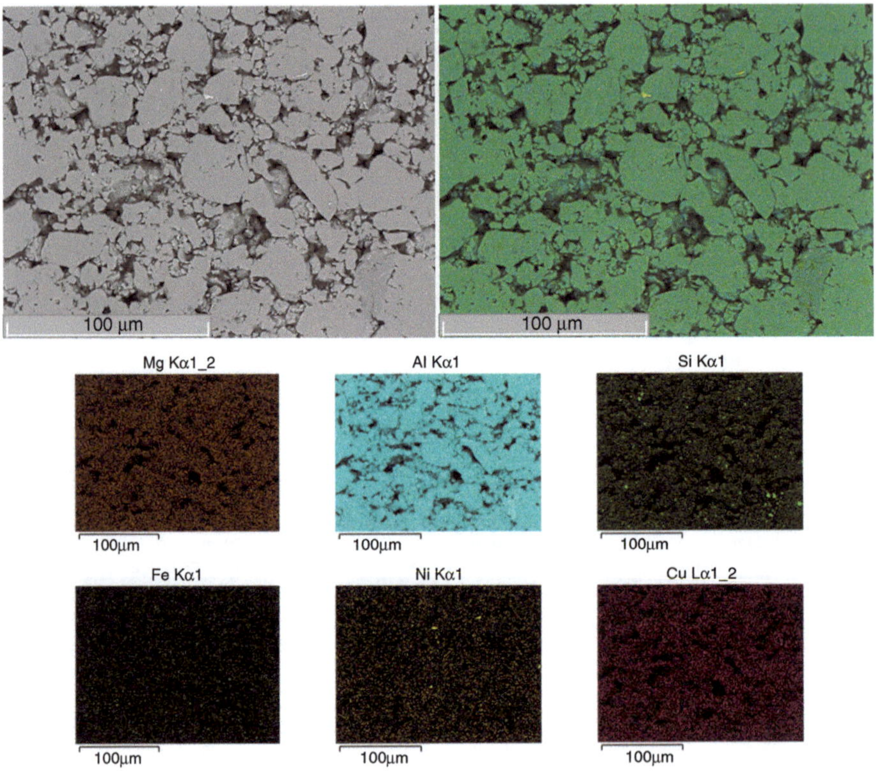

Fig. 4.16 SEM/EDS mapping analysis of the green compact MAed for 20 h

the neck formation (Fig. 4.17b). The necking coarsens with time (Fig. 4.17c) by mass transport, which is promoted by diffusion [71]. Here, it is important to state that rapid sintering systems may be diffusionless [74, 75], but in conventional sintering the driving force for sintering is diffusion. These transport phenomena take care of most of the porosity, but there are always remaining pores among the sintered particles. The particle coarsening promotes coalescence that is observed in Fig. 4.17d. A typical microstructure of a sintered product is given in Fig. 4.17e [76]. There are several aspects that lead to the final mechanical properties in the sintered products and they include temperatures and times. The temperature range for sintering is rather narrow for most alloys; however, lower temperatures are recommended as they tend to produce better sintered product, at the expense of longer times.

The graph presented in Fig. 4.18 sketches the sintering procedure followed during the sintering of the samples presented in Fig. 4.19. The sintering process was carried out at 500 °C for 3 h at a heating rate of 5 °C/min for the 393 alloy. To avoid oxidation of the consolidated samples, the process was performed under argon inert atmosphere. After sintering, samples were cooled down to room temperature inside the furnace.

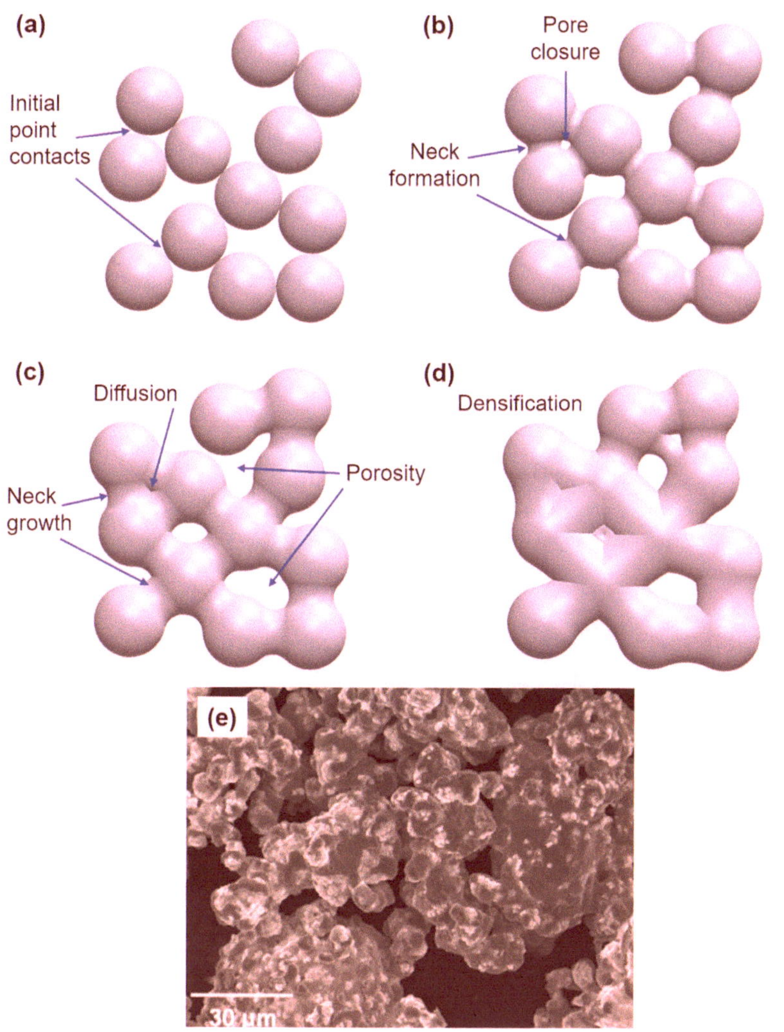

Fig. 4.17 Stages of the sintering process

In order to have a good comparison of the advantages of sintering over casting, one may need to compare the mechanical properties of the same alloy manufactured under both methods. Figure 4.20 shows the XRD results from the raw-sintered, milled-sintered, and the cast samples as a form of comparison of the samples manufactured under different conditions. This practice is common and has been reported previously [77]. The XRD results indicate that the 393 alloy has a higher crystal quality when cast, followed by the sintered product and finally the raw sample [78]. The crystal quality does not necessarily mean it is a better product since the grain

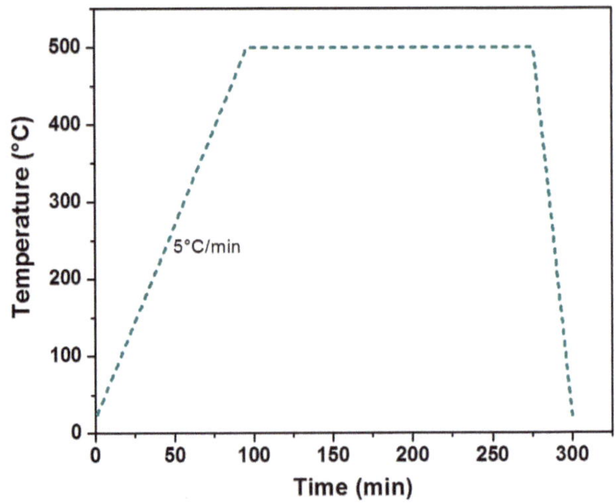

Fig. 4.18 Sintering cycle applied to 393 alloy samples

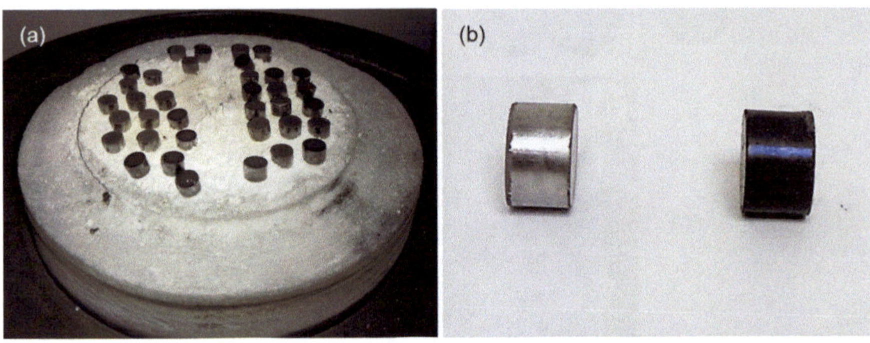

Fig. 4.19 (**a**) Sintered 393 alloy samples inside the furnace. (**b**) Comparison between a green compact (*left*) and a sintered sample (*right*)

size is a better indicator of the mechanical properties. Yet, the relative intensity of the cast sample is typical of a well-crystallized sample that is common for a casting. On the other hand, the milled and sintered samples have wider reflections and less intensity, particularly for Si, which is a brittle crystal and refines faster than other constituents that are more ductile such as Al, Cu, Ni, etc. during milling.

The mechanical milling refines the microstructure and this is observed in Fig. 4.21. Now, by comparing the microstructure characteristics of the samples produced by either processes one can conclude that the mechanical properties are different. This difference can be preestimated using the Hall-Petch model. This permits the assumption that improvements in mechanical properties are mainly attributed to a dislocation confinement by the more abundant presence of grain boundaries.

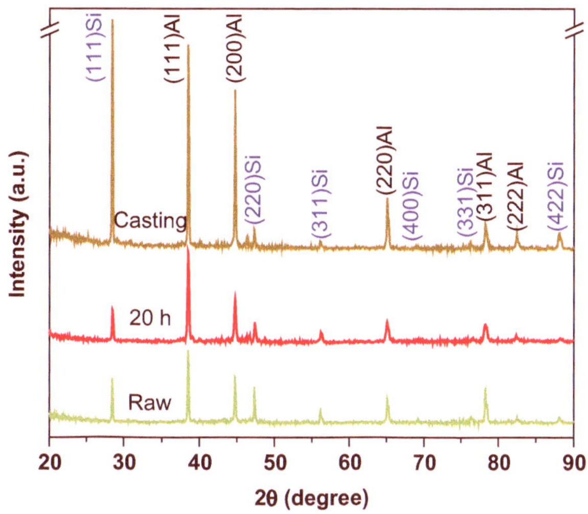

Fig. 4.20 X-ray diffraction patterns of the 393 alloy: raw-sintered sample, milled-sintered sample, and cast sample

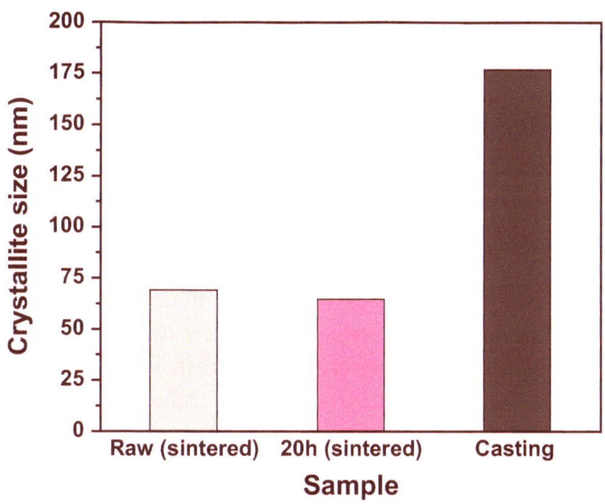

Fig. 4.21 Crystallite size of the 393 alloy: raw-sintered sample, milled-sintered sample, and cast sample

Representative microstructures of the sintered 393 alloy samples are shown in Fig. 4.22. The milling time usually improves the final products' densification; yet, there is a threshold at which the powder particles are optimized and beyond this point the densification is compromised, which is the case of the sample milled for more than 20 h. Usually a major problem in this case is the increase in porosity when compared to samples milled for shorter times such as 8 h.

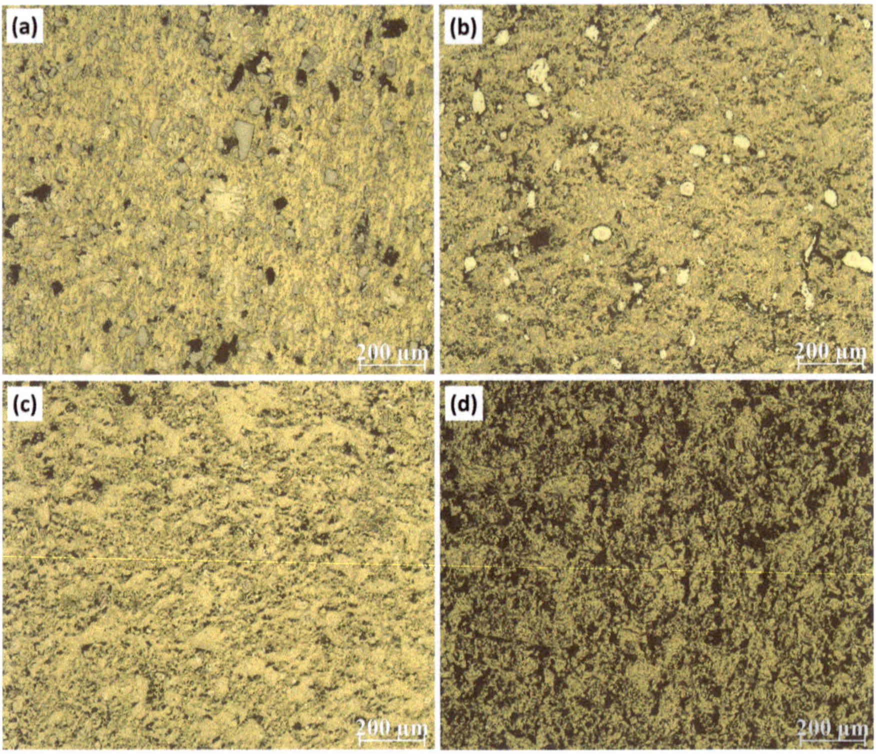

Fig. 4.22 Optical micrographs of the sintered 393 alloy samples as a function of milling time: (**a**) 0 h, (**b**) 1 h, (**c**) 8 h, and (**d**) 20 h

The homogeneity of the milled powders is usually maintained and preserved during sintering. This contributes to the isotropy and microstructure refinement, resulting in improved mechanical properties. The microstructures of the raw-sintered sample and milled-sintered sample are presented in Figs. 4.23 and 4.24, respectively. In both microstructures, the advantage of mechanical milling on homogeneity and how it is preserved even after sintering can be compared. Additional advantages over conventional castings include dendritic free structure, no segregation, and primary silicon refinement (e.g., no blocky or needle-like primary Si particles). All these benefits are contributors to the mechanical properties, including wear. Milling time is the major contributor to all these benefits; nevertheless, it can be detrimental. Therefore, it is important to balance all the processing parameters to identify the best product.

Figure 4.25 presents two micrographs of a cast 393 sample. The casting of this sample was carried out at 800 °C. In the microstructure, the segregation of the phases and constituents of the alloy are evident. The relatively coarse primary Si particles are clearly evident and they have the typical needle-like shape that is commonly found in alloys with more than 20 wt% Si [16, 18, 79–86]. Another phase that is unwelcome is the Fe-rich intermetallics that are common in cast products.

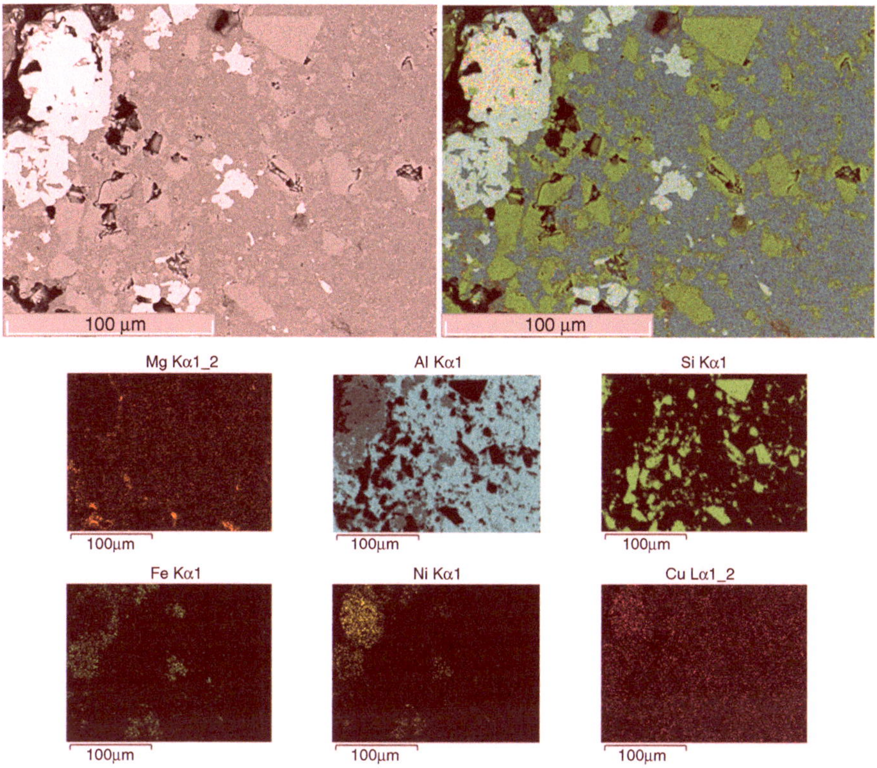

Fig. 4.23 SEM/EDS mapping analysis of the sintered elemental powder mixture (0 h of milling)

The unwelcome Fe-rich intermetallics include the phase also known as the Chinese script (inset in Fig. 4.25a). These observations are complemented by SEM/back-scattered electrons (BSE) and SEM/EDS results presented in Figs. 4.26 and 4.27, respectively. Apart from needle-like Si particles, a Ni-Fe rich phase is identified, as well as one Cu-rich phase, probably corresponding to the Al_2Cu phase or fine eutectic.

Hardness test results are presented herein as a brief example to demonstrate some of the benefits of mechanical alloying [77]. Figure 4.28 displays the results of the Vickers microhardness measurements for sintered and cast 393 alloy samples. These measurements correspond to the average of ten indentations; note that the standard deviation is low, indicating that samples are microstructurally homogeneous. As can be seen, the hardness of the sintered products is more than two times higher than that in the sample prepared by casting. It should be noted that it was not possible to perform the hardness measurement in the sample milled for 20 h due to its high porosity. Therefore, here we demonstrate that milling time is critical as it results in excessive porosity. This porosity can be observed in the optical and scanning electron micrographs in Figs. 4.22 and 4.24, respectively.

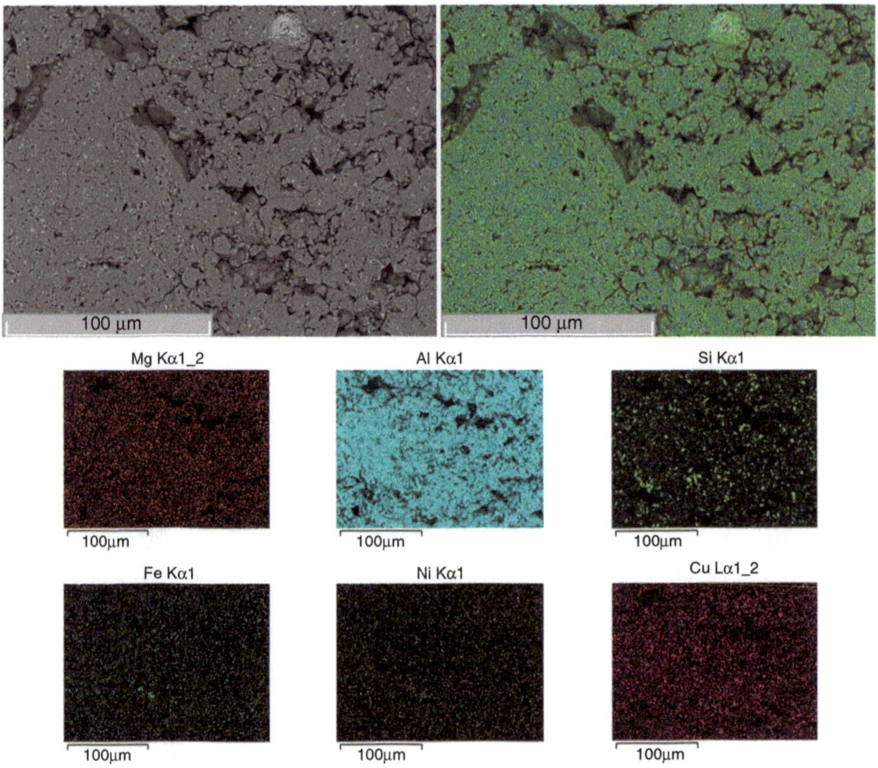

Fig. 4.24 SEM/EDS mapping analysis of the sintered 393 alloy MAed for 20 h

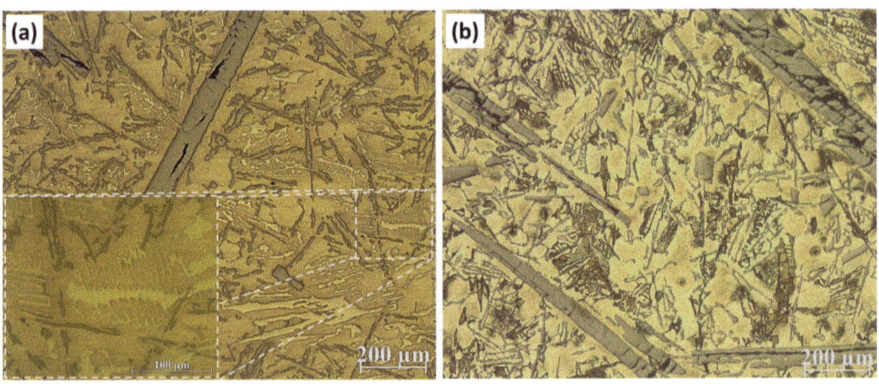

Fig. 4.25 Optical micrographs of the molten 393 alloy

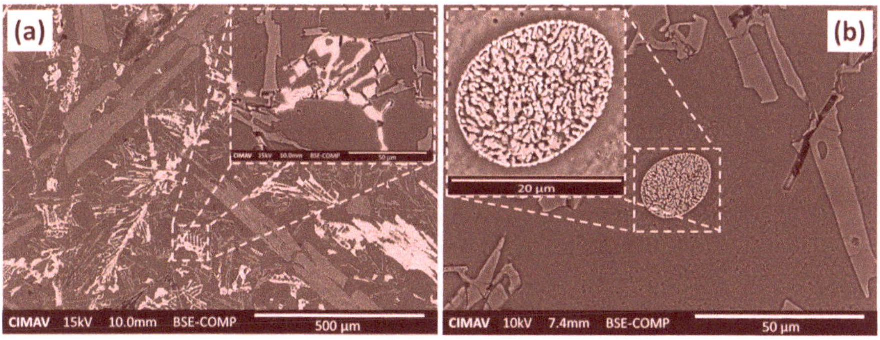

Fig. 4.26 SEM-BSE micrographs of the molten 393 alloy

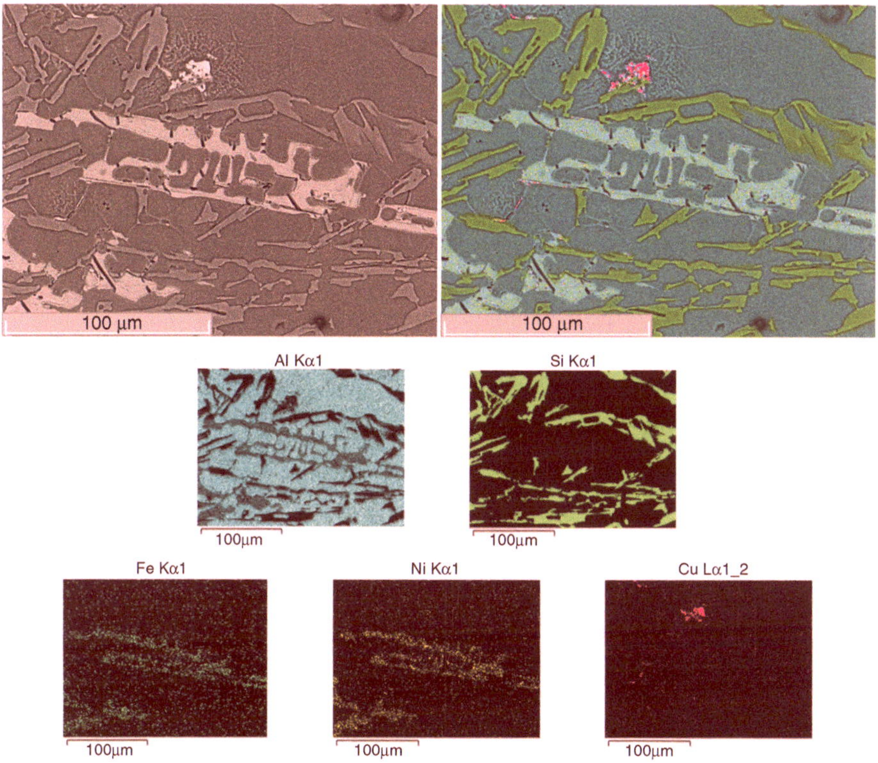

Fig. 4.27 SEM/EDS mapping analysis of the molten 393 alloy

The hardness value of the molten 393 alloy (917 MPa) corresponds well with those reported in the literature. Uzun et al. [77] found values of 602 MPa (ingot) and 1892 MPa (rapidly solidified ribbons) for an Al-16 wt% Si alloy. In the case of casting 390 alloys, values from 1030 MPa (as-fabricated sample) to 1500 MPa (T6 temper sample) are reported [15]. The hardness values of the sintered samples, especially

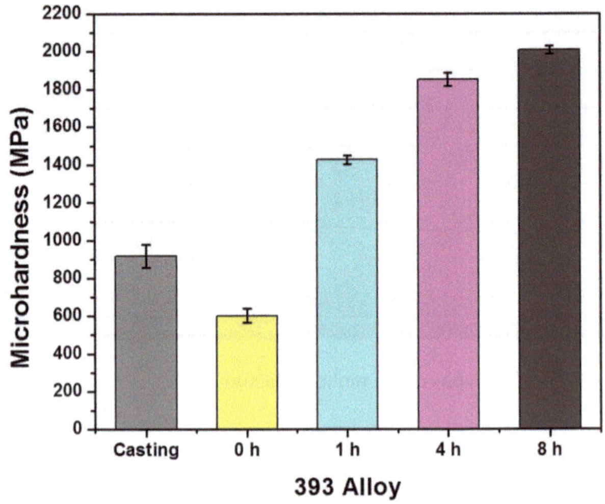

Fig. 4.28 Vickers microhardness of 393 alloys

those of samples milled for 4 and 8 h (1852 MPa and 2007 MPa, respectively), are much higher than those reported in other works for as-cast samples. We expect that further heat treatments applied to our samples could enhance their mechanical properties further.

The disadvantage of porosity in the sintered samples, in comparison to the melted one, was overcome by the hardening effect produced by the mechanical alloying. This demonstrates the effectiveness of the powder metallurgy method, particularly that of mechanical alloying combined with conventional compaction and sintering.

References

1. Suryanarayana, C. 2001. Mechanical alloying and milling. *Progress in Materials Science* 46 (1–2): 1–184. https://doi.org/10.1016/S0079-6425(99)00010-9.
2. Pickens, J.R. 1981. Aluminium powder metallurgy technology for high-strength applications. *Journal of Materials Science* 16 (6): 1437–1457. https://doi.org/10.1007/bf00553958.
3. German, R.M. 1994. *Powder metallurgy science*, 279. Princeton: Metal Powder Industries Federation.
4. Benjamin, J.S. 1970. Dispersion strengthened superalloys by mechanical alloying. *Metallurgical Transactions* 1 (10): 2943–2951. https://doi.org/10.1007/bf03037835.
5. Upadhyaya, G.S. 1997. *Powder metallurgy technology*, 160. Cambridge, UK: Cambridge Int Science Publishing.
6. Waseda, Y., and S. Suzuki. 2006. Characterization of corrosion products on steel surfaces. In *Advances in materials research*, 297. Berlin/Heidelberg: Springer.
7. McEwan, G.F. 2006. The Incas: New perspectives. In *Understanding ancient civilizations series*. Santa Barbara: Abc-Clio.
8. Guisbiers, G., S. Mejia-Rosales, S. Khanal, F. Ruiz-Zepeda, R.L. Whetten, and M. José-Yacaman. 2014. Gold–copper nano-alloy, "tumbaga", in the era of nano: Phase diagram and segregation. *Nano Letters* 14 (11): 6718–6726. https://doi.org/10.1021/nl503584q.

9. Hildeman, G.J., and M.J. Koczak. 2012. Aluminum powder metallurgy. *JOM* 38 (8): 30–32. https://doi.org/10.1007/bf03257784.
10. Voss, D.P. 1979. *Structure and mechanical properties of powder metallurgy 2024 and 7075 aluminum alloys*, AFOSR 77-3440, Final Report, 289.
11. Vijeesh, V., and K. Narayan Prabhu. 2013. Review of microstructure evolution in hypereutectic Al–Si alloys and its effect on wear properties. *Transactions of the Indian Institute of Metals* 67 (1): 1–18. https://doi.org/10.1007/s12666-013-0327-x.
12. J.R. Davis & Associates, and ASM International. Handbook Committee. 1993. Aluminum and aluminum alloys. In *ASM specialty handbook*. Vol. iii, 784. Materials Park: ASM International.
13. Campbell, J., and ScienceDirect (Online service). 2011. *Complete casting handbook metal casting processes, metallurgy, techniques and design*. Oxford, UK/Waltham: Elsevier Butterworth-Heinemann. http://www.sciencedirect.com/science/book/9781856178099.
14. Campbell, J. 2015. *Complete casting handbook metal casting processes, metallurgy, techniques and design*. Amsterdam: Butterworth-Heinemann. http://www.sciencedirect.com/science/book/9780444635099.
15. ASM International Handbook Committee. 1998. *ASM handbook, volume 02 – Properties and selection: Nonferrous alloys and special-purpose materials*. Materials Park: ASM International.
16. Prabhu, V.V.K.N. 2014. Review of microstructure evolution in hypereutectic Al-Si alloys and its effect on wear properties. *Transactions of the Indian Institute of Metals* 67 (1): 1–18. https://doi.org/10.1007/s12666-013-0327-x.
17. Eisen, W.B., B.L. Ferguson, R.M. German, R. Iacocca, P.W. Lee, D.S. Madan, K.E. Moyer, H. Sanderow, and Y. Trudel. 1998. *ASM handbook, volume 07 - Powder metal technologies and applications*. Materials Park: ASM International.
18. Jorstad, J., and D. Apelian. 2009. Hypereutectic Al-Si alloys: Practical casting considerations. *International Journal of Metalcasting* 3 (3): 13–36.
19. Jorstad, J.L. 1971. The hypereutectic aluminum-silicon alloy used to cast the Vega engine block. *Modern Casting* 60 (4): 59–64.
20. Jorstad, J.L. 1978. Low-pressure process – Companion to die casting. *Die Casting Engineer* 22 (1): 10.
21. Jorstad, J.L. 1980. Influence of metallurgical factors on machinability of aluminum casting alloys. *Die Casting Engineer* 24 (6): 26–32.
22. Jorstad, J.L. 1984. Trends in aluminum casting. Pt. 1. Automotive applications. *Modern Casting* 74 (10): 26–29.
23. Jorstad, J.L. 2008. Permanent mold casting processes. *Advanced Materials and Processes* 166 (4): 30–34.
24. German, R.M. 2005. *Powder metallurgy and particulate materials processing: The processes, materials, products, properties, and applications*. Princeton: Metal Powder Industries Federation.
25. Ebrahimi, F., G.R. Bourne, M.S. Kelly, and T.E. Matthews. 1999. Mechanical properties of nanocrystalline nickel produced by electrodeposition. *Nanostructured Materials* 11 (3): 343–350. https://doi.org/10.1016/S0965-9773(99)00050-1.
26. Volkert, C.A., and E.T. Lilleodden. 2006. Size effects in the deformation of submicron Au columns. *Philosophical Magazine* 86 (33–35): 5567–5579. https://doi.org/10.1080/14786430600567739.
27. Wyrzykowski, J.W., and M.W. Grabski. 1986. The Hall–Petch relation in aluminium and its dependence on the grain boundary structure. *Philosophical Magazine A* 53 (4): 505–520. https://doi.org/10.1080/01418618608242849.
28. Mott, N.F., and F.R.N. Nabarro. 1940. An attempt to estimate the degree of precipitation hardening, with a simple model. *Proceedings of the Physical Society* 52 (1): 86.
29. Callister, W.D. 2007. *Materials science and engineering: An introduction*. Vol. xxv. 7th ed, 122. New York: Wiley.
30. Reed-Hill, R.E., and R. Abbaschian. 1994. Physical metallurgy principles. In *PWS series in engineering*. Vol. xv. 3rd ed., 926. Boston: PWS Pub.

31. Ortner, S.R., C.R.M. Grovenor, and B.A. Shollock. 1988. On the structure and composition of G-P zones in high-purity AlZnMg alloys. *Scripta Metallurgica* 22 (6): 839–842. https://doi.org/10.1016/S0036-9748(88)80060-7.
32. Benjamin, J.S., and T.E. Volin. 1974. The mechanism of mechanical alloying. *Metallurgical Transactions* 5 (8): 1929–1934. https://doi.org/10.1007/BF02644161.
33. Benjamin, J.S. 1976. Mechanical alloying. *Scientific American* 234 (5): 40–48.
34. Benjamin, J.S., and R.D. Schelleng. 1981. Dispersion strengthened aluminum-4 pct magnesium alloy made by mechanical alloying. *Metallurgical Transactions A* 12 (10): 1827–1832. https://doi.org/10.1007/bf02643766.
35. Gilman, P.S., and J.S. Benjamin. 1983. Mechanical alloying. *Annual Review of Materials Science* 13 (1): 279–300.
36. Benjamin, J.S. 1990. Mechanical alloying—A perspective. *Metal Powder Report* 45 (2): 122–127. https://doi.org/10.1016/S0026-0657(10)80124-9.
37. Benjamin, J.S. 1992. Mechanical alloying-history and future potential. *Advances in Powder Metallurgy and Particulate Materials* 7: 155–155.
38. Soni, P.R., and ebrary Inc. 2001. *Mechanical alloying fundamentals and applications*. Cambridge, UK: Cambridge International Science Publishing. http://site.ebrary.com/lib/umich/Doc?id=10064335, http://libproxy.umflint.edu:2048/login?url=http://site.ebrary.com/lib/umich/Doc?id=10064335.
39. Suryanarayana, C., E. Ivanov, and V.V. Boldyrev. 2001. The science and technology of mechanical alloying. *Materials Science and Engineering A-Structural Materials Properties Microstructure and Processing* 304: 151–158. https://doi.org/10.1016/S0921-5093(00)01465-9.
40. Soni, P.R., and Books24×7 Inc. 2001. *Mechanical alloying fundamentals and applications*. Cambridge, UK: Cambridge International Science Pub. http://www.books24x7.com/marc.asp?bookid=14854.
41. Morsi, K., and A. Esawi. 2007. Effect of mechanical alloying time and carbon nanotube (CNT) content on the evolution of aluminum (Al)-CNT composite powders. *Journal of Materials Science* 42 (13): 4954–4959. https://doi.org/10.1007/s10853-006-0699-y.
42. Guerrero-Paz, J., F.C. Robles-Hernandez, R. Martínez-Sánchez, D. Hernandez-Silva, and D. Jaramillo-Vigueras. 2001. Particle size evolution in non-adhered ductile powders during the mechanical alloying. *Materials Science Forum* 360: 317–322.
43. Lee, W., and S.I. Kwun. 1996. The effects of process control agents on mechanical alloying mechanisms in the Ti-A1 system. *Journal of Alloys and Compounds* 240 (1–2): 193–199. https://doi.org/10.1016/0925-8388(96)02224-4.
44. Shaw, L., M. Zawrah, J. Villegas, H. Luo, and D. Miracle. 2003. Effects of process-control agents on mechanical alloying of nanostructured aluminum alloys. *Metallurgical and Materials Transactions A-Physical Metallurgy and Materials Science* 34 (1): 159–170. https://doi.org/10.1007/s11661-003-0217-7.
45. Madavali, B., J.H. Lee, J.K. Lee, K.Y. Cho, S. Challapalli, and S.J. Hong. 2014. Effects of atmosphere and milling time on the coarsening of copper powders during mechanical milling. *Powder Technology* 256: 251–256. https://doi.org/10.1016/j.powtec.2014.02.019.
46. Lin, F., D.M. Jiang, and X.M. Ma. 2009. The effect of milling atmospheres on photocatalytic property of Fe-doped TiO2 synthesized by mechanical alloying. *Journal of Alloys and Compounds* 470 (1–2): 375–378. https://doi.org/10.1016/j.jallcom.2008.02.067.
47. Ong, T.S., and H. Yang. 2000. Effect of atmosphere on the mechanical milling of natural graphite. *Carbon* 38 (15): 2077–2085. https://doi.org/10.1016/S0008-6223(00)00064-6.
48. Ito, M., K. Majima, S. Katsuyama, and H. Nagai. 1996. Effect of milling atmosphere on the preparation of Sm2Fe17Nx powder by mechanical grinding (Reprinted from J Japan Inst Metals, vol 60, pg 427, 1996). *Materials Transactions JIM* 37 (11): 1704–1709.
49. Suryanarayana, C., G.H. Chen, and F.H.S. Froes. 1992. Milling maps for phase identification during mechanical alloying. *Scripta Metallurgica et Materialia* 26 (11): 1727–1732. https://doi.org/10.1016/0956-716x(92)90542-M.
50. Fogagnolo, J.B., F. Velasco, M.H. Robert, and J.M. Torralba. 2003. Effect of mechanical alloying on the morphology, microstructure and properties of aluminium matrix composite pow-

ders. *Materials Science and Engineering A-Structural Materials Properties Microstructure and Processing* 342 (1–2): 131–143. https://doi.org/10.1016/S0921-5093(02)00246-0.
51. Koch, C.C. 1989. Materials synthesis by mechanical alloying. *Annual Review of Materials Science* 19: 121–143. https://doi.org/10.1146/annurev.ms.19.080189.001005.
52. Pérez-Bustamante, R., F. Pérez-Bustamante, I. Estrada-Guel, C.R. Santillán-Rodríguez, J.A. Matutes-Aquino, J.M. Herrera-Ramírez, M. Miki-Yoshida, and R. Martínez-Sánchez. 2011. Characterization of Al2024-CNTs composites produced by mechanical alloying. *Powder Technology* 212 (3): 390–396. https://doi.org/10.1016/j.powtec.2011.06.007.
53. Ahamed, H., and V. Senthilkumar. 2010. Role of nano-size reinforcement and milling on the synthesis of nano-crystalline aluminium alloy composites by mechanical alloying. *Journal of Alloys and Compounds* 505 (2): 772–782. https://doi.org/10.1016/j.jallcom.2010.06.139.
54. Sivasankaran, S., K. Sivaprasad, R. Narayanasamy, and V.K. Iyer. 2010. An investigation on flowability and compressibility of AA 6061100 – x-x wt.% TiO2 micro and nanocomposite powder prepared by blending and mechanical alloying. *Powder Technology* 201 (1): 70–82. https://doi.org/10.1016/j.powtec.2010.03.013.
55. Zheng, R., H. Yang, T. Liu, K. Ameyama, and C. Ma. 2014. Microstructure and mechanical properties of aluminum alloy matrix composites reinforced with Fe-based metallic glass particles. *Materials and Design* 53: 512–518. https://doi.org/10.1016/j.matdes.2013.07.048.
56. Estrada-Ruiz, R.H., R. Flores-Campos, J.M. Herrera-Ramírez, and R. Martínez-Sánchez. 2016. Mechanical properties of aluminum 7075–Silver nanoparticles powder composite and its relationship with the powder particle size. *Advanced Powder Technology* 27 (4): 1694–1699.
57. Garibay-Febles, V., H.A. Calderon, F.C. Robles-Hernández, M. Umemoto, K. Masuyama, and J.G. Cabanas-Moreno. 2000. Production and characterization of (Al, Fe)-C (graphite or fullerene) composites prepared by mechanical alloying. *Materials and Manufacturing Processes* 15 (4): 547–567.
58. Robles, F. 2004. Production and characterization of Fe-C graphite and Fe-C fullerene composites produced by different mechanical alloying techniques. *Metalurgija* 10 (2): 107–118.
59. Brysch, C., E. Wold, F.C. Robles Hernandez, and J.F. Eberth 2012. Sintering of chitosan and chitosan composites. In *ASME 2012 international mechanical engineering congress and exposition*. Houston: American Society of Mechanical Engineers.
60. Savio, A.K., D. Starikov, A. Bensaoula, R. Pillai, L.L. de la Torre García, and F.C. Robles Hernández. 2012. Tunable TiO_2 (anatase and rutile) materials manufactured by mechanical means. *Ceramics International* 38 (5): 3529–3535.
61. Fals, A.E., J. Quintero, and F.C. Robles Hernández. 2010. Manufacturing of hybrid composites and novel methods to synthesize carbon nanoparticles. *MRS Proceedings* 1276: 3. Cambridge University Press.
62. Santana-García, I., F. Hernandez-Robles, V. Garibay-Febles, and H.A. Calderon. 2010. Metal (Fe, Al)-fullerene nanocomposites: Synthesis and characterization. *Microscopy and Microanalysis* 16 (S2): 1688–1689.
63. Robles Hernández, F.C., and H.A. Calderon. 2012. Nanostructured Al/Al 4 C 3 composites reinforced with graphite or fullerene and manufactured by mechanical milling and spark plasma sintering. *Materials Chemistry and Physics* 132 (2): 815–822.
64. Fals, A.E., V.G. Hadjiev, and F.C.R. Hernandez. 2012. Multi-functional fullerene soot/alumina composites with improved toughness and electrical conductivity. *Materials Science and Engineering: A* 558: 13–20.
65. Fals, A.E., V. Hadjiev, and F.R. Hernández. 2013. Porous media reinforced with carbon soots. *Materials Chemistry and Physics* 140 (2): 651–658.
66. Robles Hernández, F.C., H.A. Calderon, D. Barber, A. Okonkwo, R. Ordoñez Olivares and V. Hadjiev. 2014. Characterization of all carbon composites reinforced with in situ synthesized carbon nanostructures. *MRS Proceedings* 1611: 145–151. Cambridge University Press.
67. Estrada-Guel, I., O. Anderson-Okonkwo, and F. Robles-Hernandez. 2016. In-situ transformation of amorphous soot into carbon-nanostructures by high-energy ball milling. *Microscopy and Microanalysis* 22 (S3): 1902–1903.
68. Raming, T.P., W.E. van Zyl, E.P. Carton, and H. Verweij. 2004. Sintering, sinter forging and explosive compaction to densify the dual phase nanocomposite system Y2O3-

doped ZrO2 and RuO2. *Ceramics International* 30 (5): 629–634. https://doi.org/10.1016/j.ceramint.2003.09.006.
69. Hatch, J.E., A. Association, and A.S. Metals. 1984. *Aluminum: Properties and physical metallurgy*. Metals Park: American Society for Metals.
70. Robles Hernandez, F.C. 2004. *Improvement in functional characteristics of aluminum-silicon cast components through the utilization of a novel electromagnetic treatment of liquid melts*. Ph.D Thesis, University of Windsor (Canada).
71. Masuda, H., K. Higashitani, and H. Yoshida. 2006. *Powder technology: Fundamentals of particles, powder beds, and particle generation*. Boca Raton: CRC Press.
72. Matsuura, K., M. Kudoh, H. Kinoshita, and H. Takahashi. 2003. Precipitation of Si particles in a super-rapidly solidified Al–Si hypereutectic alloy. *Materials Chemistry and Physics* 81 (2–3): 393–395. https://doi.org/10.1016/S0254-0584(03)00030-0.
73. Ejiofor, J.U., and R.G. Reddy. 1997. Developments in the processing and properties of particulate Al-Si composites. *JOM* 49 (11): 31–37. https://doi.org/10.1007/s11837-997-0008-5.
74. Omori, M. 2000. Sintering, consolidation, reaction and crystal growth by the spark plasma system (SPS). *Materials Science and Engineering: A* 287 (2): 183–188. https://doi.org/10.1016/S0921-5093(00)00773-5.
75. Munir, Z.A., U. Anselmi-Tamburini, and M. Ohyanagi. 2006. The effect of electric field and pressure on the synthesis and consolidation of materials: A review of the spark plasma sintering method. *Journal of Materials Science* 41 (3): 763–777. https://doi.org/10.1007/s10853-006-6555-2.
76. Gómez-Esparza, C.D., F.J. Baldenebro-López, C.R. Santillán-Rodríguez, I. Estrada-Guel, J.A. Matutes-Aquino, J.M. Herrera-Ramírez, and R. Martínez-Sánchez. 2014. Microstructural and magnetic behavior of an equiatomic NiCoAlFe alloy prepared by mechanical alloying. *Journal of Alloys and Compounds* 615 (Supplement 1): S317–S323. https://doi.org/10.1016/j.jallcom.2014.01.233.
77. Uzun, O., T. Karaaslan, M. Gogebakan, and M. Keskin. 2004. Hardness and microstructural characteristics of rapidly solidified Al–8–16 wt.% Si alloys. *Journal of Alloys and Compounds* 376 (1): 149–157.
78. Cullity, B.D. 1956. *Elements of X-ray diffraction*. Reading: Addison-Wesley Pub. Co.
79. Chen, M., and A.T. Alpas. 2008. Ultra-mild wear of a hypereutectic Al-18.5 wt.% Si alloy. *Wear* 265 (1–2): 186–195. https://doi.org/10.1016/j.wear.2007.10.002.
80. Hernandez, F.C.R., and J.H. Sokolowski. 2005. Identification of silicon agglomerates in quenched Al-Si hypereutectic alloys from liquid state. *Advanced Engineering Materials* 7 (11): 1037–1043.
81. Hernandez, F.C.R., and J.H. Sokolowski. 2006. Thermal analysis and microscopical characterization of Al-Si hypereutectic alloys. *Journal of Alloys and Compounds* 419 (1–2): 180–190. https://doi.org/10.1016/j.jallcom.2005.07.077.
82. Hernández, F.C.R., and J.H. Sokolowski. 2005. Novel image analysis to determine the Si modification for hypoeutectic and hypereutectic Al-Si alloys. *JOM* 57 (11): 48–52.
83. Hernandez, F.C.R., J.H. Sokolowski, and J.D.C. Rivera. 2007. Micro-Raman analysis of the Si particles present in Al-Si hypereutectic alloys in liquid and semi-solid states. *Advanced Engineering Materials* 9 (1–2): 46–51. https://doi.org/10.1002/adem.200600173.
84. Kyffin, W., W. Rainforth, and H. Jones. 2001. Effect of phosphorus additions on the spacing between primary silicon particles in a Bridgman solidified hypereutectic Al-Si alloy. *Journal of Materials Science* 36 (11): 2667–2672. https://doi.org/10.1023/a:1017904627733.
85. Lu, D.H., Y.H. Jiang, G.S. Guan, R.F. Zhou, Z.H. Li, and R. Zhou. 2007. Refinement of primary Si in hypereutectic Al-Si alloy by electromagnetic stirring. *Journal of Materials Processing Technology* 189 (1–3): 13–18. https://doi.org/10.1016/j.jmatprotec.2006.12.008.
86. Robles Hernandez, F.C., and J.H. Sokolowski. 2006. Comparison among chemical and electromagnetic stirring and vibration melt treatments for Al-Si hypereutectic alloys. *Journal of Alloys and Compounds* 426 (1–2): 205–212. https://doi.org/10.1016/j.jallcom.2006.09.039.

Chapter 5
Liquid and Semisolid Melt Treatment: Electromagnetic Stirring

5.1 Introduction to Electromagnetic Stirring

In the literature, there are a number of reports on microstructure refining using electromagnetic [1–27] and mechanical stirring [28, 29] methods. Some of these processes have been patented in order to protect the intellectual property of the inventions [30–43]. Most of those methods are effective in the semisolid state, except one [34] that is applied for Al-Si hypoeutectic alloys in the liquid state. The number of publications covering electromagnetic stirring and vibration for Al-Si hypereutectic alloys is rather limited [23–25, 44–46]. The main reason for the effectiveness of liquid-state treatments for Al-Si hypereutectic alloys is the presence of the Si agglomerates that are susceptible to modification in the liquid state.

Metal treatment by means of stirring has been documented since the 1960s. The applications are vast and have a wide range of purposes, such as degassing, grain refinement, mixing, desulfuring (e.g., iron), removing inclusions, and iron passivation. The most common stirring technologies are mechanical (MS), electromagnetic (ES), and electromagnetic stirring and bibration (ESV) [2–20, 47–49]. Both, ES and ESV have been well accepted and implemented for melt treatments of ferrous and nonferrous alloys. The electromagnetic fields formed within the melt are due to the presence of the Lorentz forces. Those forces induce a rotational field with well-defined shear components that cut through the coarsening dendrites in their early growth. When the electromagnetic forces in the melt are larger than the rigidity or strength of the solidifying alloy, the shear components cut through the dendrites and refine the microstructure by increasing the number of nucleating sites. The refining effect results from the breakage of dendrites, forming new nuclei and causing the formation of new grains as sketched in Fig. 5.1. In the end, the final microstructure possesses a larger number of grains that are smaller having shorter dendrites. The secondary dendrite arm spacing (SDAS) is in most cases unchanged due to the stirring effects. In some cases, the

© Springer International Publishing AG 2017

F.C. Robles Hernandez et al., *Al-Si Alloys*, DOI 10.1007/978-3-319-58380-8_5

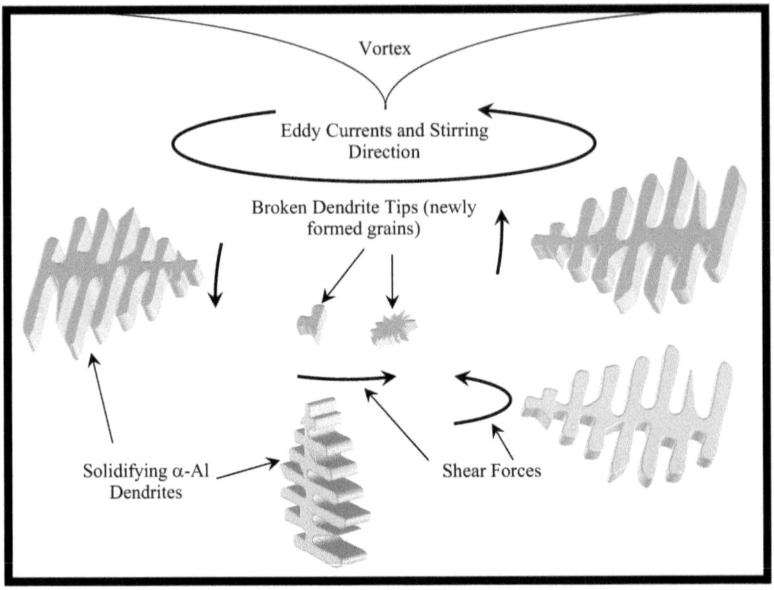

Fig. 5.1 Sketch of the refining mechanism of electromagnetic stirring in solidifying dendrites

resulting castings are dendrite-free, or they have a fully globular microstructure. Additionally, stirring promotes interdendritic feeding and reduces porosity [15, 47].

Most electromagnetic stirring apparatus consist of coil(s), a crucible, cooling system [for both, coil(s) and sample(s)], heating system (for the sample), computer-based thermal analysis (optional), electrical (voltage/current) controllers, and insulation(s). Figure 5.2 sketches a standard electromagnetic stirring workstation. In the case of mechanical stirring systems, the induction coils are replaced by an impeller that is introduced into the melt. The severity of the stirring is controlled by changing the frequency, voltage, and/or amplitude of the current, as well as time and the treatment temperature. This results in different levels of microstructure modification, and porosity reduction.

5.2 Electromagnetic Stirring (ES) Principle

In the literature, the intensity of the electromagnetic stirring is usually reported as a function of current or voltage. However, none of those parameters are the true intensity parameters. In addition to those parameters, the geometry and number of turns in the coil(s) have a direct influence on the intensity of the magnetic field. Therefore, a more meaningful way to report the severity of electromagnetic stirring method is by using magnetic field intensity per mass of treated sample.

Another important parameter is the depth of penetration that indicates the depth where the eddy currents are produced and interact with the melt. This parameter can be used to assess if the ES setup is rotational or nonrotational. The intensity of the

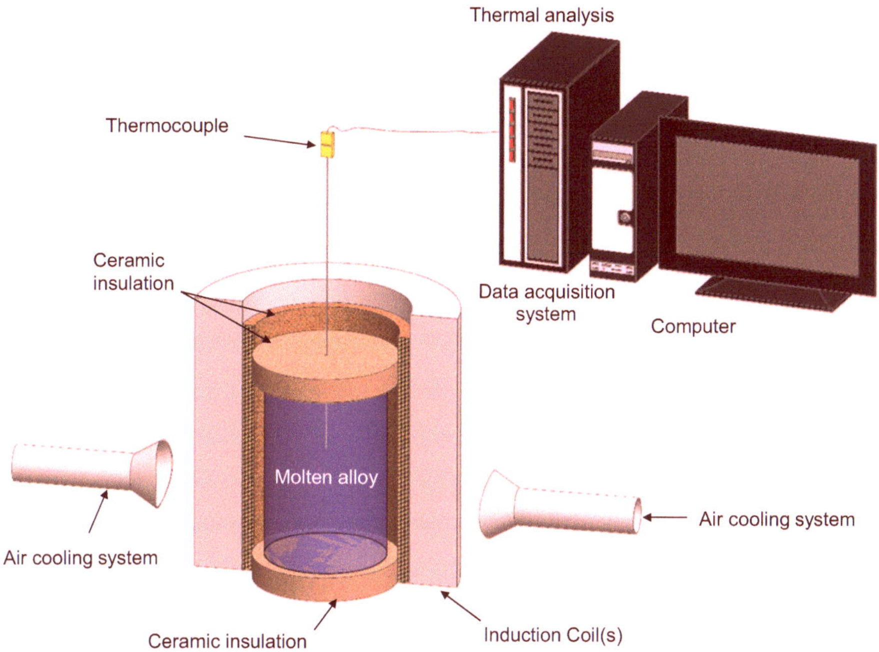

Fig. 5.2 Schematic drawing for a standard electromagnetic stirring

eddy currents, together with the intensity of the magnetic field, provides the severity of the electromagnetic stirring (Lorentz forces). Usually, an ES system generates a vortex on the liquid metal, and the depth of the vortex can be directly related to the intensity of the ES treatment. Unfortunately, the vortex promotes gas entrapment and thus porosity and oxide entrapment. The main equations required to design the magnetic field intensity, the depth of penetration, and the density of the eddy currents for an ES or ESV system are as follows.

$$B = \frac{\mu_0}{4\pi} \int_C \frac{I \cdot \overrightarrow{dL_W} \times \vec{l_r}}{r^2} \tag{5.1}$$

Magnetic flux density of a coil as a function of the coil's geometry:

$$B_X = \frac{\mu_0 \cdot I \cdot N}{h(b-a)} a \cdot \beta \cdot \ln\left(\frac{\alpha + \sqrt{\alpha^2 + \beta^2}}{1 + \sqrt{1 + \beta^2}}\right)$$

$$where: \tag{5.2}$$

$$\beta = \frac{L}{2a}$$

$$\alpha = \frac{b}{a}$$

Depth of penetration (δ):

$$\delta = \sqrt{\frac{2\rho}{\mu_r \cdot \mu_0 \cdot f}} \tag{5.3}$$

Density of eddy currents (J_m):

$$J_m = \frac{\sqrt{2}}{\delta} \cdot \frac{H_m}{\sqrt{\frac{r}{r_2}}} \cdot \exp\left(\frac{r - r_2}{\delta}\right)$$

where :

$$H_m = \sqrt{2} \cdot I \cdot f \tag{5.4}$$

where B_X is the magnetic flux density;
B and B_0 are the magnetic flux intensities for AC and DC currents, respectively.
μ_0 is the absolute permeability
μ_r is the relative permeability
I is the AC or DC current
L is the length of the coil
α and β are parameters that depend on the coils' geometry
δ is the depth of penetration
f is the frequency of the AC current
J_m is the density of eddy currents
r_1 and r_2 are the inside and outside diameters of the coil, respectively
dL_W is the infinitesimal length of a conductor (wire) carrying an I current
l_r is a unit vector that indicates the direction of the vector distance r from the current to the field point

The basic principle of the ES process is the rotation of the melt in the presence of a magnetic field that develops Lorentz forces. These forces are due to the fluid motion in the presence of a magnetic field forming rotational vectors that cause stirring of the melt. This results in a combination of centrifugal, electromagnetic, and gravitational fields. In order for the system to work effectively, the Lorentz forces must be larger than the incipient strength of the solidifying microstructure. The Lorentz forces are usually in shear and cut through the solidifying dendrites that result in microstructure refinement [50].

At the industrial level, the most common apparatus used for electromagnetic stirring are linear and rotary stirrers, which are fundamentally electromagnetic coils (similar to those used in an electric motor). The coils are usually installed around the crucible, the ladle, or furnace to induce a rotating magnetic field and spin the semisolid alloy [51]. Equation 5.5 is used to determine the Lorentz forces created during the ES treatment. B can be determined using Eq. 5.2, or if the geometry of the coil is too intricate, it can be measured at particular locations along the coil.

$$F_L = J \times B \tag{5.5}$$

where F_L are the Lorentz forces (N/m^3),
J is the current density (A/m^2), and
B is the magnetic field intensity [Tesla (T)]

Lu et al. [52] used an electromagnetic stirring system to treat Al-Si hypereutectic alloys containing 20 wt% Si. They found that using currents between 4 and 32 A (equivalent to 5×10^{-3}–25×10^{-3} T) is more beneficial for microstructure refinement and phase modification than higher currents (16–32 A). Higher currents usually promote primary Si agglomeration. The agglomeration effect is not the result of the current itself but rather an effect of the stirring motion and the difference in density between Al and Si. The agglomeration effect can be determined using Stokes' law.

5.3 Mechanical and Ultrasonic Stirring

Kang et al. [53] used a method called droplet emulsion technique (DET) that consists of emulsification of liquid Al-Si hypereutectic alloy with different amounts of Si (from 13 to 25 wt%) in a bath of liquid salts (LiCl-KCl). The emulsification is assisted by an impeller. The liquid salts bath is a massive environment that allows to control the rate of heat released from the Al-Si alloy droplet, forcing its solidification at different undercooling temperatures (from 14 to 124 K). This method induces significant modification of the Al-Si eutectic. The mechanism consists of the development of forces that change the solidification mechanisms of the Si crystals from a layer pattern to a radial pattern. Radial growth favors coarsening in less favorable directions. This slows down the coarsening. In this case, the undercooling can be used to predict the dendrite by means of the LKT growth theory [54]. A disadvantage of this method is that the final Al-Si product is obtained as droplets that require further treatment (e.g., thixocasting, rheocasting, sintering) to be used in the final components.

Mechanical pulses are an alternative way to treat Al-Si liquid melts [47–49]. This method introduces pulses with frequencies that can be modulated between 20 and 50 kHz [26]. High frequencies need to be closely controlled since they may result in unexpected cavitation. In some cases, frequencies above 5 kHz do not seem to have the grain-refining benefits and may promote grain coarsening [26, 55]. In contrast, low frequencies (<5 Hz) [26] promote the transformation of grains from equiaxed to globular. Cavitation can be beneficial for degassing purposes when applied to liquid melts. In semisolid or solid states, cavitation can be responsible for oxide formation and crack formation. Both are detrimental to melts and compromise the integrity of the castings components.

5.4 Electromagnetic Stirring and Vibration

The most extensive and complete work published on electromagnetic stirring and vibration was done by Vivès [19, 26, 27, 30–33, 41]. Vivès' work mainly consists of combining AC and DC magnetic and/or electric fields along a pool of liquid metal and/or semisolid alloy and in some cases combinations with piezoelectric transducers. Additional works found in the literature focus on Al-Si hypereutectic alloys [45, 46, 56]. Figure 5.3a, b show the ESV principle when both, AC and DC, fields are electric and/or electromagnetic. A sketch of a typical ESV apparatus is shown in Fig. 5.3c. The depth of penetration of eddy currents is a function of the frequency of the AC current (Eq. 5.3). This generates a skin effect that is a function of the inverse of the AC current frequency. The depth of penetration needs to be determined well to have a rotational system. The higher the frequency, the thinner the skin effect, which may result in excessive shear forces that may induce heat and raise the melt temperature instead of only stirring the melt.

In the ESV principle two magnetic fields are used. One is a stationary field (B_0) and the other one a variable field (B). The variable field is usually AC, and it does travel along the vicinity of the outer wall of the molten metal. The stationary field, DC, covers the entire sample. Both the magnetic and the electric fields are parallel to the vertical axis of the ingot. Therefore, there is a region near the outer wall where both fields interact. In this region two shear components are formed, one parallel to the DC field and the other one perpendicular. The interaction of both fields result in vibration. As seen in Fig. 5.3c, d, this interaction occurs in the outer region of the sample.

The DC field can be conveniently oriented to form a convex meniscus at the top of the liquid metal avoiding the formation of a vortex that can entrap gases (i.e., hydrogen and oxygen). The interaction of the magnetic fields produces two components, one vertical (rotational) and the other radial (nonrotational). The interaction of both components is responsible for vibration during the stirring process. The sinusoidal electric current used can be of low or high frequency (N) with a respective current intensity $i = I \quad \sin t$ (where I is the intensity of the current applied, t is time, and i is the intensity of the current at time t). Similar vibration effect can be generated by passing AC current along the sample in the presence of a DC magnetic field (Fig. 5.3b) [26].

Stirring with only DC fields is not recommended because a DC field covers the entire sample, and its severity is reduced when compared to that of the AC field which is concentrated in the skin zone. The ESV can be used to prevent or minimize the potential of gas entrapment in the melt; on the contrary, the ES method usually forms a deep vortex and, in most cases, a turbulent flow.

The interaction between the external direct magnetic field (B_o) and the variable electrical current of density "j" gives rise to a vertical periodic vibration electromagnetic force ($j \times B_o$). The vibration force has a frequency N that forces the molten alloy to vibrate. The vertical electromagnetic vibration force (expressed as pressure) covers the area where the eddy currents are formed and the pressure of the electromagnetic pulses is given by:

$$P = \frac{B_0 \cdot I \cdot L}{a} \sin \varpi \cdot t \quad \left(\mathrm{N/m^2}\right) \tag{5.6}$$

Electric vibration using AC electric fields require currents in the order of 0–3500 A. The requirements for an electromagnetic system can be reduced considerably (100 A or less) by using a well-designed electromagnetic coil(s) system. This is the result of the coil's effect that increases the intensity of the magnetic field as a function of the number of turns in the coil. In the literature, intensities of the magnetic

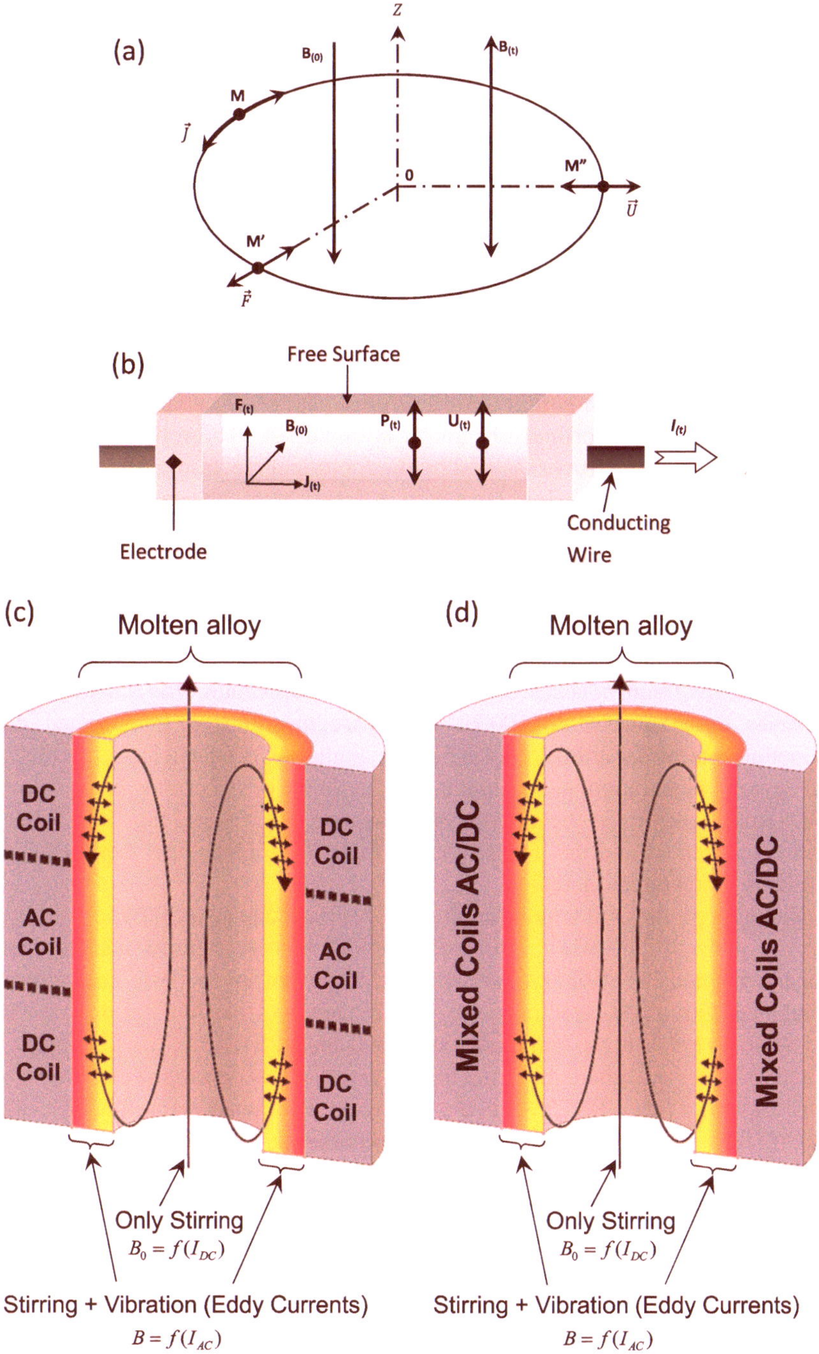

Fig. 5.3 Principle of electromagnetic vibration in molten metals using (**a**) a noncontact method and (**b**) mechanoacoustic irradiation and (**c**, **d**) vibration taking place in the regions where the AC and DC fields interact. In (**d**) both currents are forced to pass through the same conductor resulting in a more homogeneous field. The eddy currents are the result of the AC field and are responsible for the vibration effect. Figures (**a**) and (**b**) were made based on sketches found in [26]

field(s) between 0 and 0.9 T [19, 26, 27, 30–33, 41, 56] have been reported, but supermagnets of 10 T or more have been used for liquid metal treatments for Al-Si hypereutectic alloys [45, 46].

On ESV apparatus the intensity of the AC electromagnetic field can generate eddy current densities of $J = 4.38 \times 10^6$ A/m^2. The resultant pulse pressures are in the order of $P = 0.1$ to 0.5 MPa. These pulses are formed along the external section of the sample (*skin effect*) and are the result of the collisions between the AC and DC waves. The collisions build local pulses in the liquid melt that in some cases generate cavitation. The pulses break the Si agglomerates promoting a refinement on Al-Si eutectic and hypereutectic liquid melts.

It is possible to refine the microstructure in semisolid alloys from the liquid state by refining the primary Si and α-Al dendrites. The refining effect is the result of the shearing forces formed by the rotational electromagnetic fields similar to electromagnetic stirring. However, in some cases, the use of vibration is not mandatory. Instead, for semisolid state one of the magnetic and/or electric fields (the nonrotational) can be turned off allowing a pure electromagnetic stirring in the absence of vibration. This can result in energy savings restricting the primary Si modification to the liquid state since the stirring motion reduces the formation of the Si-rich channels.

Ideally, the frequencies used for an ES or ESV apparatus are relatively low frequencies, in the range of 20–300 Hz. Otherwise, the system is transformed into an induction heating apparatus. This is because higher AC frequencies result in a thinner depth of penetration, increasing the density of the eddy currents in the skin zone to a narrower section in the exterior of the sample (Eq. 5.4 and Fig. 5.3). When the skin is too narrow and the intensity is too high, the Joulean heat increases, which results in temperature increase.

Radjai et al. [45, 46] conducted an ESV research using Al–Si hypo- and hypereutectic alloys. In this work, a superconducting magnet was used to create the stationary magnetic field with a variable flux density of up to 10 T. This setup was combined with an AC electric field. The frequency of the AC current was varied from 0 to 10 kHz and the amplitude between 0 and 4.25×10^6 A/m^2. The interaction of the DC electromagnetic field and the AC electric field developed electromagnetic pressures from 0.1 to 0.225 MPa, depending on the intensity. For this type of test, the AC field promotes the stirring motion improving homogeneity of the sample. Here one has to understand that the effect of this system on refining is almost negligible in the liquid state. In contrast, by using both fields in the semisolid state, it is possible to eliminate the columnar dendritic structure. Due to the intensity of the stationary field, it is important to have a good control over the AC field; otherwise, the Joulean heat will build up and the stirring and vibration effects are minimized.

Radjai et al. [45, 46] reported that short holding times (120 s) at the superheat temperature are not sufficient to "*dissolve*" Si in the hypereutectic alloy. However, other reports [57–84] mention the presence of the Si agglomerates, and they are independent of the holding time [56, 84]. The presence of the Si agglomerates is independent of holding times at the liquidus or casting temperatures. One of the major concerns for the refinement of the Si agglomerates and the primary Si particles is their macrosegregation that is sponsored with lack of homogeneity in the sample as well as the presence of dead stirring zones along the sample. The primary Si particles can freely coarse, and due to their lower density than Al and the remaining liquid metal, they segregate in the upper sections of the sample.

A simple approach to determine the segregation effect of the primary Si particles is Stokes' law. The law states that a spherical drop or object (e.g., Si agglomerates or primary Si particles in stage 1) in a viscous, but laminar, fluid must travel up or down proportionally to the difference in specific gravity with respect to the fluid. According to this definition, Stokes' law cannot be applied for a stirring or stirring and vibration system. However, here we are proposing a hypothetical case where the system is laminar in the absence of motion. A more sophisticated and reliable approach is that of Navier-Stokes [85–88]. The segregation rate is proportional to the square of the diameter of the particle(s) and inversely proportional to the viscosity of the fluid. Therefore, the Si agglomerates and primary Si particles segregate faster as they coarse. In the following section, Stokes' law, together with a sketch where the primary Si particles ideally segregate, is discussed. While this approach may not be for valid ES and/or ESV, it is perfectly valid for a still sample allowed to solidify under natural heat exchange conditions that take place when the treatment(s) is completed.

$$v = \frac{1}{18}\left[\frac{(\rho_S - \rho_F) g d^2}{\mu}\right] (cm/s) \tag{5.7}$$

where: ν the settling velocity (cm/s)

g the gravitational acceleration (cm/s^2)

ρ_S the density of the dropped object (g/cm^3)

ρ_F density of the fluid (g/cm^3)

d the diameter of the dropped object (cm)

μ the viscosity of the fluid ($Dyne{\cdot}s/cm^2$)

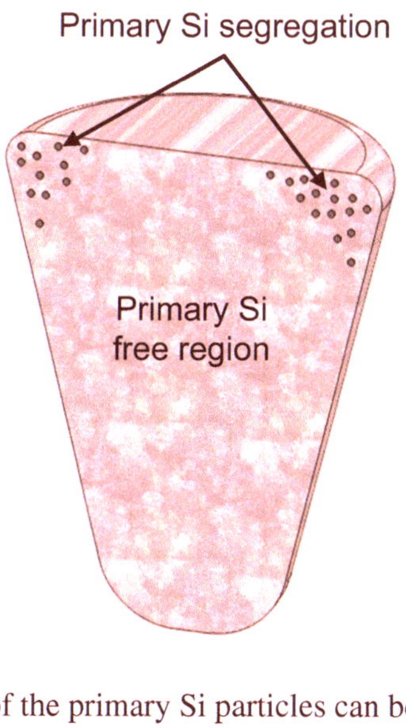

The average diameter of the primary Si particles can be determined using image analysis [89] from quenched samples [76, 77] and the Silicon Modification Level

(SiML) method [89]. With these methods, the average size of the Si (agglomerates or primary) particles can be determined precisely, and these results can be used for the particle's setting velocity (Eq. 5.7). The setting velocity can be translated into a direct measurement of the macrosegregation effect of Si. This macrosegregation rate can easily be up to three to five times higher for untreated samples than for ESV melt-treated samples [56].

Reference [56] reports the application of ESV melt treatments using 14 different Al-Si hypo- and hypereutectic alloys compositions. The chemical composition of the investigated alloys varied from 5 to 28 wt% Si and 1, 2, and 4 wt% Cu [56]. The ESV melt treatments were carried out using AC and DC electromagnetic fields. Figure 5.3 shows a schematic representation of a typical ESV melt treatment apparatus. In order to modify primary Si, the ESV treatments can be applied at superheat temperatures of up to 150 °C or as low as ±10 °C from the liquidus. The problem in applying the melt treatments at lower temperature is the rapid increase in strength by the solidifying melt, particularly after the dendrite coherency point [90, 91]. The magnetic field intensity was varied between 0 and 0.9 T and AC currents with frequencies from 0 to 100 Hz. The DC and AC fields were always generated using the same amplitude (from 0 to 100 A) for both currents, and the ESV treatment times varied from 0 to 560 s. Figure 5.4 sketches the optimum temperature ranges where the ESV melt treatments should be applied.

Figure 5.4 is a cooling curve showing the solidification reactions of a 390 Al-Si hypereutectic alloy. This curve is selected because of its similarity with an Al-Si hypoeutectic (e.g., 319, 356) with the extra reaction that is the primary Si. The benefits for the ESV treatment are better observed when applied between points 1 and 3. Ideally, if the intention is to modify only the primary Si, ESV should be focused between points 1 and 2. For modifications of the dendritic structure, ESV treatment can be used at temperatures as low as the Al-Si eutectic (e.g., point 4). ES treatments should be applied only between points 3 and 4, and they are not recommended for Al-Si hypereutectic alloys.

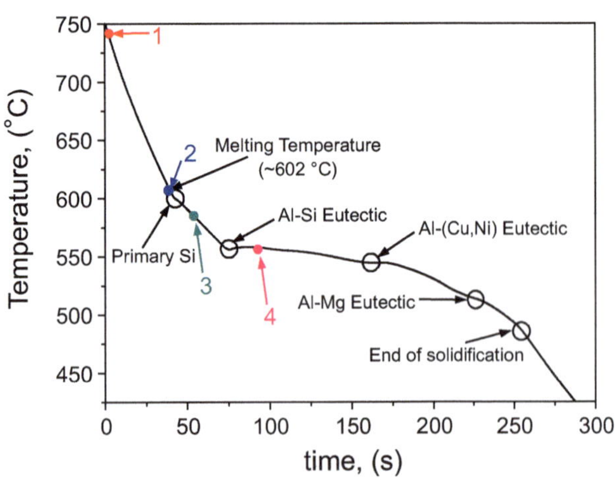

Fig. 5.4 Cooling curve of a typical Al-Si 390 alloy showing the locations where each of the solidification reactions occurs. Points 1, 2, 3, and 4 indicate the locations where the ESV and ES treatments are usually applied. Treatments below the dendrite coherency point are not recommended. Point 2 is where the primary Si start precipitating

Si agglomerates coarse in a quasilinear manner, and this coarsening can be hindered with ES in the semisolid state because the Si-rich channels that cause the coarsening of the primary Si particles cannot be formed. However, in the case of ES, it is necessary to use a quite powerful system because the solidifying melt builds up its viscosity. In contrast, the ESV treatment generates local pulses of significantly higher intensity that can easily break the Si channels (Fig. 5.5a) resulting in the multiplication of Si agglomerates, hence Si nuclei. The primary Si particles usually

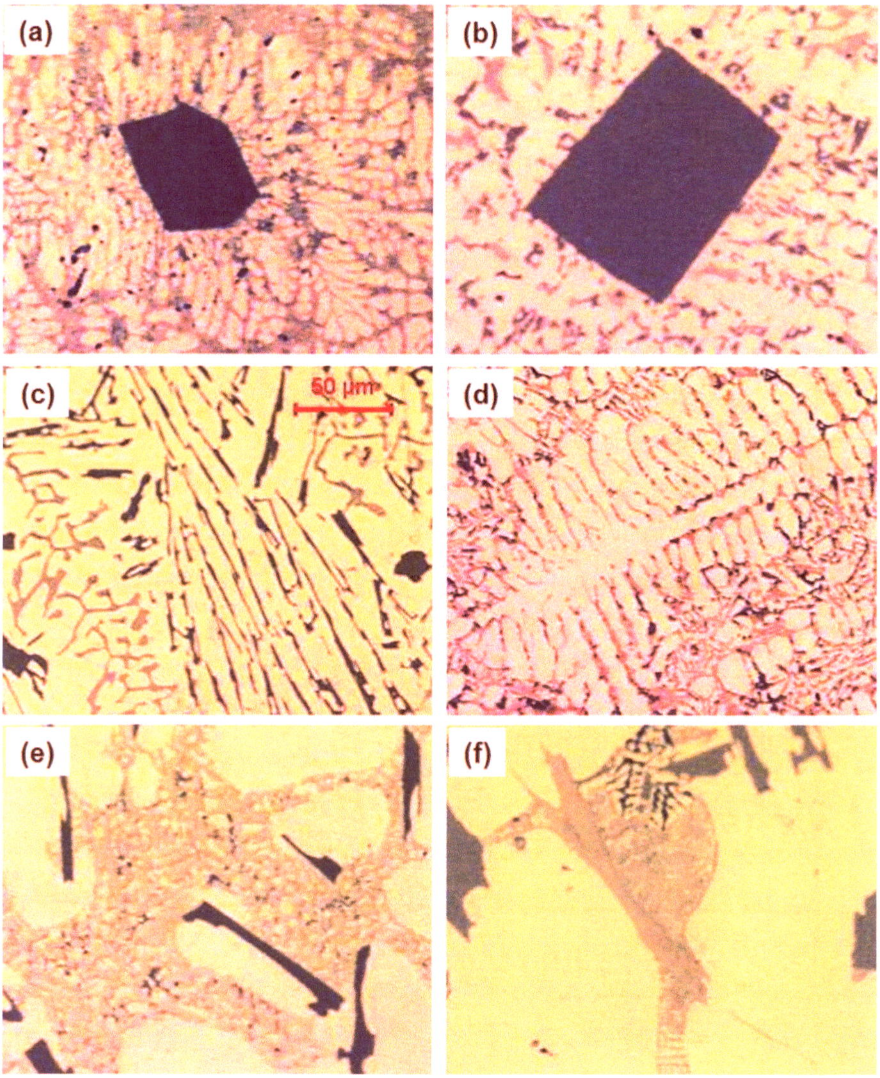

Fig. 5.5 Microstructures of the 390 Al-Si alloy. The (**a**–**f**) micrographs are representative of all the important solidification phases present in the melt: (**a**) a frozen primary Si particle and the respective "Si channels", (**b**) a fully formed primary Si particle, (**c**) Al-Si eutectic, (**d**) dendrite, (**e**) Al-(Cu, Ni) eutectic, (**f**) Si-Mg eutectic. The scale in (**c**) is the sample for all micrographs

stop coarsening when the precipitation of the Al-Si eutectic (Fig. 5.5c) starts. An ESV-treated sample can develop a dendritic structure that is unusual in Al-Si hypereutectic alloys and rather common in the hypoeutectic ones (Fig. 5.5d).

The effects of ESV or ES treatments at temperatures below the dendrite coherency point (slightly below the Al-Si eutectic) are not recommended. Therefore, the effects of ES or ESV in the Al-(Cu, Ni) and Mg-Si eutectics are rather negligible.

Some of the advantages of ESV treatments include the full refinement of the primary Si and Al-Si eutectic. Al-Si hypereutectic alloys that are ESV treated are usually free from primary Si. The melt has up to a 40% higher fluidity and a 300% higher feedability when the alloy was used for gravity die castings. Hardness, Vickers microhardness, and Brinell microhardness increased considerably. Heat-treated samples (using the T6 heat treatment) showed a further increase in mechanical properties of the material, reaching a maximum of 35% higher microhardness and 25% higher Brinell hardness.

Combinations of the ESV melt treatment and thermal treatments using cooling rates of 1.6 °C/s or higher are fundamental to fully refine primary Si. Those cooling rates are comparable to die-casting conditions. Castings produced under those conditions showed primary Si-free microstructures in a wide range of compositions (Al with 12–25 wt% Si). Figures 5.6 and 5.7 show the typical microstructure of a

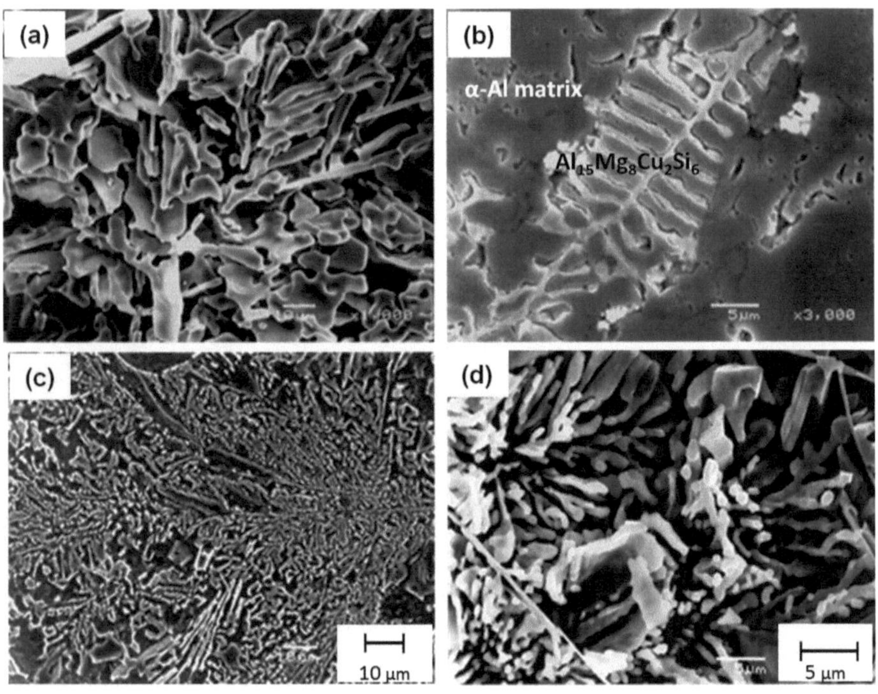

Fig. 5.6 SEM/SE micrographs of samples of the 390.1 [56] alloy deep-etched microstructures ESV melt treated using a $I_{AC/DC}$ = 100 A, a f_{AC} = 100 Hz, T_{ESV} = 610 °C for 3 min solidified at a cooling rate of 0.08 °C/s, (**a**) Al-Si eutectic particles, microstructures solidified at a cooling rate of ~1.6 °C/s (**b**) highly refined Fe-rich phases and, Keller's Reagent deep etched Al-Si eutectic at (**c**) 1000× and (**d**) 3000× [56].

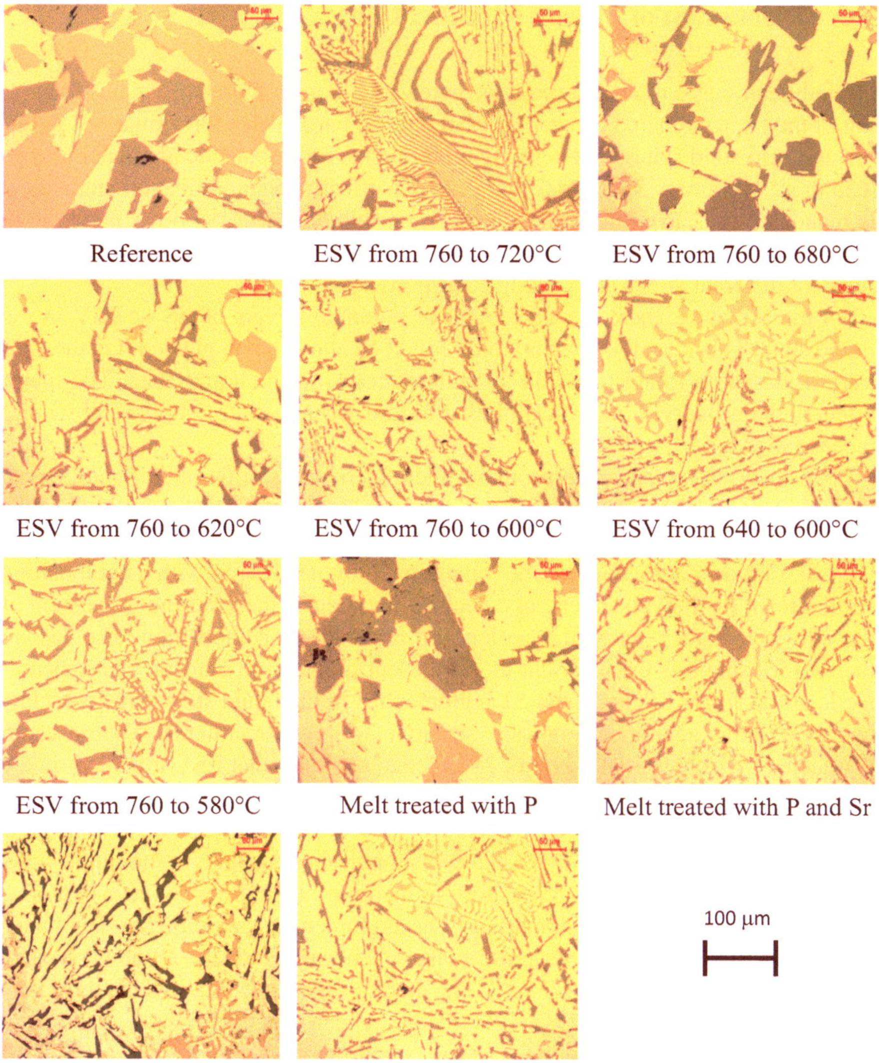

Fig. 5.7 Representative microstructures samples treated under ESV and chemical refinement for primary and eutectic Si. In all cases, for the chemical treatments 60 ppm of P and Sr were added

casting produced under the above conditions in some of the microstructures chemical refinement is used to enhance the ESV effect. Other microstructures, including deep-etched, Al-Si alloys that were ESV treated and solidified under natural heat exchange conditions, are also included in Fig. 5.6 for reference. It is important to notice the improved refining effect of the combined effect of the ESV melt treatment with the higher cooling rate.

5.5 Discussion of Silicon Modification

Si modification is a topic that requires review and more comprehensive understanding of the nucleation and coarsening phenomena, particularly for primary Si in liquid and semisolid states. The refining mechanisms for Si eutectic are better known than those of primary Si. Therefore, to develop effective Si modification methods, it is mandatory to carry out further research on this field. On the other hand, grain refinement is a well-known practice that has been used and implemented successfully for different alloys and casting conditions.

In the literature, there are no reports of potential effects of ES or ESV in the liquid state. This is well known not only for Al-Si alloys but in general. On the contrary, it has been found that ESV is quite effective in the liquid state. The major effect is the primary Si modification that has not been well accomplished by other means [45, 46, 56].

Two theories explain the reasons for the effectiveness of the melt treatments in the liquid stage. In the first one [45, 46], it is attributed to the lack of sealing of the melting system and the electromagnetic field that forced the liquid alloy, particularly Si, out; as a consequence, the remaining alloy had less Si with a composition similar to an Al-Si hypoeutectic or eutectic alloy. Hence, there was no refinement of the alloy, but rather macrosegregation and a change in composition. The second theory [56] supports the idea that the Si present in the Al-Si eutectic and hypereutectic alloys forms Si clusters or Si agglomerates in the liquid state, and such clusters or atomic agglomerations are sensitive to melt treatments. The research conducted by Robles et al. [49, 56, 76–77] demonstrates that ESV treatment in the liquid state is the most effective method to refine primary Si and its benefits are enhanced when the alloys is rapid cooled (C.R. > 1.6 °C).

Several researchers [57, 76–83, 92–94] support the idea that, in the liquid state, some elements, including Si, form short- to medium-range-order colonies of atoms. Those colonies are called agglomerates. The agglomerates have either broken bonds or unbonded atoms. This "weakens" makes them vulnerable to refinement when subjected to melt treatments [76, 95]. The methods that have reported the most successful melt treatments are ESV [76, 77] and electromagnetic vibration [45, 46]. This success can be attributed to the severe conditions created by vibration along the molten alloy and the weak internal energies holding the Si agglomerates together.

Some of the advantages of melt treatments applied in the liquid state are their potential to preserve the fluidity of the melt and casting temperatures. Yet, treatments such as ESV are quite effective in improving the fluidity even when applied in semisolid states. ES and ESV methods are contact-free and eliminate the potential for contamination. Likewise, ES or ESV systems are usually set away from any contact or interaction with the melt avoiding harsh conditions (e.g., high temperatures). This prevents or eliminates potential damage to the electromagnetic coils. ESV treatments in the liquid state needless power when compared to a system used for the semisolid state. This is because in the semisolid state, the alloy builds up strength, particularly below the dendrite coherency point. The dendrite coherency point is more common for Al-Si hypoeutectic and eutectic alloys. But well-treated (primary Si-free) Al-Si hypereutectic alloys are also dendritic.

In the liquid state, primary Si does not exist; instead, clusters of Si with limited or broken Si-Si bonding have been identified. The clusters are known as Si agglomerates. The primary Si particles initiate their growth below the liquidus temperature and it is concluded at the Al-Si eutectic temperature [56, 76, 84, 96–98]. At this point, the remaining Si initiates the Al-Si eutectic nucleation and coarsening. The above stages of the Al-Si hypereutectic alloys are key for the development of successful melt treatment apparatus and conditions. At temperatures above the formation of the Si agglomerates, melt treatments are not effective. Therefore, it is mandatory to use special tools (e.g., characterization, rapid quenching, etc.) to identify the exact moment when Si agglomerates nucleate. Melt treatments above this temperature will only mix the melt with no modification benefits.

High cooling rates from temperatures above the formation of the Si agglomerates is a potential solution to refine primary Si microstructure. Unfortunately, the hydrogen concentration increases sharply above 750 °C. As a result, the affinity of hydrogen increases dissolving larger amounts and promoting excessive porosity. Porosity is usually unwelcome due to its effects on the casting's integrity. Below the Al-Si eutectic or the dendrite coherency point, the solidifying alloy forms a skeleton (dendrite coherency point) with strength high enough to eliminate any possible benefit of the ES or ESV treatments. Any refining treatment, including mechanical, below this temperature is not recommended because it will introduce significant amounts of oxygen into the melt. Therefore, any treatment below the Al-Si eutectic temperature (~577 °C) is not recommended. Furthermore, for Al-Si hypereutectic alloys melt treatments applied at temperatures close to the liquidus are the most effective. Figure 5.6 shows the effectiveness of melt treatments and the effect of cooling rate during solidification on Al-Si hypereutectic alloys at temperatures above the liquidus.

An excellent example of an overall investigation of the effects of different melt treatments (chemical and ESV) and their combinations is given in [99]. Image analysis [89] and thermal analysis methods are potential tools to predict the characteristics of the castings including those treated with ES or ESV. This approach allowed a precise analysis of the level of modification of Al-Si alloys. This analysis can be used for a wide range of composition including melt-treated samples. It also demonstrates that temperatures, times, and chemical modifiers contribute to the refinement of primary Si. Furthermore, the effectiveness of thermal analysis is demonstrated by the cooling curve and first-derivative analysis. In this work, a thermal analysis methodology is presented, where it is demonstrated that the analysis of the undercooling of the Al-Si eutectic reaction can be used to assess the level of refinement of the Si-rich particles. The respective cooling curves and first derivatives are given in Chap. 2: "Thermal Analysis" and Fig. 5.6.

Microstructure refinement in Al-Si hypereutectic alloys can be achieved using ESV methods in the semisolid state. Conversely, in the semisolid state, stirring has limited, if any, effect on mirostructure refinement (Fig. 5.1). The primary Si particles should be refined in the liquid state. In addition, microstructure refinement can be further enhanced using chemical agents (Ti, B, etc.). By modifying primary

Si in Al–Si hypereutectic alloys can help its implementation for various applications, including automotive components, and potentially promote the ultramild wear proposed by Chen et al. [100]. Therefore, the greatest potential of ESV treatments is their application above the liquidus temperatures [101] of the respective Al-Si hypereutectic alloy.

Figure 5.8 presents cooling curves of Al-Si hypereutectic alloys treated under different conditions. The analysis of the undercooling for the Al-Si eutectic ($\Delta T^{Al\text{-}Si}_{E}$) is a good indicator of the effects of ESV and chemical treatments on microstructure. The narrower the $\Delta T^{Al\text{-}Si}_{E}$, the better the refinement of the microstructure. For example, when the 390 Al-Si alloys is allowed to solidified naturally, no treatment, the $\Delta T^{Al\text{-}Si}_{E}$ = 2.9 °C. Applying the ESV treatment from 760 to 680 °C, the $\Delta T^{Al\text{-}Si}_{E}$ changes to 2.6 °C. If the treatment is applied from 700 to 600 °C, the $\Delta T^{Al\text{-}Si}_{E}$ reduces to 0.43 °C and when chemical modifiers are added, the $\Delta T^{Al\text{-}Si}_{E}$ is almost negligible ($\Delta T^{Al\text{-}Si}_{E}$ = 0.08 °C). From these results, we can conclude that the level of modification of an Al-Si alloy can be monitored with a high-precision thermal analysis system. This type of system can be a great tool for industrial applications.

Figure 5.9 shows a hypothetical sketch of the effects of Si particle modification for reference, chemical, ESV, and ESV + P treatments. This image shows a refinement mechanism for the Si particles as a function of the treatment. In the literature, it has been proposed that primary Si coarsening is a linear function of temperature [76, 102]. In this graph, it is assumed that primary Si particles suppress their coarsening when the Al–Si eutectic starts nucleating. The primary Si particles were measured by optical microscopy in size of less than 200 μm².

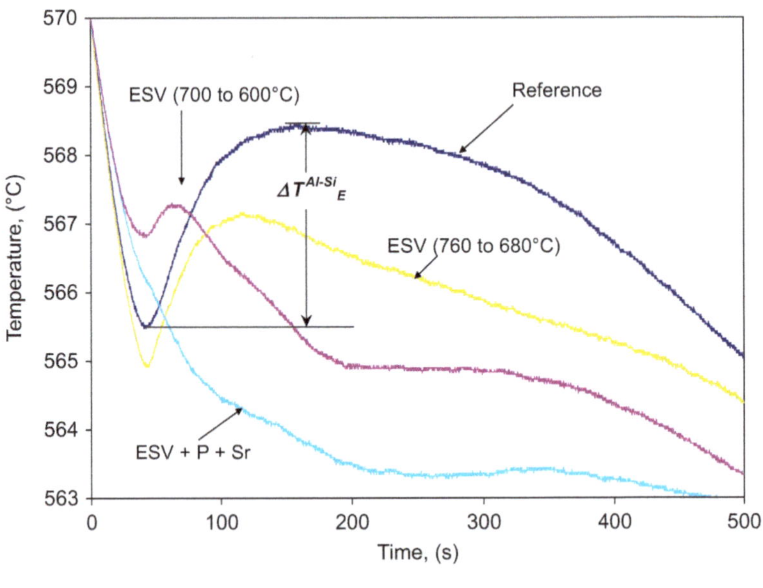

Fig. 5.8 Effect of the undercooling of the Al-Si eutectic ($\Delta T^{Al\text{-}Si}_{E}$) for samples treated using ESV and chemical treatments. Note: the narrower the undercooling, the higher the refinements of the microstructure; compare with microstructures in Fig. 5.7

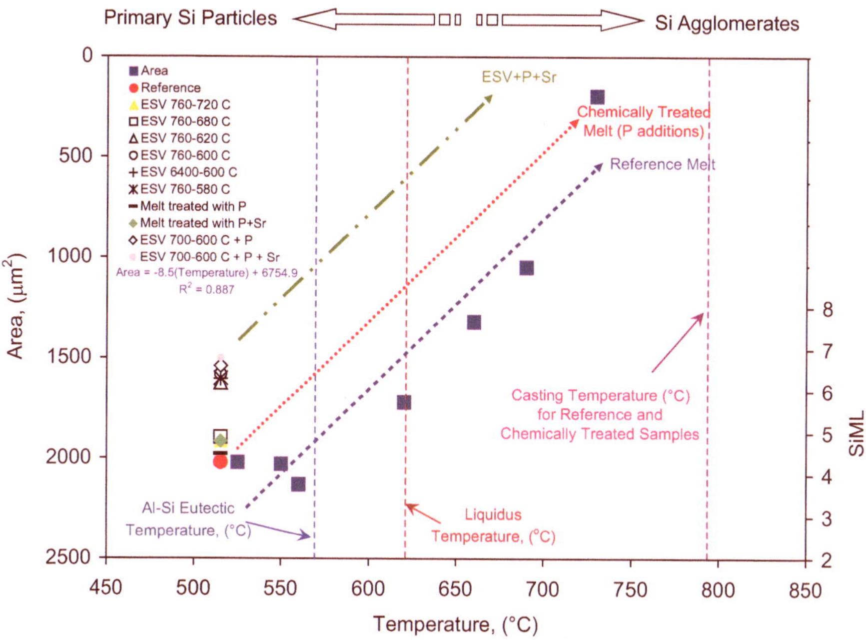

Fig. 5.9 Sketch of the coarsening of the Si agglomerates and the primary Si particles for Al-Si hypereutectic alloys solidified under different melt treatments, including additions of chemical agents such as P and Sr

The "hypothetical" coarsening presented in Fig. 5.8 for the primary Si particles assumes no changes in the diffusion rate due to the ESV. Therefore, the "hypothetical" coarsening pathways are expressed with parallel lines for the reference, ESV, and ESV + P melt-treated samples.

Use of a chemical modifier for the primary Si (e.g., P) promotes the nucleation of Si by causing Si–Si bonding. This phenomenon refines the Si particles as proposed in Fig. 5.9. Yet, only ESV limits its effect to size reduction of the Si agglomerates. When Al-P is added, it acts as a heteronucleant with an initial size that depends on the size of the AlP "seed" particle. The Al-P modifier needs to be added preferably at superheat temperatures of 200 °C or less. Chemical treatments not only reduce the Si particles but also lower the superheat temperatures.

Implementation of the ESV methods can be expensive, but mass production and energy savings can make them cost-effective. The major cost savings using this method relate to lower casting temperatures, closer to liquidus, less severe degassing requirements, elimination of different machining stages (where applicable), reduction of risers, and feeding system. Most importantly, the soundness and service characteristics of ES- and/or ESV-treated castings are improved noticeably. The 390 ESV melt treated with primary Si microstructures are ideal candidates to substitute cast iron for wear-resistant applications such as pistons, cylinder liners, breaks, etc.

The mechanism under which the chemical agents work is by forming new nuclei along the molten or semisolid alloy sponsoring the formation of new grains and/or particles. This obviously reduces their effectiveness with agglomerates because the size of the agglomerates depends on the melt's temperature. Consequently, at a particular temperature the agglomerates have a well-established size that chemical modifiers cannot reduce. In some cases, the heteronuclei can be as big as or bigger than the Si agglomerate at the respective superheat temperature limiting their effectiveness as primary Si modifiers.

The above arguments do not mean to say that chemical agents are not effective, but rather to point out some of their limitations when compared to other microstructure-refinement and phase-modification techniques. In fact, chemical modifiers are the most commonly used melt treatment methods and it is attributed to their low price and easy implementation together with the fact that there are not known better methods, exept for ESV. Unfortunately, the AlP has limited effect, but in most cases their marginal effect is enough for some casting applications.

References

1. Griffiths, W.D., and D.G. McCartney. 1997. The effect of electromagnetic stirring on macrostructure and macrosegregation in the aluminium alloy 7150. *Materials Science and Engineering A* 222 (2): 140–148. https://doi.org/10.1016/S0921-5093(96)10527-X.
2. Vives, C. 1992. Elaboration of semisolid alloys by means of new electromagnetic rheocasting processes. *Metallurgical Transactions B* 23 (2): 189–206.
3. Easton, M.A., and D.H. StJohn. 2013. The effect of alloy content on the grain refinement of aluminium alloys. In *Essential readings in light metals*, 393–399. Springer.
4. Quested, T.E., A.T. Dinsdale, and A.L. Greer. 2005. Thermodynamic modelling of growth-restriction effects in aluminium alloys. *Acta Materialia* 53 (5): 1323–1334. https://doi.org/10.1016/j.actamat.2004.11.024.
5. Quested, T.E., and A.L. Greer. 2003. Growth-restriction effects in grain refinement of aluminium. In *Light metals 2003*, ed. P. Crepeau, 827. Warrendale: TMS.
6. Kearns, M.A., and P.S. Cooper. 1997. Effects of solutes on grain refinement of selected wrought aluminium alloys. *Materials Science and Technology* 13 (8): 650–654.
7. Johnsson, M. 1994. Influence of Si and Fe on the grain refinement of aluminium. *Zeitschrift für Metallkunde* 85 (11): 781–785.
8. Desnain, P., Y. Fautrelle, J.L. Meyer, J.P. Riquet, and F. Durand. 1990. Prediction of equiaxed grain density in multicomponent alloys, stirred electromagnetically. *Acta Metallurgica et Materialia* 38 (8): 1513–1523. https://doi.org/10.1016/0956-7151(90)90119-2.
9. Vives, C., and C. Perry. 1986. Effects of electromagnetic stirring during the controlled solidification of tin. *International Journal of Heat and Mass Transfer* 29 (1): 21–33. https://doi.org/10.1016/0017-9310(86)90031-1.
10. Mondolfo, L.F., and J.G. Barlock. 1975. Effect of superheating on structure of some aluminum alloys. *Metallurgical Transactions B* 6 (4): 565–572. https://doi.org/10.1007/BF02913849.
11. Yamada, Sen-ichi, Keitarou Kobayashi, Hong Gang Ji, and Shigeo Tsukahara. 2002. Morphology of primary silicon crystals during isothermal holding at solid-liquid co-existent temperature on hyper-eutectic Al–Si alloys. *Journal of Japan Institute of Light Metals* 52 (4): 174–178. https://doi.org/10.2464/jilm.52.174.
12. Lojen, G., A. Krizman, and I. Anzel. 1997. Spheroidization of silicon in hypereutectic Al-Si alloys. *Livarski Vestnik* 44: 1–8.
13. Bergsma, S.C., et al. 1997. Semi-solid thermal transformations of Al−Si alloys and the resulting mechanical properties. *Materials Science and Engineering A* 237 (1): 24–34. https://doi.org/10.1016/S0921-5093(97)00112-3.
14. Perry, T. 1965. Ladle degassing with induction stirring and high vacuum. *Iron Steel Engineering* 42 (10): 89–94.

15. Johnston, W. C., G. R. Kotler, S. Ohara, H. V. Ashcom, and W. A. Tiller. 1965. Grain refinement via electromagnetic stirring during solidification. *Transactions of the Metallurgical Society of AIME* 233: 1856–1860.
16. Pakhomov, A.I. 1965. Effect of electromagnetic mixing on the content on gases in electric smelting of steel. *Leningradskogo Politekhnicheskogo Instituta* 235: 54–68.
17. Chekin, B.V. 1965. How electromagnetic forces help to remove non-metallic inclusions from molten metals. *Trudy Donetsk N-I Inst. Chern. Met.* 2: 127–136.
18. Nawakowskij, W.M., E.F. Lapschina, and M.S. Bloch. 1965. Importance of liquid stirring in passivation of iron and cast iron in sulfuric acids. *The Journal of Physical Chemistry* 230: 313–326.
19. Vives, C. 1990. Hydrodynamic, thermal and crystallographical effects of an electromagnetically driven rotating flow in solidifying aluminium alloy melts. *International Journal of Heat and Mass Transfer* 33 (12): 2585–2598. https://doi.org/10.1016/0017-9310(90)90194-Y.
20. Vives, C. 1989. Electromagnetic refining of aluminum-alloys by the crem process. Part I. Working principle and metallurgical results. *Metallurgical Transactions B: Process Metallurgy* 20 (5): 623–629. https://doi.org/10.1007/Bf02655919.
21. Dahle, A.K., and L. Arnberg. 1997. Development of strength in solidifying aluminium alloys. *Acta Materialia* 45 (2): 547–559. https://doi.org/10.1016/S1359-6454(96)00203-0.
22. Vivès, C. 1996. Effects of forced electromagnetic vibrations during the solidification of aluminum alloys: Part II. Solidification in the presence of colinear variable and stationary magnetic fields. *Metallurgical and Materials Transactions B* 27 (3): 457–464. https://doi.org/10.1007/BF02914910.
23. Adam, F., and C. Vives. 1998. Net shape forming of electromagnetically elaborated semisolid metal matrix composites. *Science and Engineering of Composite Materials* 7 (3): 223–229.
24. Yang, Y.S., and C.Y.A. Tsao. 1997. Viscosity and structure vibrations of Al-Si alloy in the semisolid state. *Journal of Materials Science* 32 (8): 2087–2092.
25. Haoran, G., M. Jiaji, and B. Xiufang. 1997. Thermal rate treatment and its effect on modifi cation of Al-Si alloy. *Transactions of Nonferrous Metals Society of China* 7 (1): 137–141.
26. Vivès, C. 1998. Grain refinement in aluminum alloys by means of electromagnetic vibrations including cavitation phenomena. *JOM–e* 50 (2).
27. Chen, J.S.J., K. Zhang, M.N.O. Sadiku, and A.A. Tseng. 1998. The electromagnetic enhancement of melt spinning and strip casting. *JOM–e* 50 (2).
28. Gabalthular, J.P., S. Steeb, and P. Laparner. 1979. Über die Struktur von Aluminium-Silizium-Schmelzen. *Zeitschrift für Naturforschung* 34a: 1305–1315.
29. Ward, P.J., H.V. Atkinson, P.R.G. Anderson, L.G. Elias, B. Garcia, L. Kahlen, and J.M. Rodriguez-ibabe. 1996. Semi-solid processing of novel MMCs based on hypereutectic aluminium-silicon alloys. *Acta Materialia* 44 (5): 1717–1727. https://doi.org/10.1016/1359-6454(95)00356-8.
30. Zavaras, A.A., R. Sobolewski, and C.B. Griffith. 1977. Continuous casting of metal using electromagnetic stirring. Google Patents: USA. p. 10.
31. Dantzig, J.A., and D.E. Tyler. 1984. Mold for use in metal or metal alloy casting systems. Google Patents: USA. p. 15.
32. Dussart, B. 1979. Continuous-casting mould provided with an electromagnetic stirring device. Google Patents: USA. p. 4.
33. Dussart, B.E.A.. 1978. Process for electromagnetic stirring. Google Patents: USA. p. 5.
34. Kasprzak, M.S., W. Kasprzak, and J.H. Sokolowski. 2009. Electromagnetic method and apparatus for treatment of engineering materials, products, and related processes. Google Patents: Canada. p. 16.
35. Long, L.J., and J.A. Henrickson. 1989. Counterflow electromagnetic stirring method and apparatus for continuous casting. Google Patents: USA. p. 22.
36. Metz, P. 1994. Electromagnetic stirring process for continuous casting. Google Patents: USA. p. 7.
37. Mori, T., K. Ayata, J. Miyazaki, T. Fujimoto, and H. Nakata. 1985. Electromagnetic stirring method in horizontal continuous casting process. Google Patents. p. 8.
38. Mulcahy, J.A., and J. Szekely. 1990. Magnetic control of molten metal systems. Google Patents: USA. p. 6.

39. Saeki, M., T. Nishio, and H. Omura. 1989. Electromagnetic stirring method. Google Patents: USA. p. 6.
40. Schmid, M. 1992. Electromagnetic stirrer for continuous casting. Google Patents: USA. p. 6.
41. Vives, C. 1985. Process for casting metals in which magnetic fields are employed. Google Patents: USA. p. 6.
42. Vives, C. 1987. Process for the electromagnetic casting of metals involving the use of at least one magnetic field which differs from the field of confinement. Google Patents: USA.
43. Whittington, K.R. 1984. Electromagnetic stirring. Google Patents: USA. p. 9.
44. Radjai, A., and K. Miwa. 2002. Structural refinement of gray iron by electromagnetic vibrations. *Metallurgical and Materials Transactions A* 33 (9): 3025–3030. https://doi.org/10.1007/s11661-002-0287-y.
45. Radjai, A., K. Miwa, and T. Nishio 1998. An investigation of the effects caused by electromagnetic vibrations in a hypereutectic Al-Si alloy melt. *Metallurgical and Materials Transactions A* 29 (5): 1477–1484. https://doi.org/10.1007/s11661-998-0363-z.
46. Radjai, A., and K. Miwa. 2000. Effects of the intensity and frequency of electromagnetic vibrations on the microstructural refinement of hypoeutectic Al-Si alloys. *Metallurgical and Materials Transactions A* 31 (3): 755–762. https://doi.org/10.1007/s11661-000-0017-2.
47. Spencer, D.B. 1971. *Rheology of liquid-solid mixtures of lead-tin*. Cambridge, MA: Massachusetts Institute of Technology.
48. Moon, D.B. 1990. *Massachusetts Institute of Technology: Cambridge, MA*.
49. Robles Hernández, F.C., J.H. Sokolowski, and W. Kasprzak. 2004. Electromagnetic refinement of the 319 aluminum structure. *Science Forum Light Metals*: 583–592.
50. Zhang, W.Q., Y.S. Yang, Q.M. Liu, Y.F. Zhu, and Z.Q. Hu. 1998. Structural transition and macrosegregation of Al-Cu eutectic alloy solidified in the electromagnetic centrifugal casting process. *Metallurgical and Materials Transactions A* 29 (1): 404–408. https://doi.org/10.1007/s11661-998-0193-z.
51. Li, B.Q. 1998. Solidification processing of materials in magnetic fields. *JOM–e* (2): 50.
52. Lu, D., Y. Jiang, G. Guan, R. Zhou, Z. Li, and R. Zhou. 2007. Refinement of primary Si in hypereutectic Al–Si alloy by electromagnetic stirring. *Journal of Materials Processing Technology* 189 (1–3): 13–18. https://doi.org/10.1016/j.jmatprotec.2006.12.008.
53. Kang, H.S., W.Y. Yoon, K.H. Kim, M.H. Kim, and Y.P. Yoon. 2005. Microstructure selections in the undercooled hypereutectic Al–Si alloys. *Materials Science and Engineering A* 404 (1–2): 117–123. https://doi.org/10.1016/j.msea.2005.05.041.
54. Lipton, J., W. Kurz, and R. Trivedi. 1987. Rapid dendrite growth in undercooled alloys. *Acta Metallurgica* 35 (4): 957–964. https://doi.org/10.1016/0001-6160(87)90174-X.
55. Abramov, V., O. Abramov, V. Bulgakov, and F. Sommer. 1998. Solidification of aluminium alloys under ultrasonic irradiation using water-cooled resonator. *Materials Letters* 37 (1–2): 27–34. https://doi.org/10.1016/S0167-577X(98)00064-0.
56. Robles-Hernández, F.C. 2004. *Improvement in functional characteristics of Al-Si cast components through the utilization of a novel electromagnetic treatment of liquid melts*, 251. Windsor: Mechanical Engineering, University of Windsor.
57. Bian, X., and W. Wang. 2000. Thermal-rate treatment and structure transformation of Al–13 wt.% Si alloy melt. *Materials Letters* 44 (1): 54–58. https://doi.org/10.1016/S0167-577X(00)00011-2.
58. Djurdjevic, M.B., W. Kasprzak, C.A. Kierkus, W.T. Kierkus, and J.H. Sokolowski. 2001. Quantification of Cu enriched phases in synthetic 3XX aluminum alloys using the thermal analysis technique. *AFS Transactions* 16: 1–12.
59. Kapranos, P., D.H. Kirkwood, H.V. Atkinson, J.T. Rheinlander, J.J. Bentzen, P.T. Toft, C.P. Debel, G. Laslaz, L. Maenner, S. Blais, J.M. Rodriguez-Ibabe, L. Lasa, P. Giordano, G. Chiarmetta, and A. Giese. 2003. Thixoforming of an automotive part in A390 hypereutectic Al–Si alloy. *Journal of Materials Processing Technology* 135 (2–3): 271–277. https://doi.org/10.1016/S0924-0136(02)00857-9.
60. Gruzleski, J.E. 2000. *Microstructure development during metal casting*, 238. Des Plaines: American Foundrymen's Society.
61. McDonald, S.D., K. Nogita, and A.K. Dahle. 2004. Eutectic nucleation in Al–Si alloys. *Acta Materialia* 52 (14): 4273–4280. https://doi.org/10.1016/j.actamat.2004.05.043.

62. Callister, W.D. 2007. *Materials science and engineering: An introduction*. 7th ed. 832. New York: Wiley.
63. Porter, D.A., and K.E. Easterling. 2001. *Phase transformations in metals and alloys*. 2nd ed. Boca Raton: CRC Press.
64. Evans, E.L., C.B. Alcock, and O. Kubaschewski, eds. 1967. *Metallurgical thermochemistry*. 4th ed. Oxford, UK/New York: Pergamon Press.
65. Wang, F., Z. Zhang, Y. Ma, and Y. Jin. 2004. Effect of Fe and Mn additions on microstructure and wear properties of spray-deposited Al–20Si alloy. *Materials Letters* 58 (19): 2442–2446. https://doi.org/10.1016/j.matlet.2004.02.027.
66. Narayanan, L.A., F.H. Samuel, and J.E. Gruzleski. 1992. Thermal analysis studies on the effect of cooling rate on the microstmcture of 319 aluminum alloy. *AFS Transactions* 100: 383–391.
67. Kierkus, W.T., and J.H. Sokolowski. 1999. Recent advances in cooling curve analysis: A new method of determining the 'Base Line' Equation. *AFS Transactions* 66: 161–167.
68. Polishchuck, V.P., and N.R. Aranova. 1965. *Trudy Donetsk N-I Inst. Chern. Met.* 2: 146–152.
69. Nafisi, S., et al. 2004. Advances in lightweight automotive castings and wrought aluminum alloys. In *SAE world congress*. Detroit: Society of Automotive Engineers.
70. Barclay, R.S., P. Niessen, and H.W. Kerr. 1973. Halo formation during unidirectional solidification of off-eutectic binary alloys. *Journal of Crystal Growth* 20 (3): 175–182. https://doi.org/10.1016/0022-0248(73)90001-8.
71. Sadigh, B., M. Dzugutov, and S.R. Elliott. 1999. Vacancy ordering and medium-range structure in a simple monatomic liquid. *Physical Review B* 59 (1): 1–4.
72. Spittle, J.A., J.M. Keeble, and M.A. Meshhedani. 1997. The grain refinement of Al-Si foundry alloys. *TMS Light Metals*: 795–800.
73. Baez, J.C., C. Gonzalez, M.R. Chavez, M. Castro, and J. Juarez. 2004. Fourier thermal analysis of the solidification kinetics in A356/SiCp cast composites. *Journal of Materials Processing Technology* 153–154: 531–536. https://doi.org/10.1016/j.jmatprotec.2004.04.119.
74. Cruz, H., C. Gonzalez, A. Juárez, M. Herrera, and J. Juarez. 2006. Quantification of the microconstituents formed during solidification by the Newton thermal analysis method. *Journal of Materials Processing Technology* 178 (1–3): 128–134. https://doi.org/10.1016/j.jmatprotec.2006.03.152.
75. Backerud, L., P. Gustafson, and M. Johnsson. 1991. Grain refining mechanisms in aluminium as a result of additions of titanium and boron, Part I. *Aluminum* 67 (9): 910–915.
76. Robles Hernández, F.C., and J.H. Sokolowski. 2005. Identification of silicon agglomerates in quenched Al-Si hypereutectic alloys from liquid state. *Advanced Engineering Materials* 7 (11): 1037–1043.
77. Robles Hernandez, F.C., J.H. Sokolowski, and J.D.J. Cruz Rivera. 2007. Micro-Raman analysis of the Si particles present in Al-Si hypereutectic alloys in liquid and semi-solid states. *Advanced Engineering Materials* 9 (1–2): 46–51. https://doi.org/10.1002/adem.200600173.
78. Yu, L., X. Liu, H. Ding, and X. Bian. 2007b. A new nucleation mechanism of primary Si by peritectic-like coupling of AlP and TiB2 in near eutectic Al–Si alloy. *Journal of Alloys and Compounds* 432 (1–2): 156–162. https://doi.org/10.1016/j.jallcom.2006.06.005.
79. Yu, L., X. Liu, H. Ding, and X. Bian. 2007a. A new nucleation mechanism of primary Si by like-peritectic coupling of AlP and Al4C3 in near eutectic Al–Si alloy. *Journal of Alloys and Compounds* 429 (1–2): 119–125. https://doi.org/10.1016/j.jallcom.2006.04.011.
80. Xiufang, B., and Q. Jingyu. 2004. Aluminium alloys: Their physical and mechanical properties. In *9th International Conference on Aluminium Alloys (ICAA9)*. North Melbourne: Institute of Materials Engineering Australasia.
81. Xiunanfang, B., Z. Guohua, Z. Shengxu, and M. Jiaji. 1992. *Cast Metals* 5 (1): 39–42.
82. Wang, W., X. Bian, J. Qin, and S.I. Syliusarenko. 2000. The atomic-structure changes in Al-16 pct Si alloy above the liquidus. *Metallurgical and Materials Transactions A* 31 (9): 2163–2168. https://doi.org/10.1007/s11661-000-0134-y.
83. Wang, W., X. Bian, J. Qin, and T. Fan. 2000. Study on atomic density changes in the liquid Al-Si alloys by X-ray diffraction method. *Journal of Materials Science Letters* 19 (17): 1583–1585. https://doi.org/10.1023/A:1006741626728.

84. Robles Hernández, F.C., and J.H. Sokolowski. 2006. Thermal analysis and microscopical characterization of Al–Si hypereutectic alloys. *Journal of Alloys and Compounds* 419 (1–2): 180–190. https://doi.org/10.1016/j.jallcom.2005.07.077.
85. Ashurst, W.T., A.R. Kerstein, R.M. Kerr, and C.H. Gibson. 1987. Alignment of vorticity and scalar gradient with strain rate in simulated navier-stokes turbulence. *Physics of Fluids* 30 (8): 2343–2353. https://doi.org/10.1063/1.866513.
86. Lee, P.D., A. Chirazi, R.C. Atwood, and W. Wang. 2004. Multiscale modelling of solidification microstructures, including microsegregation and microporosity, in an Al-Si-Cu alloy. *Materials Science and Engineering A-Structural Materials Properties Microstructure and Processing* 365 (1–2): 57–65. https://doi.org/10.1016/j.msea.2003.09.007.
87. Bhat, M.S., D.R. Poirier, J.C. Heinrich, and D. Nagelhout. 1994. Permeability normal to columnar dendrites at high fraction liquid. *Scripta Metallurgica et Materialia* 31 (3): 339–344. https://doi.org/10.1016/0956-716x(94)90293-3.
88. Fu, H., M. Dehsara, M. Krivilyov, S.D. Mesarovic, and D.P. Sekulic. 2016. Kinetics of the molten Al-Si triple line movement during a brazed joint formation. *Journal of Materials Science* 51 (4): 1798–1812. https://doi.org/10.1007/s10853-015-9550-7.
89. Robles Hernández, F.C., and J.H. Sokolowski. 2005. Novel image analysis to determine the si modification for hypoeutectic and hypereutectic Al-Si alloys. *JOM* 57 (11): 48–53. https://doi.org/10.1007/s11837-005-0027-z.
90. Veldman, N.L.M., A.K. Dahle, D.H. St John, and L. Arnberg. 2001. Dendrite coherency of Al-Si-Cu alloys. *Metallurgical and Materials Transactions A-Physical Metallurgy and Materials Science* 32 (1): 147–155. https://doi.org/10.1007/s11661-001-0110-1.
91. Chai, G.C., L. Backerud, T. Rolland, and L. Arnberg. 1995. Dendrite coherency during equiaxed solidification in binary aluminum-alloys. *Metallurgical and Materials Transactions A-Physical Metallurgy and Materials Science* 26 (4): 965–970. https://doi.org/10.1007/Bf02649093.
92. Qiao, J., X. Song, X. Bian, L. Zhu, and Q. Zhang. 2002. Application of AI-P master alloy to AI-Si piston alloy. *Special Casting & Nonferrous Alloys*: 43–45.
93. Xiufang, Bian, Wang Weimin, and Qin Jingyu. 2001. Liquid structure of Al–12.5% Si alloy modified by antimony. *Materials Characterization* 46 (1): 25–29. https://doi.org/10.1016/S1044-5803(00)00089-9.
94. Eskin, G.I. 1998. *Ultrasonic treatment of light alloy melts*. 1st ed. CRC Press Taylor and Francis Group.
95. DeWolf, I. 1996. Micro-Raman spectroscopy to study local mechanical stress in silicon integrated circuits. *Semiconductor Science and Technology* 11 (2): 139–154.
96. Djurdjevic, M., H. Jiang, and J. Sokolowski. 2001. On-line prediction of aluminum–silicon eutectic modification level using thermal analysis. *Materials Characterization* 46 (1): 31–38. https://doi.org/10.1016/S1044-5803(00)00090-5.
97. Djurdjevic, M.B., W.T. Kierkus, G.E. Byczynski, and J.H. Sokolowski. 1998. Calculation of liquidus temperature for aluminum 3XX series of alloys. *American Foundry Society Transactions* 47: 143–147.
98. Djurdjevic, M.B., J.H. Sokolowski, and T.J. Stockwell. 1998. Control of the aluminum-silicon alloy solidification process using thermal analysis. *Journal of Metallurgy* 4: 237–248.
99. Robles Hernández, F.C., and J.H. Sokolowski. 2006. Comparison among chemical and electromagnetic stirring and vibration melt treatments for Al–Si hypereutectic alloys. *Journal of Alloys and Compounds* 426 (1–2): 205–212.
100. Kyffin, W.J., W.M. Rainforth, and H. Jones. 2001. Effect of treatment variables on size refinement by phosphide inoculants of primary silicon in hypereutectic Al–Si alloys. *Materials Science and Technology* 17 (8): 901–905. https://doi.org/10.1179/026708301101510870.
101. Robles Hernandez, F.C., M.B. Djurdjevic, W.T. Kierkus, and J.H. Sokolowski. 2005. Calculation of the liquidus temperature for hypo and hypereutectic aluminum silicon alloys. *Materials Science and Engineering A* 396 (1–2): 271–276. https://doi.org/10.1016/j.msea.2005.01.024.
102. Hernandez, F.C.R., J.H. Sokolowski, and J.D.C. Rivera. 2007. Micro-Raman analysis of the Si particles present in Al-Si hypereutectic alloys in liquid and semi-solid states. *Advanced Engineering Materials* 9 (1–2): 46–51. https://doi.org/10.1002/adem.200600173.

Chapter 6
Mechanical Properties

Acronyms

AA	The Aluminum Association
ACARE	Advisory Council for Aeronautics Research in Europe
AFS GFN	American Foundry Society Grain Fineness Number
ASTM	American Society for Testing and Materials
CAFÉ	Corporate Average Fuel Economy
FCC	Face-Centered Cubic
GSPM	Gravity Semipermanent Molding process
GTAW	Gas Tungsten Arc Welding
HAZ	Heat-Affected Zone
HPDC	High-Pressure Die Casting
LFC	Lost foam casting
LPPM	Low-Pressure Permanent Mold
NHTSA	National Highway Traffic Safety Administration
OES	Optical Emission Spectrometer
PPAP	Production Part Approval Process
PSCP	Precision Sand Casting Process
RPT	Reduced Pressure Test
SPM	Semipermanent Mold Process
TPRE	Twin Plane Reentrant Edge
USCAR	United States Council of Automotive Research
VALF	Vacuum Assist Variant for Lost Foam

© Springer International Publishing AG 2017

F.C. Robles Hernandez et al., *Al-Si Alloys*, DOI 10.1007/978-3-319-58380-8_6

6.1 Mechanical Strength and Durability Assessment

This section will focus on the standard methods to assess aluminum castings' functionality in their intended application. Since most aluminum castings will encounter both static and dynamic loading, both monotonic (e.g., tensile, compression, and shear) and fatigue mechanical testing are part of the initial validation of a casting process and/or become part of a regular auditing scheduled testing regime.

Other testing regimes that are conducted to assess functionality involve corrosion testing, cryogenic cycling, permanent growth dimensional testing, machined surface strain gauge testing, casting dimensional (layout) testing, and debris (internal cavity cleanliness) and fluoroscopy auditing. Some of these will be discussed in a later section of this chapter.

6.2 Tensile Testing Properties

The first aspect before reviewing mechanical testing methods of aluminum castings is to discuss the greater approach toward how mechanical testing is treated and researched. Here we recommend internationally established standards and protocols, and in the general undergraduate level review of the determination of mechanical strength. Wrought materials have a texture established from either the forming or the rolling process. For wrought aluminum components the recrystallization to equiaxed grain structure due to the thermal processing after rolling, or to be present throughout the tensile test bar extracted. Thus, the geometric confines of the established tensile test bar shape dictate that failure will occur in the reduced gauge section and will thereby reflect the intrinsic mechanical state of the material. This is not the case in many cast aluminum components made today, and this issue will review some of the considerations that must be taken in order to establish the targeted mechanical properties required for casting components.

The factors that affect mechanical strength are, among others, the chemistry of the alloy, solidification rate, phase size, homogeneity, and finally heat treat regime. These would generally be considered inherent metallurgical factors. There are secondary factors that affect mechanical strength making more difficult the assessment of the inherent metallurgy of an aluminum casting that is being manufactured. The two most common secondary factors are pores and oxide damage. Both factors are discontinuities that have resulted in stress concentration. The size and distribution of pores and oxides can affect the number of test samples required to make a sound metallurgical determination of a process development. Statistical methods can be used to help with the data population determination (e.g., -3σ, Weibull modulus [1]) and is discussed later in this chapter. These secondary factors as described by the effect of pores and oxides can be considered noninherent metallurgical factors.

Inherent metallurgical factors that affect the tensile strength of many cast aluminum alloys are limited by the size and distribution of the primary phase constituents such as the secondary dendrite arm spacing (λ_2). The Si particle morphology is another important parameter having direct influence on mechanical properties of the Al-Si eutectic. Al-Si-Cu sand castings will range in elongation between 1% and 8%, rarely below 1%, and seldom exceeding 8% [2, 3]. There is also a contribution by the secondary phase constituents such as β-Al_5FeSi, α-Al_8Fe_2Si, α-$Al_{12}Fe_3Si_2$, and α-$Al_{15}(Fe,Cr,Mn)_3Si_2$ located in the interdendritic regions [4–14]. Typically, the heat treatment T6 (peak aging) and T7 (over-aging), solution, quenching, and then artificial aging used in most 3xx series alloys modify the properties within the face-centered cubic (FCC) aluminum structure. This modification is carried out via precipitation, which elevates the matrix microhardness, elongation, and yield point. Effects of the T7 treatment do not modify the morphology of the phases that grow via the twin plane reentrant edge (TPRE) growth method (i.e., β-Al_5FeSi and Si platelet crystals), due to the low solution temperature (470–499 °C) used for Cu-containing alloys [4–6]. Alloys such as 356 and 357 have solution temperatures around 530 °C and are usually heat-treated to spheroidize the β-Al_5FeSi and Si platelet crystals [12–14].

As mentioned, tensile properties for castings can have more scatter than what is typically found for wrought aluminum. This scatter in properties is what can limit aluminum casting usage in several dynamic applications. Wrought aluminum products can achieve nearly 20% elongation. This is due to the fact that cast alloys will have a larger volume of secondary phases, higher volume fraction of porosity, and the absence of crystalized texture.

An important factor toward the understanding of tensile properties includes both inherent and noninherent metallurgical conditions. The fractography analysis is used as a complement for tensile tests. Figure 6.1 shows the fracture surface of cylindrical tensile test samples from different aluminum casting processes. All tests were conducted using ASTM B557 [15] with gauge diameter 0.25″ (6.35 mm). This ASTM standard is common and is cited in many casting prints.

The stereomicrograph shown in Fig. 6.1a is the fracture surface of a tensile test sample extracted from a sand casting and has no secondary artifacts such as oxides or porosity. Figure 6.1b shows another fracture surface of an identical test sample from the identical casting process but this time with an oxide present. The oxide would have contributed to a lower elongation and/or lower ultimate tensile stress. Figure 6.1c, d shows the stereomicrographs of fracture surfaces of tensile test bars extracted from an High Pressure Die Casting (HPDC), and both gas voids and oxides are seen. Both secondary artifacts contributed to lower elongation and ultimate tensile stress.

The critical point being made here is that when trying to develop an alloy composition and/or heat treatment regimen, the tensile testing results may have a follow-up examination of the tensile fracture surfaces if scatter of test results is large. If oxides and gas voids are found to be the contributing factors to unfavorable tensile test results, then focus on the casting process itself (metal filling turbulence tendencies) or metal treatment practices may have to be examined further. Perhaps in cases where modification is performed, the component needs a review before the alloy chemistry or heat treatment is done.

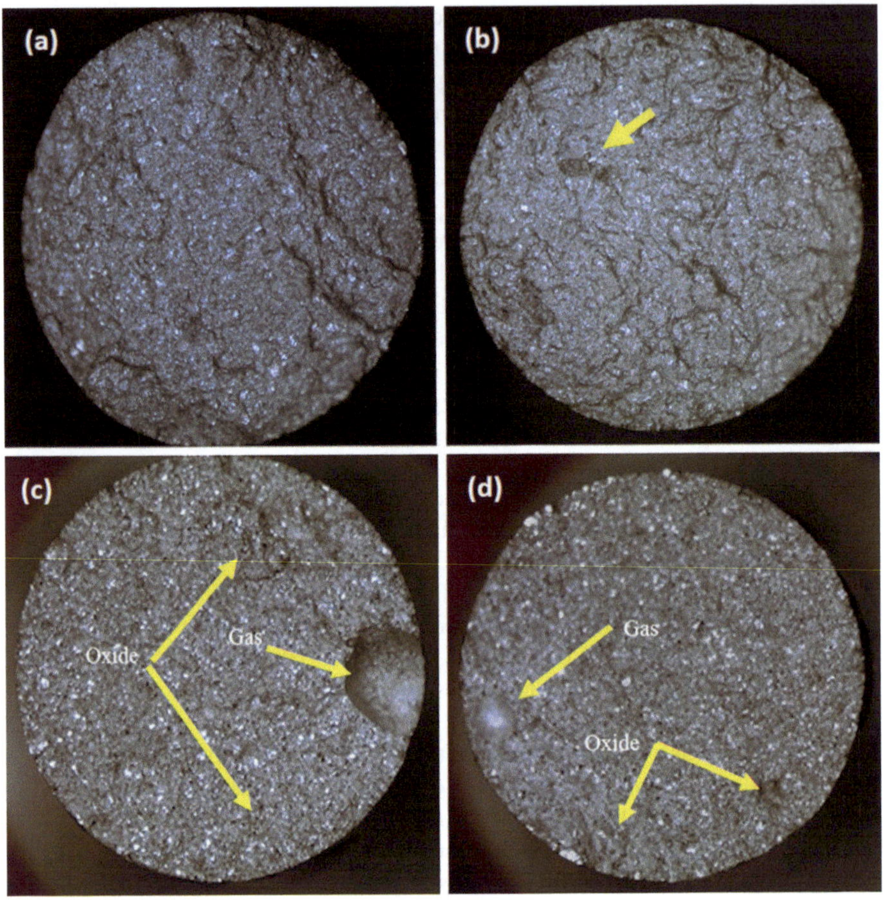

Fig. 6.1 (**a**) Sand casting with no defects. (**b**) Sand casting with oxide defect (*arrow*). (**c**) HPDC with gas void and oxides. (**d**) HPDC with smaller gas void and oxides

The scanning electron image seen in Fig. 6.2 reveals the oxide damage seen near the edges of the surface. These oxides contribute to the failure during plastic deformation. The oxides seen are the result of liquid metal fronts that coalesced inside the mold during filling. The quiescent filling process attempts to not only deter splashing but also allow for oxide skins on the front of advancing liquid metal fronts to fold outward. Inward folding contributes to coalescence and compromises the properties. Now, the presence of these oxides near the surface of the tensile test bar has most likely contributed to the drop in plastic elongation, more profoundly if the same two oxides identified were in the center of the fracture. This can be related simply due to the stress concentration effect of a discontinuity having one sharp edge versus one having two. Thus, simply by chance, a mold-filling-related issue could be disguised unless a sufficient number of tensile tests from many castings are conducted. This is why most automotive and aerospace aluminum castings

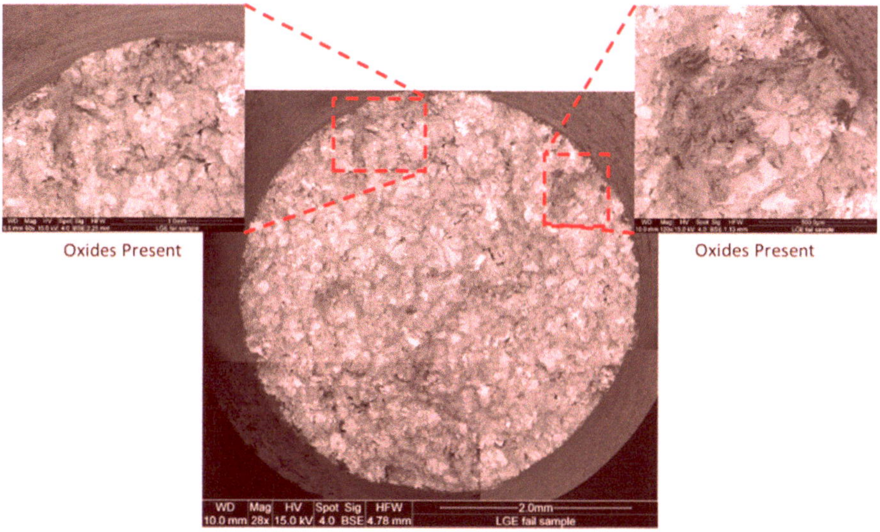

Fig. 6.2 Scanning electron micrograph of fracture surface of a tensile test sample extracted from a sand casting process

undergoing a process development require that a statistically based number of tests be conducted to validate process parameters.

In the case with semipermanent mold and sand mold casting processes, oxide damage is more difficult to determine unless SEM analysis is used. For HPDC process, the oxide damage becomes more evident. However, if process modifications have been conducted and some oxides and gas voids are still present, it may be just considered that this will be the inherent limitations of the casting process. Statistical methods will have to be used each time to check that the process is being established as part of an audit. In practice, most literature reports show a large number of tensile tests due to the scatter mentioned [16–19]. The results present the average, and a standard deviation is used for a final result. Boileau [20] reported that a minimum of 16 test samples were needed for 98% confidence interval for a true mean in coarse structures typical of sand castings. The parameters that tend to scatter the most are elongation and UTS. For a similar heat treatment condition and cooling rate, the yield strength should have a much lower level of scatter [21–23].

For cast component development, however, there will be a desired mechanical property requirement determined from stress analysis modeling. The casting developer must then have a process to achieve the mechanical test results in a well-established setup or production environment. The problem that may exist with an average or a mean of testing results with a standard deviation (σ) is that tensile results being presented in this fashion are implied as a Gaussian (also known as normal) distribution or a "bell-shaped" curve. A single $\pm\sigma$ (sigma) represents 68% of the true scatter of results and $\pm 3\sigma$ represents 99.73%. However, the true scatter of tensile test result data sets is not symmetrically Gaussian but rather is skewed [24].

The skewed distribution can arise due to the fact that a tensile test sample can have a low elongation or UTS due to a large defect. Opposite is the case when a test sample is essentially defect free and represents the highest result possible. Therefore, the method of Weibull statistics has been used, which can determine a single number (modulus) representing the reproducibility of data. This method has been growing in use in the metal casting industry.

6.2.1 Tensile Test Sample Geometries for Casting

As with fatigue test sample considerations, microstructural gradients and the physical size of a cast section in sand castings with chills provide limitations. They could affect the quality of tensile results and the predictive value. In HPDC components, where turbulence-based oxides are high, a situation is reached where tensile results are a guide, but are not necessarily indicative of casting performance. Here a major limitation is the presence of oxides that are not located near the cast surface. When a manufacturer outlines on a casting print a location for a tensile test sample along with its dimensions, it is imperative to follow those considerations as outlined in this section.

6.2.2 Precision Sand with Implemented Chill

In the case with precision sand process with implemented chill, microstructural gradients become an inevitable realty. Unlike the case with fatigue sample testing, tensile failure will generally occur in the weakest location of the reduced gauge section, and this may always occur near the same end of the test bar. Figure 6.3 shows schematic of a test bar with a total of eight documented fracture locations along the reduced gauge section. Also shown in the figure is the representation of the value of λ_2 along the test bar, reflecting the impact of the chill on the casting structure. If the ASTM E557 protocol is specified (commonly listed for automotive and aerospace casting prints), failure must occur in the middle two third of the reduced gauge section; otherwise, the test must be repeated with two more test samples. A total of four of the indicated fracture locations in the reduced gauge section are made with a nontapered geometry, which is typically the norm in tensile testing. In this schematic representation, two of the four fractures fall outside the middle two third of the reduced gauge section and thus must be repeated. When considering a much larger tensile test sample population, the time and cost impacts (e.g., each tensile test will range from USD 70 to 100/test) could be significant when attempting to gauge the mechanical integrity of a specified casting location and following the listed ASTM E566 standard on the print.

Customer specifications at times define a tensile test bar location based on stress modeling or a safety factor result. In many cases, little consideration is given

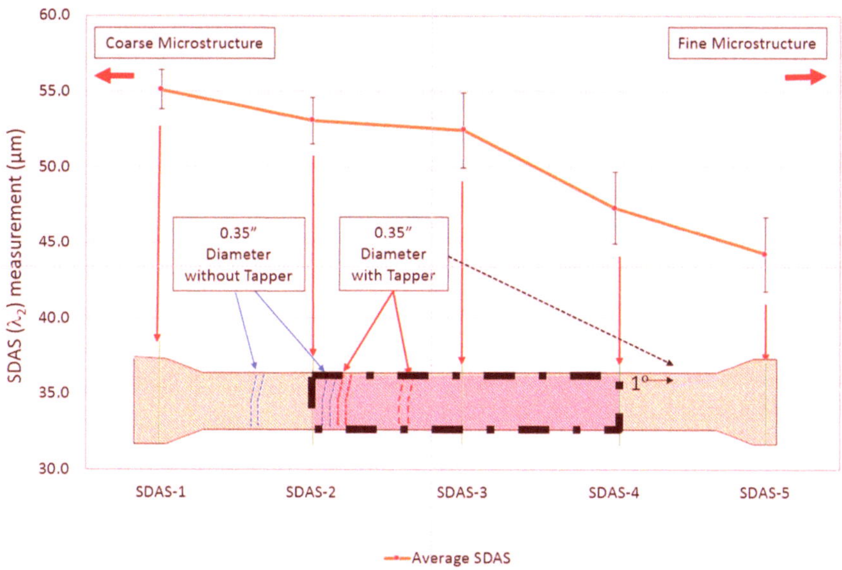

Fig. 6.3 Effect of fracture location in reduced gauge section of an ASTM E557 tensile test bar with a gradient in λ_2. *Dashed box* represents the ASTM E557 compliance fracture permissible

to the fact that a defined test bar geometry and dimension will encompass a large range of microstructure coarseness in terms of secondary dendrite arm spacing (λ_2) and porosity.

There are limited options that exist to get around issues where tensile test bars fail near an end of the reduced gauge section due to λ_2 gradients. The ASTM E557 does have a provision that allows a small tapper in the fabrication of the reduced gauge section (ASTM E557, Sect. 6.1, note 1). This clause in the standard indicates that a tapper of no more than 1° from the grip radius to the center of the reduced gauge section can be permitted. In conditions where a gradient in microstructural coarseness exists as in Fig. 6.3, and chronic shoulder or grip end failures occur, the tapper allowance may provide the compliance to ASTM E557 required (e.g., failure occurring in the middle two third of the reduced gauge section).

Referring back to Fig. 6.3, the additional four fracture points represented are from the identical test bar size (length and grip end diameter), and from the same casting location having the same λ_2 gradient, but now containing the 1 degree tapper fabricated in the reduced gauge section. In this case, all fractures occur within the middle two third of the reduced gauge section, and thus now all four samples are compliant with the ASTM E577 standard.

Without the tapper, 50% of the tensile test samples fail outside the middle two third and thus are not compliant. In order to maintain the required test sample number protocol, two more castings must be sacrificed, which is added cost and added time. To put a cost implication associated to this, consider that a 30-piece tensile test requirement is used to assess a casting process optimization. Using the fallout example seen in Fig. 6.3, consider the cost of the casting (an average of USD 200) and the cost of the tensile test (USD 70, which includes extraction, fabrication, and the test itself) that would cost USD 16 K without the tapper or USD 8 K with the tapper. The cost differential has an impact on metallurgical lab budgets.

While using a tapper in tensile test bar fabrication may reduce the number of repeat testing, it does also mean that larger-sized discontinuities with high stress concentration not located near the center of the reduced gauge section are not revealed. On the other hand, tensile testing may be viewed strictly on developing a process based on solidification rate, chemistry, and heat treatment.

6.2.3 Tensile Properties of Casting Sections Where Cylindrical Samples Are Not Possible

There are casting sections where tensile properties need to be assessed in geometrically small cross-sections. An example that will be reviewed is the need to assess the mechanical integrity of the inter-cylinder bore cast cross-section which is 4 mm thick. The smallest diameter test bar (reduced gauge section diameter of 6.35 mm) is not possible and therefore flat tensile test samples are the options. However, there can be limitations in terms of the quality of the tensile results, particularly in relation to a given defect size with respect to the cross-section area of the tensile test sample. Flat tensile bars originated from the need to test wrought metal components and the inherent stress effects of a noncylindrical cross-section are mitigated either by a texture in the microstructure or by the recrystallization that occurs after thermal process.

Figure 6.4 illustrates one of the most common reasons flat tensile test samples may be considered, and it is related to the inter-bore region of an aluminum cast cylinder block. Inter-cylinder bore cast structure generally sustains the highest operating temperatures and may have some of the lowest mechanical property strengths. It has become imperative during cast component development to have at least some gauge to understand inter-cylinder bore crack susceptibility. Cryogenic cycling can be used to assess its fracture susceptibility; however, no metric is achieved other than crack/no crack observations. Flat tensile bar testing can at least provide a measurable that then could be used for the casting process developers to modify a process parameter (e.g., chemistry, heat treatment, the use of Sr modifiers) to improve inter-cylinder bore strength.

In an example shown in Figs. 6.5 and 6.6, flat tensile test bars were used to understand the impact of the heating time (cast iron liner induction heating) on the inter-cylinder bore casting strength of the casting seen in Fig. 6.4. The cast iron liner inductor heating times vary from 12, 13, 14, and 15 s and occur prior to the casting process. As mentioned previously in the chapter, cast iron liners are heated to deter cold shut and misrun defects from forming. The unfortunate side effect of a higher liner temperature is that the inter-bore structure will become coarse and thus open the engine block to a higher susceptibility to inter-cylinder bore crack failure.

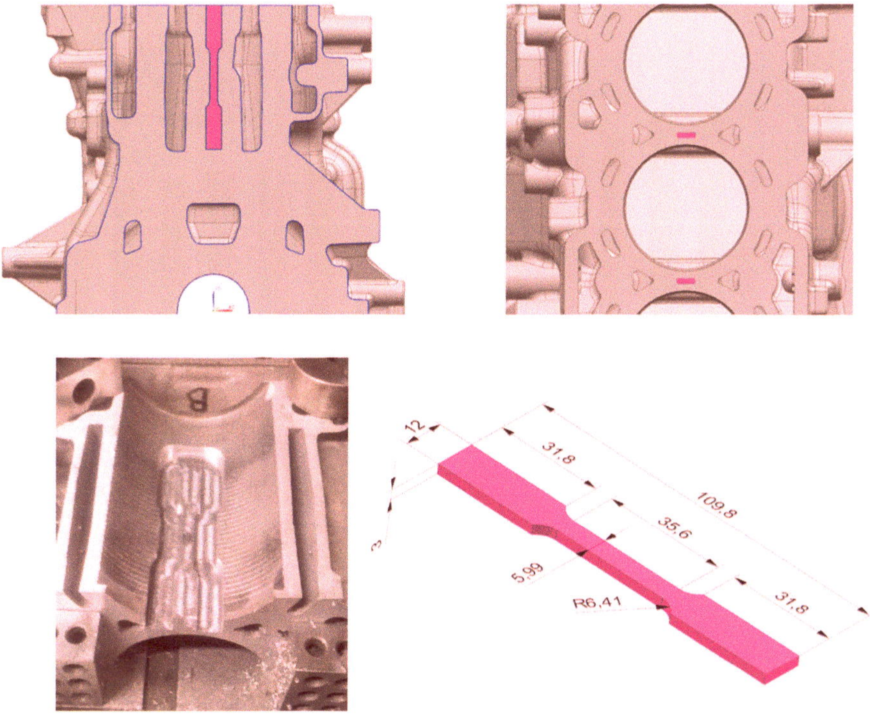

Fig. 6.4 Illustration of the extraction of a flat tensile bar from the inter-bore cast structure of an engine block casting

Figure 6.5 shows the UTS and total elongation results respectively with progressively longer cast iron inductor heating times. The 12 s inductor time dramatically provided improvements in inter-bore strength and thus points to a lower susceptibility to inter-bore crack formation. In this case, the castings made with the 12 s inductor heating time will also have to be examined to ensure that aluminum bore walls are free of misrun and cold shut defects, which would result in water jacket leak test failures at the customer's engine plant.

The reason for the improved mechanical strength was determined by conducting metallographic examination of the flat test bars and is shown in Fig. 6.6. The highest total elongation was found in samples with a lower λ_2 and porosity. Now it should be pointed out that the scatter of the test results was high, and this most likely is due to the effect of discontinuities. The discontinuities are inherent and noninherent with respect to a rectangular reduced gauge cross-section of the test bar geometry shown in Fig. 6.4. Recall that flat tensile test sample fabrication is best suited for the metallurgical testing of components which either have an equiaxed grain structure (rolled and thermally processed wrought products) or contain a uniformly directional texture (rolled wrought products). In fact, one may argue that inherent metallurgical assessment could not be made in this exercise; however, it was able to

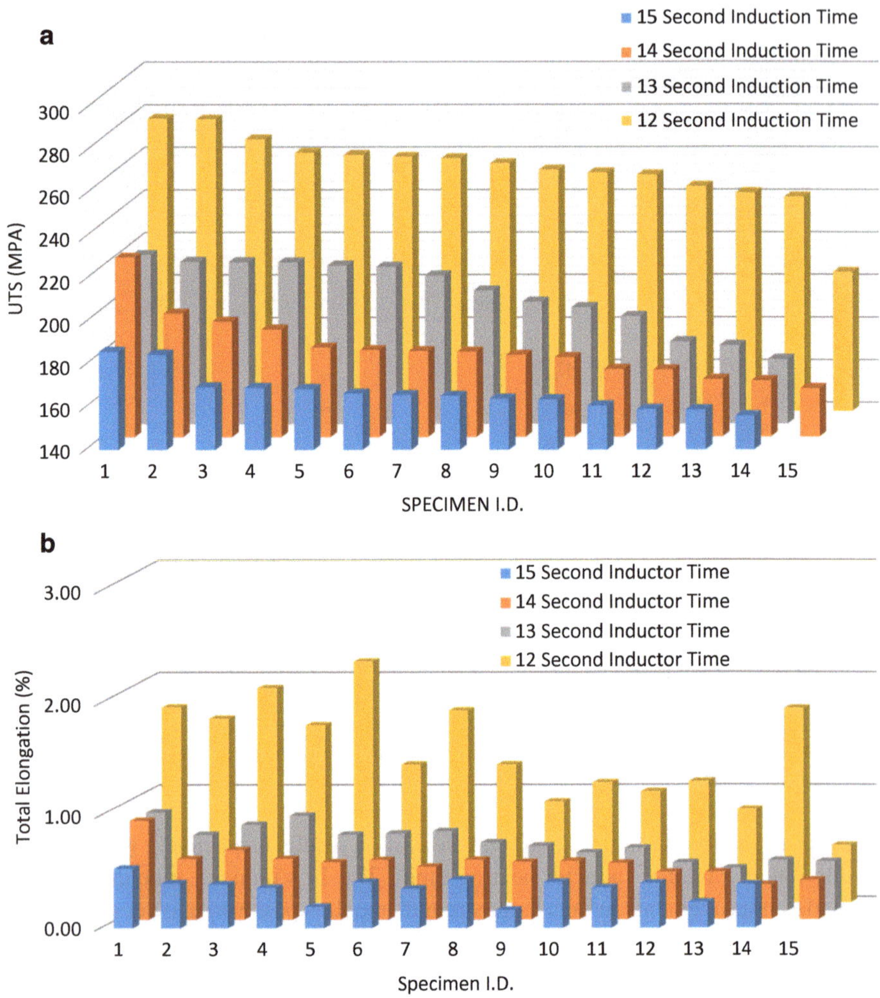

Fig. 6.5 Effect of liner inductor time on (**a**) UTS and (**b**) total elongation

provide directional information needed to confirm that reducing inductor heating times improved mechanical integrity of the inter-bore structure and thus reduced inter-bore crack susceptibility in an engine block casting.

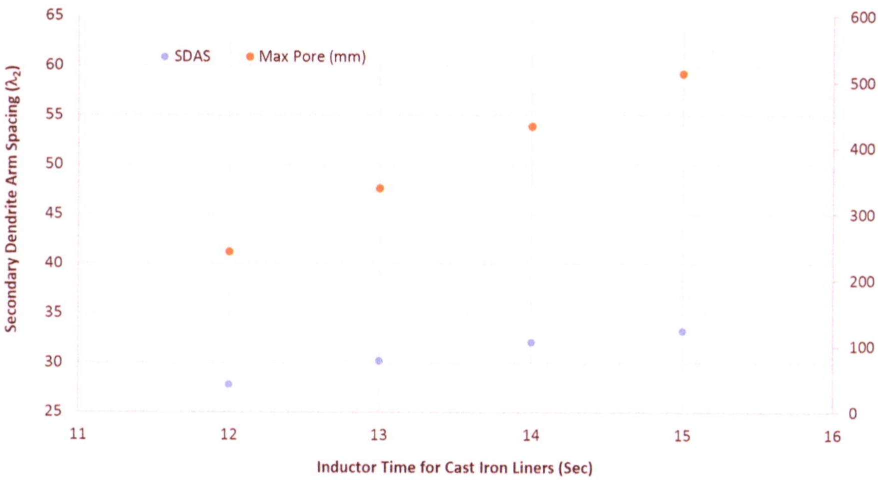

Fig. 6.6 Effect of inductor time on SDAS (λ_2)

6.3 Weibull Statistical Method for Treating Tensile Test Data

The approach used in Weibull analysis was proposed by Weibull [1]. Although this type of analysis has been developed with the investigation of mechanical properties of ceramics in mind, it has been successfully applied in the previous analysis of the tensile test data for aluminum-based cast components [24, 25]. Its usefulness lies in the fact that it accounts for the cast structural defects during the analysis of the tensile properties.

Since yield stress (YS) and yield strain (ε_{YS}) are rather insensitive to casting defects, they tend to follow a normal distribution. In that case, the statistical analysis tools are the mean, the median, and the mode, in combination with single/double/triple standard deviations (σ/2σ/3σ). However, the ultimate tensile stress (UTS) and the total elongation (El.$_{TOT}$) are casting defect sensitive and do not follow the Gaussian distribution. Instead they have a skewed distribution. As a result, Weibull statistics lends itself as a useful alternative when analyzing large quantities of UTS and El.$_{TOT}$ tensile data.

The analysis begins with a rearrangement of the data in ascending order. A frequency plot can then be constructed where the occurrence of a specific data point is quantified within narrow range sets. In this work, both the ultimate tensile strength (UTS) and total elongation (El.$_{TOT}$) have been treated using this approach. A cumulative frequency distribution of a random variable x can be represented using the three-parameter form of the Weibull (Eq. 6.1):

$$P(x) = 1 - \exp\left[-\left(\frac{x - x_o}{\theta - x_o}\right)^m\right] \tag{6.1}$$

P(x) is the cumulative frequency distribution of a random variable *x*, in this case either UTS, or $El._{TOT}$. x_o is the minimum allowable value of the analyzed variable, which in this case is 0 for both stress and strain data. θ is the scale parameter, also referred to as the characteristic value, and *m* is the shape parameter, also known as the Weibull modulus. Equation 6.1 can be converted into a linear representation by applying the "ln" operator to both sides of the equation to yield the following form:

$$\ln\left(\ln\left(\frac{1}{1-P(x)}\right)\right)=m\ln(x)-m\ln(\theta) \tag{6.2}$$

In this form, the equation can be interpreted by the linear form $y = mx + b$, where a variable *y* is plotted in the domain *x* to give a line with slope *m* and a *y*-intercept at *b*. By applying this approach, a plot of (*y* vs. *x*) was constructed for both the UTS data and the $El._{TOT}$ data sets. The results are a liner best-fit to calculate the slope *m* in each case. In order to interpret the cumulative frequency (*y*-axis) of the variable *x*, the *P(x)* is analyzed using Eq. 6.3:

$$P(x)=\frac{j-0.5}{n} \tag{6.3}$$

In Eq. 6.3, the reorganized *n* data points (ascending order) are assigned a ranking of *j* in the range of (*1-n*), such that each point has a ranking *P(j)* on the cumulative frequency plot. A combination of this ranking for *P(x)* data set with the previously described Eq. 6.2 yields the Weibull plot.

The usefulness of this analysis comes from the realization that different distributions of data of variable *x* will yield different amounts of scatter in the data sets. The Weibull modulus *m* in Eq. 6.2 is a quantitative measure of this scatter. The greater the value of *m*, the steeper the slope of the Weibull plot, and the smaller the scatter in the data. This translates into a narrower range of the analyzed property. The smaller the value of *m*, on the other hand, the shallower the slope of the Weibull plot is, which implies a large scatter in the analyzed data. This in turn makes more difficult the accurate predictions of the material property. As a result, it becomes evident that the *m* parameter in the Weibull analysis is a powerful tool in the tensile data analysis, material reliability, and the number of test samples in order to have reliable data.

6.3.1 Examples of Weibull Statics Used in Aluminum Castings

To illustrate the usefulness of the Weibull statistical method, three different aluminum casting processes used to manufacture engine blocks are studied and compared in terms of mechanical test results. The three processes are the HPDC-T5, sand casting with zircon sand and no chill (PSCP-Zircon-T5), and finally sand casting with chill (PSCP-Silica/Chill-T7). The T5 and T7 designations refer to the heat treat process used which was explained in an earlier section of this chapter. These are

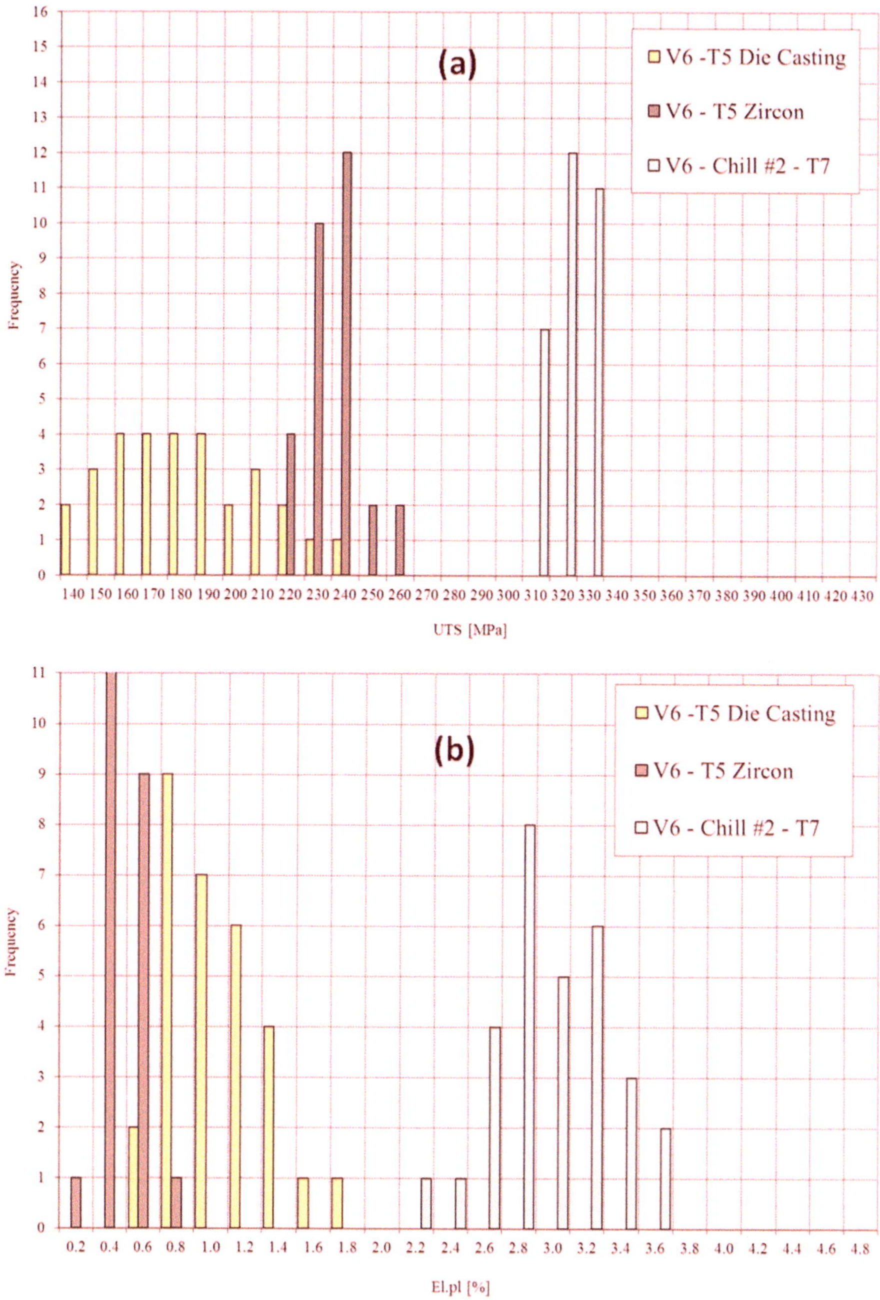

Fig. 6.7 Histogram plots for (**a**) UTS and (**b**) elongation

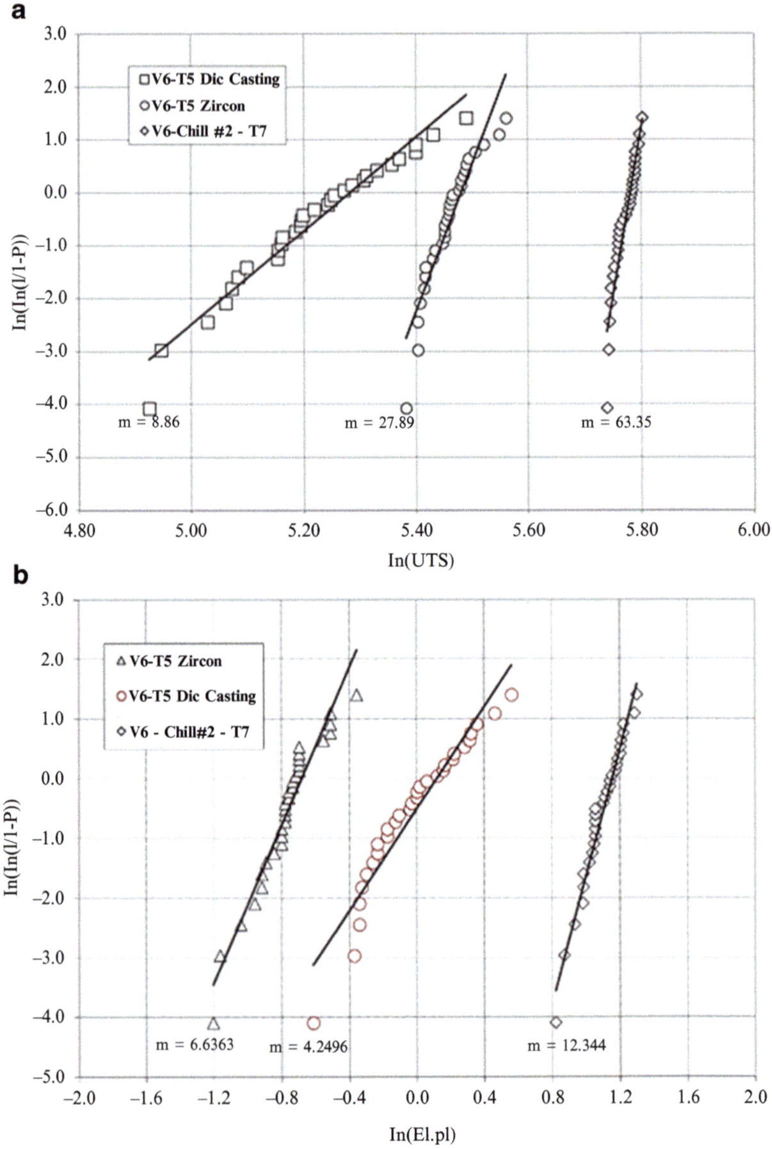

Fig. 6.8 Reliability plots of (**a**) UTS and (**b**) elongation

essentially the most common casting processes used to make engine blocks today. The histogram plots are shown in Fig. 6.7 for UTS and elongation, and the reliability plots are shown in Fig. 6.8.

Before discussing the results in Figs. 6.7 and 6.8, it is worth reviewing the differences of the casting processes first described earlier in this chapter. The turbulent fill process associated with HPDC produces a high density of oxides that leads to high porosity formation and, as a consequence, leads to a larger scatter in tensile test results. This scatter in tensile test results decreases with the PSCP-Zircon; however, the histogram plots show an overlap in UTS and elongation with HPDC. PSCP-Silica/Chill has a quiescent filling process that leads to a lower oxide generation, followed by rapid solidification imposed by the chill. This results in a low porosity level and consequently to a consistent tensile test value range.

There are secondary factors that contribute to the difference in reliability seen in Fig. 6.8. HPDC and PSCP-Zircon are thermally processed in a T5 condition, where PSCP-Silica/Chill is thermally processed in a T7 condition, and HPDC has elevated Fe concentrations in the melt. The results lead to Fe-based sludge formation (see light optical micrograph image in Fig. 6.9) that can be found in the interdendritic regions of the cast structure [10, 12, 26]. Nonetheless, considering the overlap in HPDC and PSCP-Zircon seen in the histogram plots of Fig. 6.7, it is possible to observe an HPDC sample that has a low enough density. The density is the result of the oxides/porosity formation. Under those conditions, it is possible to achieve a tensile test result similar to that found in the quiescent filled PSCP-Zircon test samples.

The similarities are the T5 thermal process and the porosity levels; the differences are the Fe concentrations (HPDC would use 380 alloy, while the PSCP-Zircon would use 319 alloy, see Table 8.3 on chapter 8) and λ_2 values as previously shown in Table 7 (see Table 8.3). Figure 6.8 also shows another difference in the reliability between the two casting processes for engine blocks. This difference is presumably due to the level of oxides. The PSCP-Silica/Chill however shows a clear separation due to low porosity, low oxides, and a refined microstructure imposed by the chill. The T7 helps develop the mechanical properties to a higher level than those achievable by the T5 alone.

As just mentioned, the value of SDAS ($\lambda 2$) is another secondary factor when comparing directly PSCP-Zircon and PSCP-Silica/Chill. Those methods usually contrast with HPDC, where there is sufficient quantity of porosity and the presence

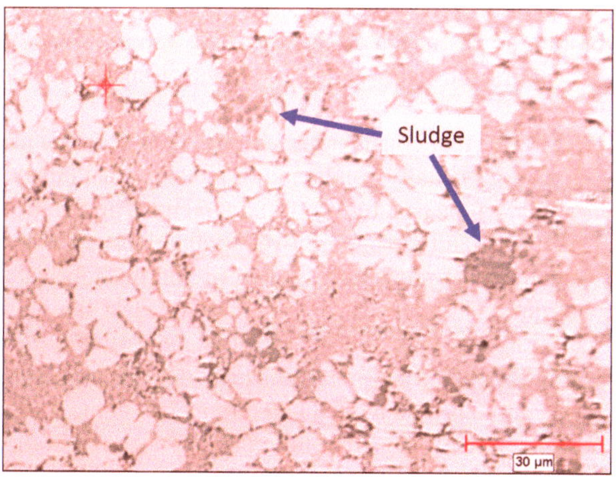

Fig. 6.9 Example of an Fe-based sludge particle in an HPDC alloy. The alloy type used in this case is a 383 alloy

of Fe-bearing sludge particles to override the effect of smaller $\lambda 2$ and finer-sized defect structure in the sand casting process. The HPDC and the PSCP-Zircon are thermally processed in a T5 condition, where PSCP-Silica/Chill is thermally processed in a T7 condition. The HPDC has elevated Fe concentrations in the melt, which leads to Fe-based sludge formation (see Fig. 6.8) that can be found in the interdendritic regions of the cast structure.

The turbulent fill process associated with HPDC produces a high density of oxides (bi-films) that consequently lead to higher amounts of porosity formation and ultimately lead to a higher scatter in tensile test results. This contrasts with PSCP process where the scatter is lower. However, when considering the values themselves, the histogram plots show notable overlap in elongation and UTS between the PSCP-Zircon and HPDC distributions. The PSCP-Silica/Chill process has a quiescent filling regime (leading to a lower oxide generation), followed by rapid solidification (shorter feeding ranges) imposed by the chill. This leads to lower porosity levels and consequently to more consistent tensile test value range.

6.4 Fatigue Durability of Aluminum Castings

The general growth of aluminum alloys in automotive powertrain systems has warranted the need to further understand the factors that affect their fatigue performance. In general, the small body of research in the fatigue of aluminum castings has historically been mainly, but not exclusively, devoted to the evaluation of aerospace component performance. The United States Council of Automotive Research (USCAR) consortium has compiled a large amount of information pertaining to the effect of porosity distribution in shape castings made from the A356, the A357, and the AM60 alloys [10]. Currently limited work has been performed for Al-Si alloys containing Cu such as the 319 [27, 28]. Nonetheless, much of the existing work on fatigue performance related to the 356 alloy is valuable and provides a sufficient understanding of the factors controlling fatigue crack growth in Al-based alloys.

In terms of this text, further consideration must be made on the usefulness of the fatigue test. Fatigue test is usually conducted in metallurgical labs when compared to the fatigue durability of the casting itself in its intended application. Fatigue testing in both, the aerospace and the automotive, applications tends to use the ASTM E466 protocol. Figure 6.10 shows the geometry of the fabricated test sample extracted from a casting. Under the ASTM protocol, the test sample is fabricated to the geometry and is then followed by a polishing stage where all circumferential machining lines are eliminated and determined under a stereomicroscope at 20X magnification. The sample is then loaded in a fatigue test frame and cycled between tension and compression, with the stress ratio defined as tested in the fully reversed condition ($R = -1$, R is defined as $\sigma_{min}/\sigma_{max}$). While there is no official standard on

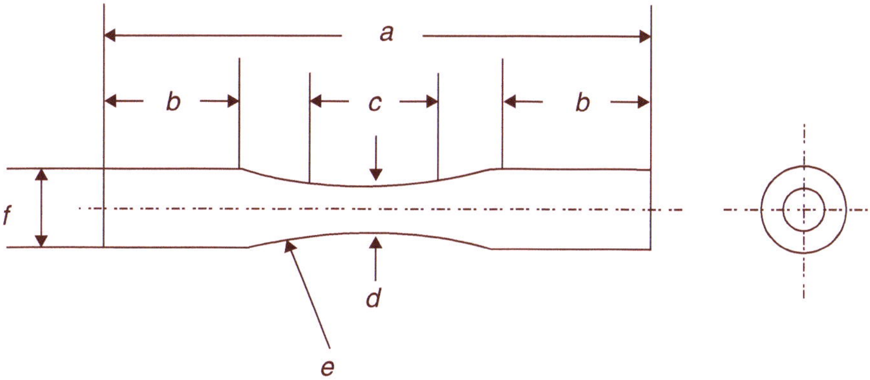

Fig. 6.10 Description of the fatigue test sample as outlined by ASTM E466–96 and commonly used in automotive and aerospace industries

the total number of cycles that need to be achieved for a successful test completion (also called Life Goal), the number generally cited on aluminum casting prints for both automotive and aerospace applications is 10^7 cycles.

The effect of having a polished surface is that it accentuates the maximum possible level of the stress concentration effect of a casting discontinuity caused by the solidification process. The reason is that surface polishing may eliminate any stress concentration leaving behind only the effect of the discontinuities in the casting. In most aluminum castings, this will be a microshrinkage defect nucleated by an oxide defect or columnar grain weakness. Figure 6.11 shows an SEM image of a fatigue test sample fracture surface with both casting defects indicated. One example is a columnar grain failure (low grain refiner content in the alloy), and the other is an oxide defect which nucleated a microshrinkage pore (turbulence upon filling). Surface roughness (nonpolished reduced gauge surface) will either suppress or exaggerate the stress concentration effect of the same discontinuity and thus skew a casting durability determination.

With the description just made using the fatigue testing method does assess the interior cast cross-section durability, it does not always reflect the way fatigue failure may manifest itself in a casting component in service. The origin of fatigue failures in aluminum castings is seen in mainly two regions. The first one is sharp edges on casting surface that have a high stress concentration under cyclical loading, and the second is exposed pores in the drilling and tapping of bolt holes. As mentioned, the fatigue testing that is commonly done for aluminum castings (fully reversed cyclic loading, $R = -1$) mainly attempts to gauge durability of the latter, as exposed pores from the drilling/tapping process provide a high stress concentration because of the one sharp edge present. Here it is important to state that internal discontinuities have two sharp edges and surface discontinuities have only one sharp edge. Stress modeling, and/or dynamometer engine testing, is the best way to reveal adverse effects of sharp edges on a casting surface.

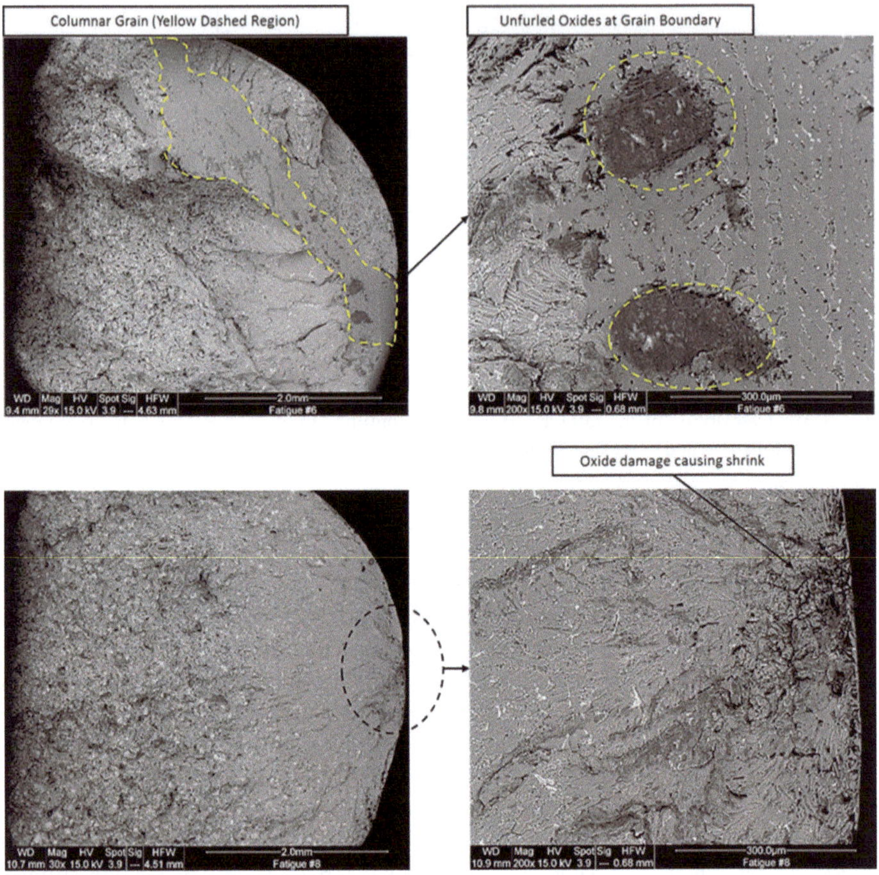

Fig. 6.11 Examples of fracture surfaces of fatigue test samples under scanning electron microscopy examination (backscattered electron mode)

There are different methods to approach improving fatigue durability in aluminum casting component once it has been identified. Nonchemistry modification methods tend to provide the fastest resolution as a requalification program does not have to be instituted for a component material change. For fatigue fractures that nucleate from a sharp edge from a casting surface (Fig. 6.12), a local radius change could be implemented. The loaded region/surface will then not have an elevated stress concentration at the sharp radius. Surface treatments may provide the same effect where shot blasting or surface penning may reduce stress concentrations at the casting surface.

For fatigue fractures that originate from an exposed pore in a final machining state, the approach can vary with casting processes used. In precision sand or semipermanent sand applications, cast in bolt holes is generally too difficult to

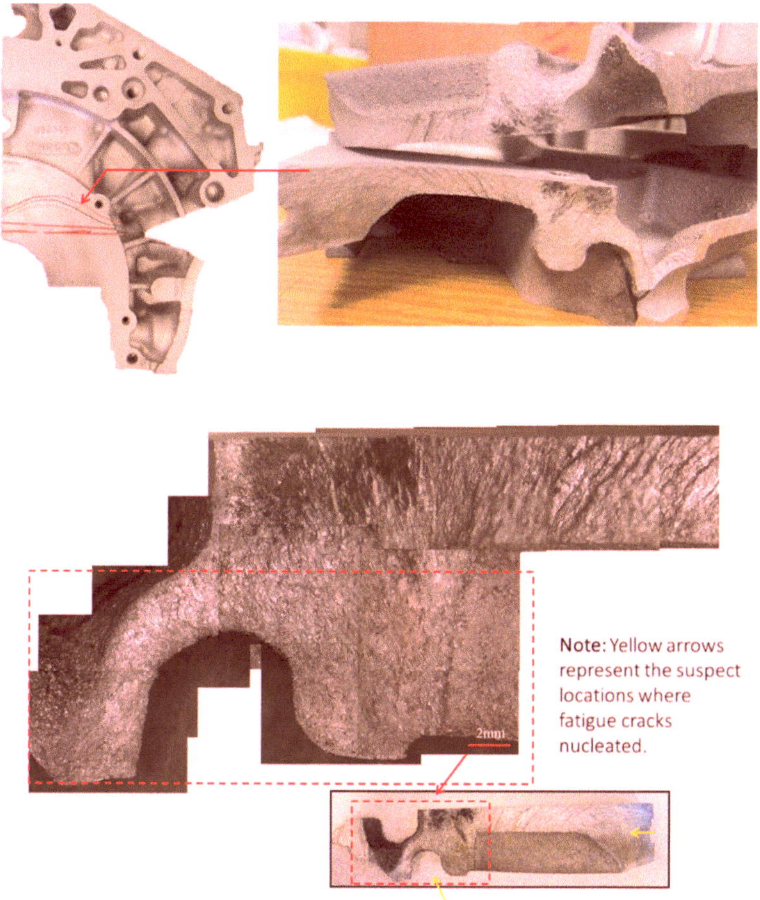

Fig. 6.12 Fatigue failure on a sharp edge of automotive casting

achieve in casting manufacturing stage, and as a result, they are drilled and then tapped out of solid metal. This of course opens up the possibility that a pore or oxide defect becomes exposed and is changed from a defect with two sharp edges to one, changing the stress concentration effect dramatically. An example of a fatigue crack originating from a drill/tapped hole in an automotive casting is shown in Fig. 6.13.

HPDC process does however have many of the casting bolt holes formed at the casting stage. Usually they are cast in the form of core pins that are protruding in the die mold, and the final tapping process is machined at the engine plant. The virtue of this process is that upon the initial stages of solidification, a rapidly solidified shell forms with the mold wall, and then the subsequent porosity and oxide damage that forms from the turbulent mold filling is forced to reside

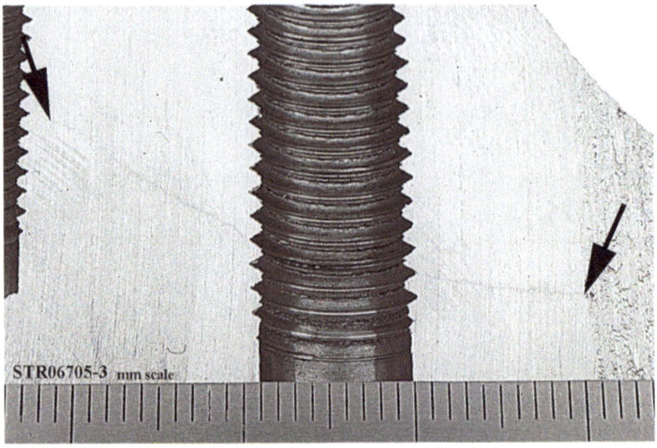

Fig. 6.13 Fatigue fracture which has nucleated from an exposed pore in a drilled and tapped bolt hole in a high cyclical stressed region of a casting component

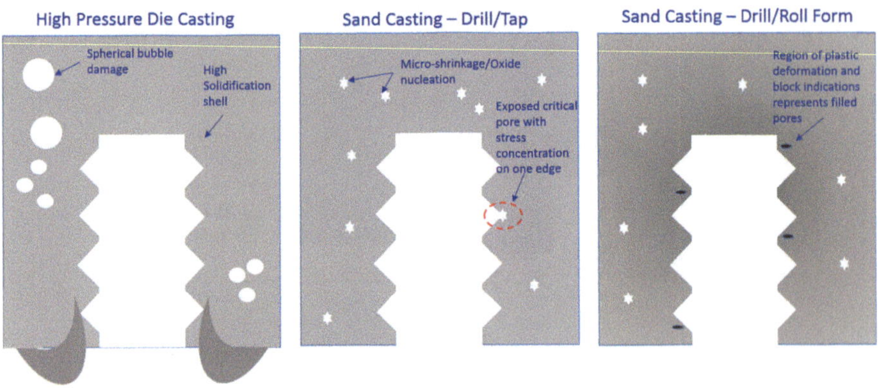

Fig. 6.14 Schematic of bolt hole with tapped threads in sand castings and HPDC. The porosity type that is seen in both processes is also included

in the center cross-section of the casting. Upon the tapping process, the oxide and pores from the metal turbulent filling process are not revealed, and thus a two sharp edge discontinuity doesn't become a one sharp edge discontinuity. This is illustrated in Fig. 6.14 for bulkhead bolt holes in the precision sand and HPDC process.

Precision sand casting process with an integrated cast iron chill, however, has a post-cast machining operation that may help further improve fatigue durability with respect to revealed pores and oxide defects. As a finishing operation in the formation of bolt holes, a *roll-forming operation* can be utilized. Roll forming is a mechanical process where the final thread geometry is achieved via a plastic deformative process in a drilled bolt hole. Roll forming not only improves yield strength, hardness, and wear resistance, but has an important impact on fatigue crack susceptibility. The plastic deformation that occurs in roll forming basically fills in pores with aluminum and reduces or eliminates its stress concentration effect. This is illustrated in Fig. 6.14. However, roll forming can be used only in bolt holes which have relatively good ductility (e.g., precision sand casting with crank bearing chill). Lower ductility regions can still be drilled and tapped, but following up with a roll-forming process could generate stress cracks and thus will result in a scraped casting.

6.4.1 Fatigue Staircase Generation for Al-Si Castings

The staircase method of determining a sustainable stress level is more amenable for product durability evaluation, which is intrinsic to the cast structure, but does not gauge sharp casting features on the casting surface. The axial fatigue test sample numbers required to generate a fatigue staircase test are generally lower than that for the stress against the number of cycles to failure in the S-N curve [29]. Before commencing a fatigue staircase study, a fatigue life cycle goal must be determined for the cast part. Normally this is defined as 107 cycles [27, 28]. The first test sample can be tested at a specific stress, and should that test sample sustain a total of 107 cycles, it would be considered a run-out or a passed test. The stress on a second test sample would then be increased by a set incremental amount ($\Delta\sigma$). This second test should fail to achieve a total of 107 cycles, and it would be a failed test. A third test sample would be stressed at the same level as the first one. Figure 6.15 illustrates the first three test samples of a staircase test.

The mean high cycle fatigue (HCF) stress (from a staircase test) is then determined for all the stress levels of the test samples, which did not fail (run-outs), or all the failed test samples, whichever number is smaller [28]. The total number of

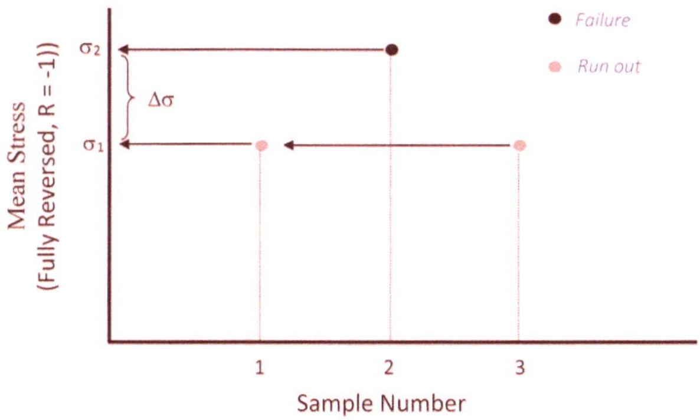

Fig. 6.15 Three fatigue test samples at the beginning of a staircase test. The sequence would continue until the desired number of tests has been completed

fatigue test samples used to support a complete staircase will vary from customer to customer, and it depends which statistical protocol is used to determine fatigue outcome. Some customers will use a 30-piece staircase where a mean stress value is determined. The fatigue mean stress ($F_{m.s}$) of the staircase study would be determined by taking the average stress of all the fatigue test samples which failed (higher value of stress) or all the fatigue test samples which passed (lower value of stress), whichever is greater in number. This is provided along with a standard deviation (σ) of the results. In some cases, a -3σ value is specified on the casting print, where the staircase mean stress minus the -3σ value is the threshold that has to be met. The logic behind using the -3σ value is that it gives design engineering a metric that indicates that component cycling in tension and compression stress that are under ($F_{m.s}$ -3σ) have a 99.999% chance of not sustaining a fatigue fracture in service.

There can be some drawbacks in using the staircase method in trying to assess cast aluminum durability. Firstly, the design of the fatigue test sample can play a significant role in the validity of the fatigue test conducted. Straight gauge fatigue test samples may sometimes suffer from shoulder failures, as per the ASTM E466 protocol. Should a shoulder failure occur, that sample must be replaced and tested at the same alternating stress. Figure 6.16 illustrates an example of a critical defect near the surface of a fatigue test sample that fails in the shoulder.

The overall effect of shoulder failures can be best illustrated in a staircase segment in Fig. 6.17. Fatigue crack location is indicated on a schematic of the fracture surface view in the perpendicular orientation. Red is used for shoulder crack location and green is for straight gauge crack location. The red circles represent the clock position where the crack occurred in the sample. The green color represents a critical defect

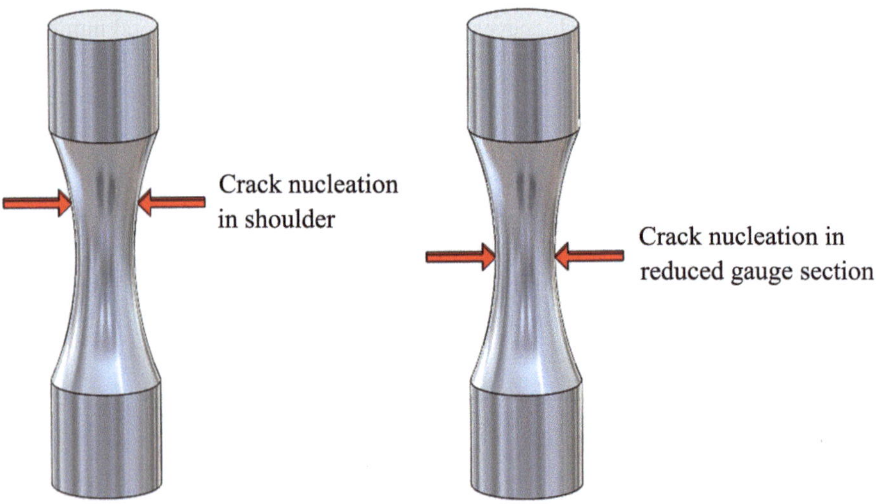

Fig. 6.16 Illustration of the fatigue crack location in the fatigue sample. Failure in the shoulder requires replacement

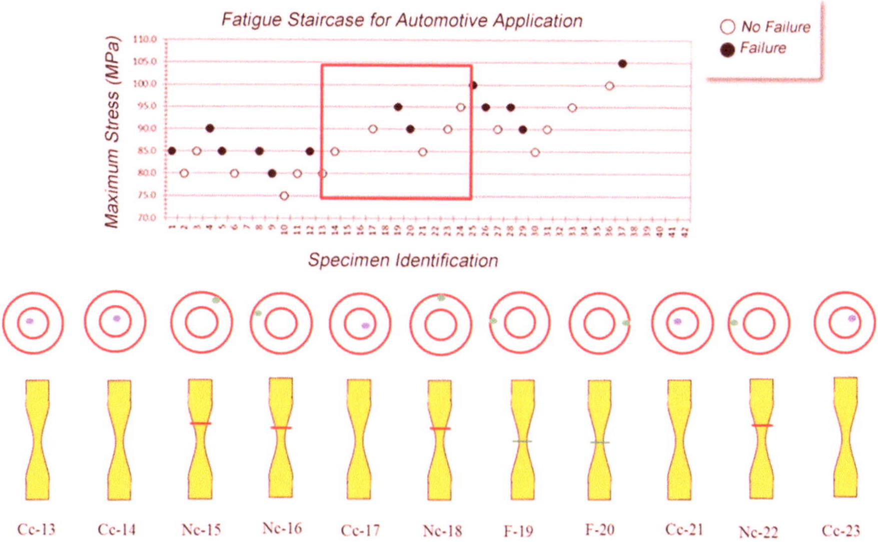

Fig. 6.17 Illustration of a fatigue staircase segment along with a representative of the fatigue test sample using the straight gauge fatigue test geometry. Fracture location indicated in the clock position (dot indicated on red cricles that denotes the top view of the fracture surface), and weather the fracture was compliant to the ASTM standard (e.g., shoulder failure or reduced gauge section failure). The nomenclature for the fatigue test samples are as follows: *Cc* compliant and completed (achieved life cycle goal of 10^7), *Nc* not compliant and not completed (shoulder failure), F failed (did not complete 10^7 life cycle goal)

located at the surface. The light purple color represents the critical defect placed near the center of the sample and was not able to result in a failure. In this case, 11 samples are tested sequentially. There were a total of five samples that reached the 10^7 cycles and thus resulted in the next sample to be tested at the 5 MPa increment. For the samples that failed to reach 10^7 cycles, four of the six experienced fracture in the shoulder. Instead of stepping the staircase down 5 MPa for the next sample, the alternating stress was repeated. This means that of the 11 fatigue test samples tested, only 2 technically failed and resulted in the 5 MPa step down for the next tested sample. Now let us consider the next example if the hourglass geometry is used. In this case, the geometric shape of the sample forces only the critical defect to fail as long as it is situated in the narrowest region of the reduced gauge section. In Fig. 6.18, considering an 11-sample segment as done for Fig. 6.17, we see that failure occurs in the mid-section of the hourglass geometry and thus complied with the ASTM 466 protocol in all test sample cases; there was no need for replacement tests.

Of the 11 test fatigue samples illustrated in Fig. 6.18, 6 samples failed, while 5 completed their 10^7 cycles and thus passed, the same number of samples as in Fig. 6.17 which only had 2 compliant and passed samples. The hourglass geometry tests the fatigue strength of a smaller cast structure than when using the straight gauge geometry, but the straight gauge geometry does suffer from necessary repeat of the alternating stress.

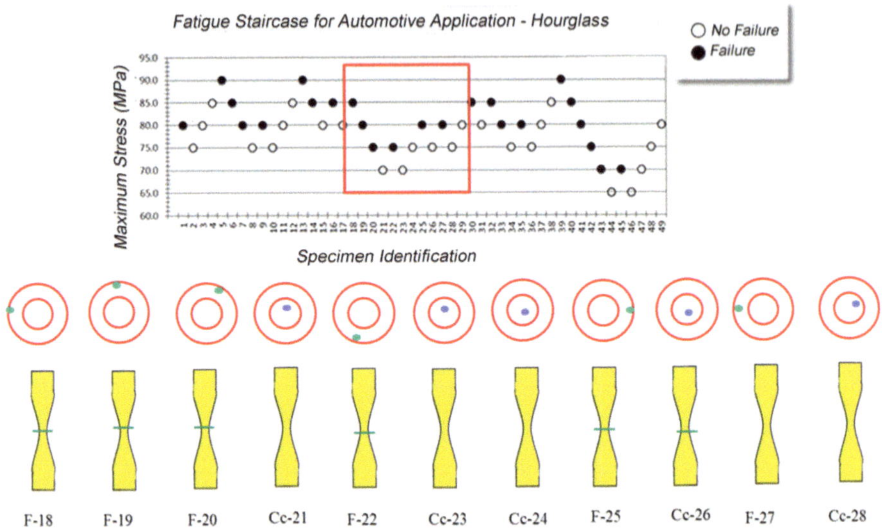

Fig. 6.18 Illustration of a fatigue staircase segment along with a representative of the fatigue test sample with hour glass fatigue test geometry. Fracture location indicated in the clock position (dot indicated on red cricles that denotes the top view of the fracture surface), and weather the fracture was compliant to the ASTM standard (e.g., shoulder failure or reduced gauge section failure). The nomenclature for the fatigue test samples are as follows: *Cc* compliant and completed (achieved life cycle goal of 10^7), *Nc* not compliant and not completed (shoulder failure), *F* failed (did not complete 10^7 life cycle goal)

The drawback with the hourglass geometry is that a smaller segment of the cast structure is truly being tested when compared to the straight gauge test bar geometry. Consider dimension C in Fig. 6.17, which covers 16 mm of the cast structure where a defect can be present and thus initiate failure. Additionally, if failure occurs in the shoulder (which actually becomes negated under the ASTM standard and requires replacement and retest), then a defect could be found in a 60 mm section of the cast structure and examined under stereomicroscopy to support feedback, even though that sample is not used in generating the staircase. Since the hourglass geometry of the reduced gauge section forces fracture in a smaller 5–10 mm cross-section, true impact of nearby defects on cast component fatigue performance may be underestimated. After all, the ultimate purpose of the fatigue staircase is not just to understand the impact of the alloy composition and structural refinement but, in some cases more importantly, to understand the role of cast defects on an aluminum cast components service life.

An additional problem that can be encountered with fatigue testing of casting sections is the effect of microstructural gradients that may exist in the fatigue geometry fabricated. Some casting processes, particularly sand castings that employ the use of chills (added to sand cores to both suppress shrinkage formation and improve fatigue strength of a casting section), will inevitably have one end that has a finer microstructure than the other. An example is shown in Fig. 6.19, where a straight gauge fatigue test bar is used. The secondary dendrite arm spacing (λ_2) is plotted

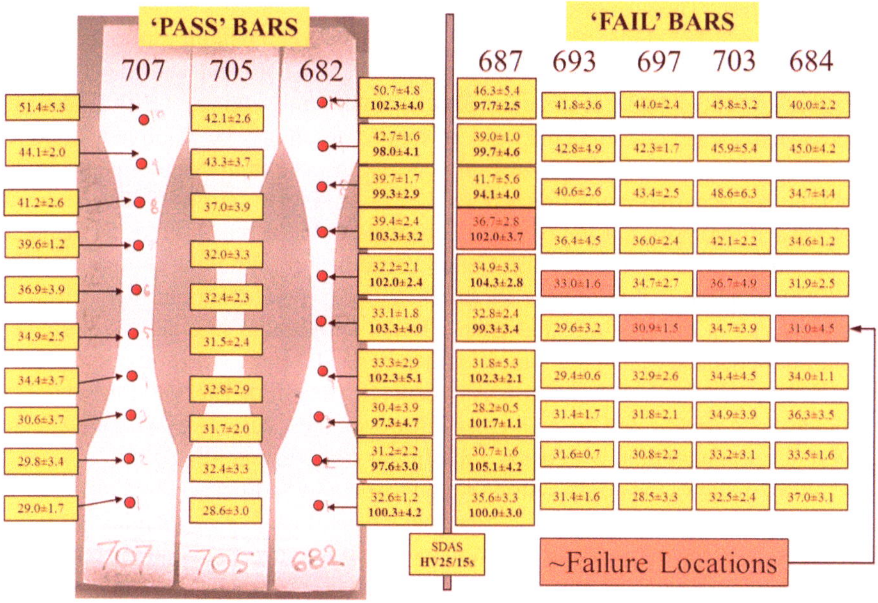

Fig. 6.19 The effect of λ_2 profile along the fatigue test bar fabricated from an Al-Si casting

along the length of the fatigue bar in both passed (again reaching 10^7 cycles) and those which did not.

The microstructural gradient can be evident near the top of the figure – coarse away from the chill (or top of the figure) and finer near the chill (bottom of the sample in the figure). What can be seen from the figure is that the fracture occurred not where λ_2 is the largest in the reduced gauge section, but where the largest critical defect was located. Using the hourglass geometry would have forced failure to occur always in the mid-section; however, if a critical defect is away from the mid-section, then the straight gauge geometry would be better to capture true fatigue performance of the cast section of interest. This distinction becomes critical in the development of print specifications, where fatigue test sample requirements are normally contained. It does become important, however, that the frequency of retesting due to shoulder failures becomes a reality, and the casting print specifications need to consider that replacement samples should be produced.

6.5 Weld Repair of Aluminum Castings

Weld repair is typically conducted to salvage scrap castings which have a defect that can be grind or cut out without damage to the casting (stress cracks). The gas-covered weld process is used mainly to provide filler metal back into the cut-out casting section. The gas tungsten arc welding (GTAW) process is used with argon being the preferred choice for the shield gas. While argon/helium mix shield gases are possible, they do provide a hotter weld, which is not ideal with most casting thicknesses. The shield gas practice used in aluminum welding is critical to deter excessive oxide generation in the weld. In some cases, it may be required to conduct fluorescent penetrant inspection or Eddy current inspection to ensure that no stress cracking has occurred in the finished weld.

First and foremost, it is important that the weld rod filler metal is closely matched to the parent metal of the casting. Using a significantly different aluminum alloy may result in a slightly different thermal expansion coefficient and may result in stress cracks in the heat-affected zone (HAZ). Differences in chemistry between weld filler chemistry and parent metal structure can be somewhat mitigated as long as a full solution and aging process is conducted afterward. The solution stage can provide some stress relief through creep. However, most customers will not want to invest development time with using a dissimilar composition filler metal.

Table 6.1 shows the composition of the most common filler weld metals used in aluminum casting repair. The 4145 filler metal is used for 319-based alloys as it has similar Si and Cu compositions to parent metal; conversely, 4043 and 5356 are used for the weld repair of 356 alloys.

Most customers that permit weld repair of aluminum castings will usually specify locations which can be welded and not welded. Machined surfaces cannot be welded, and bolt bosses which may be load bearing may not be welded. However, there may be special provisions made that are permitted after detailed metallurgical evaluation. Figure 6.20 shows a bolt boss of a sand casting which is drilled and tapped in final machining and is load bearing. The casting composition is listed in

Table 6.1 Alloy designations and compositions for weld filler rods for Al-Si castings repair

Alloy (AA)	4145	2319	4043 (for 356)	5356 (for 356)	Casting
Aluminum	Bal.	Bal.	Bal.	Bal.	Bal.
Silicon	9.3–10.7	0.2	4.5–6.0	0.25	7.5–9.0
Copper	3.3–4.7	6.3	0.3	0.1	2.5–3.5
Iron	<0.8	–	0.8	0.4	0.30–0.60
Magnesium	<0.15	–	0.05	4.5–5.5	0.32–0.4
Manganese	<0.15	0.3	0.05	0.05–0.20	0.30–0.55
Zinc	<0.2	0.18	0.1	0.1	0.4–0.8
Titanium	–	0.15	0.2	0.06–0.20	0.12–0.16
Strontium	–	–	–	–	0–0.003
Nickel	–	–	–	–	0.14
Tin	–	–	–	–	0.03
Lead	–	–	–	–	0.05
Vanadium	–	0.1	–	–	–
Beryllium	–	–	0.0003	0.0003	–
Chromium	–	–	–	0.05–0.20	0.1
Total alloy	13.9–16.7	7.23	6.0–7.5	5.5–6.9	12.5–14.3

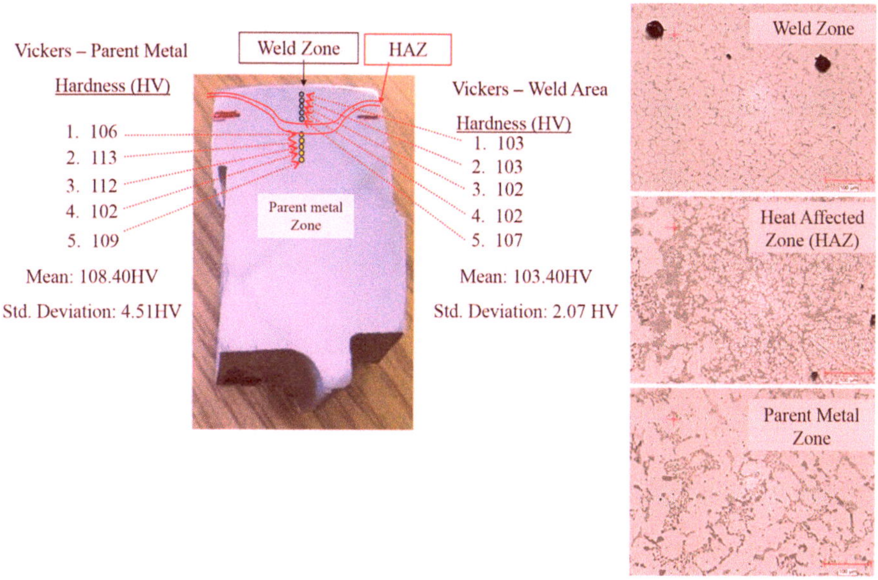

Fig. 6.20 Weld repair viewed in cross-section with microstructure images of the three

Table 6.1 and the weld filler material is 2319. Figure 6.20 includes a Vickers microhardness array conducted along the parent metal, HAZ, and the weld melt zone. In this case, there is a small drop in microhardness in the weld zone compared to the parent metal, most likely due to the fact that the total alloy content of the weld melt is lower than that of the parent metal. An additional consideration is that since 2319 filler rod is a Cu-bearing aluminum alloy, its use should be followed by a creep stress-relieving process such as a solution heat treatment. This is because Al-Cu alloys have a higher volumetric contraction on cooling than the Al-Si-Cu parent alloy composition. This leaves a high level of residual stress concentration at the transition between the weld zone and the parent metal, which is the HAZ. Should a post-weld stress-relieving process not be implemented, then the risk for a stress-induced crack to occur in the HAZ increases. The choice for a 4145 weld filler rod to be used for casting weld repair minimizes the risk associated with stress cracking in the HAZ and additionally may not need a solution treatment post-weld. The choice of which weld rod material to use will depend on the stage of the post-cast operation at which weld repair would occur. If weld repair is conducted as part of a visual inspection prior to heat treatment, then it is possible to use either the 2319 or 4145 weld material. If it is conducted after solution but prior to an age treatment, then 4145 weld material should be the stronger consideration for use.

Generally the use of either weld material type just described will have a good integrity of the HAZ in the stress-relieved condition. Figure 6.21 shows the scanning electron microscope images in backscattered electron mode for when the 4145 weld filler is used and when the 2319 weld filler is used. The weld structure differences are seen, where the 2319 weld material contains mainly α-Al phase with

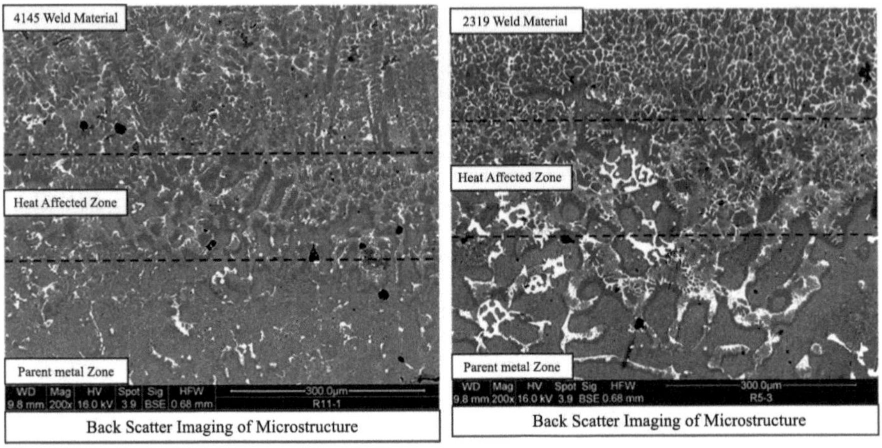

Fig. 6.21 Comparison of the HAZ and weld zone in the case for 4145 and 2319

Cu-based secondary phases, while the 4145 weld structure contains a refined version of the dendritic and Al-Si eutectic structures seen in the parent metal owing to the closer similarities of their chemistries. As mentioned, the 2319 weld material has a higher degree of volumetric contraction than for the 4145, and thus considerations on ensuring the casting component to be weld repaired is adequately heated during the weld operation and that the cool down is controlled at a slow rate. All of these prep conditions are usually conducted with some preliminary trial and error as weld crack susceptibility can be compounded by cast section architecture complexity.

The weld cast section shown in Fig. 6.20 is shown in Fig. 6.22 as a light optical micrograph mosaic. The overlay of the final machined state (drilled and tapped bolt hole) in Fig. 6.22 is placed over the weld region, HAZ, and parent metal structure. Most of the weld structure will supposedly be removed in the final machined state as done in this scaled overlay. In this case, the weld metal, after a surface-grinding operation, provides a flat casting surface to center the tool used to drill the initial bolt hole and to allow for drill torque to be targeted (measured by resistance in drilling aluminum out).

In cases where weld repair area is also a load-bearing section of the casting, the analysis shown in Fig. 6.21 can be used to assess how much of the actual weld material remains in the final machined state. Conceivably a load-bearing section which would not normally be welded (no weld repair permissible) could be repaired if the potentially discontinuous structure of the weld repair is mostly removed in the machined state, allowing the parent metal to provide the structural integrity required for load-bearing capacity in the casting end application. This of course requires customer approval using the analysis outlined to drive the case for weld repair in a nonpermissible weld repair location. The overall cost impact on both the foundry and the customer can potentially be significant, as the weld repair generally costs the time of the weld operator (USD 40/h for an in-foundry trained weld operator)

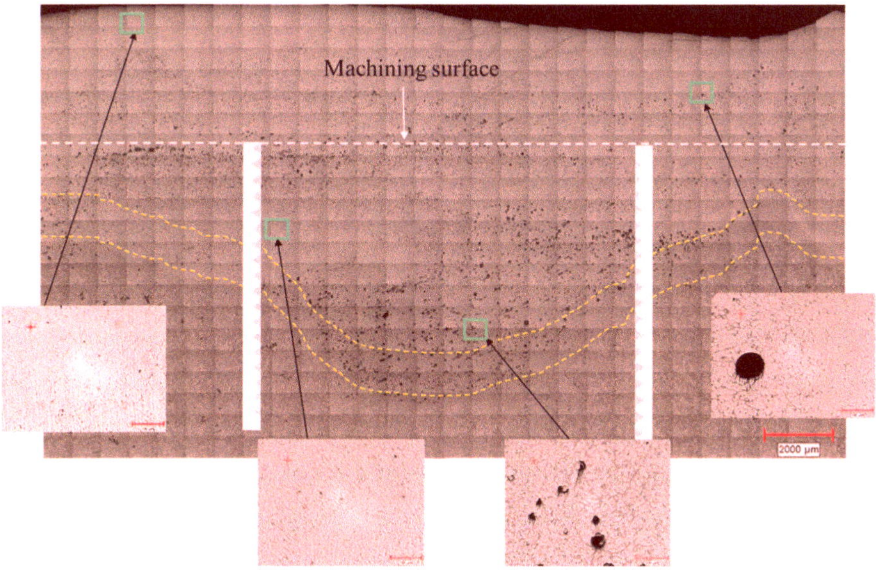

Fig. 6.22 Weld repair microstructure with the overlay of the material that is removed during the machining process

and the filler metal versus the scrap of a casting, which can range from USD 200/ unit for an engine block or several thousands of dollars per unit for an aerospace casting.

References

1. Weibull, W. 1951. A statistical distribution function of wide applicability. *Journal of Applied Mechanics* 18: 293.
2. Bourcier, G.F., J. Dickinson, J. Tessandori, and D. Schiffer. 1985. *Aluminum recycling casebook*, 56. Washington, D.C.: The Aluminum Association, Inc.
3. Gruzleski, J.E., and B.M. Closset. 1990. *The treatment of liquid aluminum-silicon alloys*, 256. Des Plaines: American Foundry Society, Inc.
4. Djurdjevic, M., H. Jiang, and J. Sokolowski. 2001. On-line prediction of aluminum–silicon eutectic modification level using thermal analysis. *Materials Characterization* 46 (1): 31–38. https://doi.org/10.1016/S1044-5803(00)00090-5.
5. Djurdjevic, M., J. Sokolowski, and T. Stockwell. 1998. Control of the aluminum-silicon alloy solidification process using thermal analysis. *Metallurgy* 4: 237–248.
6. Djurdjevic, M., T. Stockwell, and J. Sokolowski. 1997. The effect of strontium on the microstructure of the aluminum-silicon and aluminum-copper eutectics in the 319 aluminum alloy. *International Journal of Cast Metals Research* 12 (2): 67–73.
7. Hamed, Q., M. Dogan, and R. Elliott. 1993. The dependence of secondary dendrite arm spacing on solidification conditions of Al-7 Si-0.5 Mg alloys treated with TiBAl and TiBAl/Sr additions. *Cast Metals* 6: 47–53.
8. Paray, F., and J.E. Gruzleski. 1994. Microstructure-mechanical property relationships in a 356 alloy. Part I: Microstructure. *Cast Metals* 7: 29–40.

9. ———. 1994. Microstructure-mechanical property relationships in a 356 alloy. Part II: Mechanical properties. *Cast Metals* 7: 153–163. https://doi.org/10.1080/09534962.1994.11819174.
10. Zalensas, D.L. 1993. *Aluminum casting technology*, 201. Des Plaines: American Foundrymen's Society.
11. Paray, F., and J.E. Gruzleski. 1994. Factors to consider in modification. *Transactions of the American Foundrymen's Society* 102: 833–842.
12. Backerud, L., G. Chai, and J. Tamminen. 1990. *Solidification characteristics of aluminum alloys. Vol. 2. Foundry alloys*, 266. American Foundrymen's Society, Inc/Skanaluminium.
13. Paray, F., and J.E. Gruzleski. 1993. Modification-A parameter to consider in the heat treatment of Al-Si alloys. *Cast Metals* 5: 187–197.
14. Campbell, J. 1997. 10 rules for good castings. *Modern Casting* 87 (4): 36–39.
15. Wang, Y., S. Xing, X. Lu, F. Robles-Hernandez, S.-S. Pei, and J. Bao. 2014. Twisted bilayer graphene with controlled rotation angles. *American Physical Society Meeting* 1: 37003.
16. Zhao, Y.Y., E. Ma, and J. Xu. 2008. Reliability of compressive fracture strength of Mg-Zn-Ca bulk metallic glasses: Flaw sensitivity and Weibull statistics. *Scripta Materialia* 58 (6): 496–499. https://doi.org/10.1016/j.scriptamat.2007.10.052.
17. Green, N.R., and J. Campbell. 1993. Statistical distributions of fracture strengths of cast Al 7Si Mg alloy. *Materials Science and Engineering: A* 173 (1): 261–266. https://doi.org/10.1016/0921-5093(93)90226-5.
18. Wang, Q.G., D. Apelian, and D.A. Lados. 2001. Fatigue behavior of A356-T6 aluminum cast alloys. Part I. Effect of casting defects. *Journal of Light Metals* 1 (1): 73–84. https://doi.org/10.1016/S1471-5317(00)00008-0.
19. Tiryakioglu, M., and J. Campbell. 2010. Weibull analysis of mechanical data for castings: A guide to the interpretation of probability plots. *Metallurgical and Materials Transactions A-Physical Metallurgy and Materials Science* 41A (12): 3121–3129. https://doi.org/10.1007/s11661-010-0364-6.
20. Boileau, J.M., J.W. Zindel, and J.E. Allison. 1997. *The effect of solidification time on the mechanical properties in a cast A356-T6 aluminum alloy*. SAE International.
21. Campbell, J. 2003. *Castings*. 2nd ed. Oxford, UK: Butterworth-Heinemann.
22. Mohanty, P.S., F.H. Samuel, and J.E. Gruzleski. 1995. Experimental study on pore nucleation by inclusions in aluminum castings. *Transactions of the American Foundrymen's Society* 103: 555–564.
23. Rooy, E.L. 1992. Mechanisms of porosity formation in aluminum. *Modern Casting* 82 (9): 34–36.
24. MacKay, R., and D. Szablewski. 2010. The use of Weibull statistical method to assess the reliability of a development engineered automotive casting component. *International Journal of Metalcasting* 4 (4): 31–45.
25. Mackay, R., and G. Byczynski. 2011. The use of the Weibull statistical method to assess the reliability of cast aluminum engine blocks made from different casting processes. In *Shape casting: 4th international symposium*. Wiley.
26. Crepeau, P.N. 1995. Effect of iron in Al-Si casting alloys: A critical review. *Transactions of the American Foundrymen's Society* 103: 361–366.
27. Caton, M.J., J. Wayne Jones, J.M. Boileau, and J.E. Allison. 1999. The effect of solidification rate on the growth of small fatigue cracks in a cast 319-type aluminum alloy. *Metallurgical and Materials Transactions A* 30 (12): 3055–3068.
28. Mackay, R.I. 2003. *Development of a new durable Al-Si alloy for the next generation of engine block casting*. Ph.D. Thesis, University of Windsor.
29. Backerud, L., E. Krol, and T. Tamminen. 1996. *Solidification characteristics of aluminum alloys. Vol. 1. Wrought alloys,* Vol. 1, 1st ed. AFS/Skanaluminum.

Chapter 7
Applications in the Automotive and Aerospace Industries

7.1 Aluminum Alloys: A Brief History

Aluminum castings have risen steadily since the Second World War for both automotive and aerospace applications. Their growth has mainly been due to the growing availability of scrap aluminum supply and their relative ease in the manufacture of secondary-grade alloys. Prior to this, most aluminum alloys were processed from primary sources, and thus had a significant cost when compared to other alloy alternatives. Today, primary sources are still used extensively for aerospace applications, while secondary sources are predominately used in the automotive industries.

Aluminum was first publicly introduced as a metal for consideration in 1855. At that time, it was more expensive than gold and had no known applications, other than academic interest. Aluminum was first described in literature as the metal used to produce a rocket that traveled in outer space in Jules Verne's 1865 science fiction novel titled *From the Earth to the Moon* [1].

In 1888, the Pittsburg Reduction Company (reorganized as the Aluminum Company of America, Alcoa, in 1907) began the production of aluminum for sale in the United States, but had a limited application in end use (e.g., high-end tea kettles and cooking pots. The selling point was that they did not rust like cast iron and copper-based versions and thus last longer). Its daily production was only 25 kg/day, but with advancements in reduction methods, this grew to 450 kg/day in 1890. Also in 1888, the Aluminum Smelter Society started in Neuhausen, Germany, and supplied aluminum mainly in Western Europe. At inception, it made 110 kg/day and was able to grow its production to 1230 kg/day by 1895.

Aluminum use outside of opulent ornament and cookware applications did not change until military planners wanted to investigate its use in navel applications. In 1893, the Scottish shipbuilding yard Yarrow & Co created a 52 m long torpedo boat

© Springer International Publishing AG 2017
F.C. Robles Hernandez et al., *Al-Si Alloys*, DOI 10.1007/978-3-319-58380-8_7

made of aluminum and was named *Sokol*. The *Sokol* was made for the Russian Empire's Navy and achieved a speed of 32 knots, a record speed for any naval vessel at the time. Despite this accomplishment, widespread use of aluminum did not occur as most militaries could only afford a lower number of aluminum-hulled ships compared to a ferrous-hulled fleet. Navy military planners still felt numbers was better than speed at winning sea-based conflicts.

Initial price reductions did not occur till 1900, when the first signs that aluminum's application in military ordainments and survival gear began to emerge. However, aluminum was a soft metal, and thus its durability was limited in these applications. In 1909, the first cast "aluminum alloy" was made to address this issue. It was developed by the German scientist Alfred Wilm, and it was called Duralumin. Duralumin was aluminum alloyed with copper, magnesium, and manganese using a proprietary blending process. At the time, its strength and durability were superior to those of cast aluminum but had nearly the same weight.

In 1907, in a separate attempt to produce aluminum with high strength and durability, Robert Victor Neher invented the method used for continuous aluminum rolling foil production in Switzerland, and in 1910 commissioned the first production rolling mill for aluminum.

Further price reductions occurred thanks to the advent of the First World War, where aluminum saw its use expand into the emerging military aerospace industry. In fact, aluminum was nicknamed the "wing metal" because its strength and low weight literally drove design architectures possible for biplanes in terms of range and maneuverability. The driving force for aluminum use in military aerospace applications at the time may have been helped by its impact on the historical first flight that occurred in 1903. The Kitty Hawk had a 69 kg crankcase casting, which was made from aluminum alloyed with copper. The Wright brothers did consider an existing ferrous-based automotive engine for the Kitty Hawk, but was determined to be too heavy to allow liftoff.

A unique engine made with cast aluminum appeared to be the only option for the historic first flight to occur. By the time the Second World War had arrived, the need to improve the range of armored personal carriers, tanks, and aircraft led to the strategic need for higher aluminum content, as logistics for fuel supply reserves and delivery were critical in war planning. At the same time, aluminum processing and alloy development had advanced to allow mass casting production with relatively improved reliability and quality. For example, aluminum was absolutely critical for the range and maneuverability for a total of 296,000 fighter aircraft and long-range bombers produced to support the Second World War effort in the United States. Most war buffs are well aware of the long range the P-51 Mustang had and how it was able to provide the needed bomber escorts over Germany. The P-51 introduced the semi-monocoque fuselage that was constructed entirely of aluminum to save weight. Further weight reductions were achieved using a derivative version of the Duraluim alloy to cast the cylinder block, crankcase, wheel-case, and pistons of the P-51's 12-cylinder Merlin engine.

Aluminum introduction into the automotive industry occurred nearly at the same time as for aerospace industry, but grew at a much slower rate over time due mainly to its cost. Aluminum was first reported in an automotive application in the 1899

Berlin International Motor Show. Carl Benz, future cofounder of Mercedes Benz, introduced the first aluminum engine in 1901. However, civilian use of aluminum in applications such as automotive would have to wait till after the Second World War, where a glut of recycled aluminum, wrought, and cast components, left over from retired war aircraft, etc., gave way for a possible cost-effective use as secondary sourced aluminum.

Increased aluminum production volumes globally and demand for the metal resulted in aluminum becoming an exchange commodity in the 1970s. In 1978, the exchange trade for aluminum contracts started on the London Metal Exchange. Today aluminum casting components are playing a critical role in the cost-effective manufacture in both the automotive and aerospace industries as a whole. The sources of aluminum and casting techniques have advanced significantly over the years.

7.2 Casting Quality Requirements

In a typical car, minivan, crossover or light truck, aluminum castings can be found as a roof bracket or door handle in its smallest size, and cast aluminum block, the most weighted aluminum component, will range from 23 to 45 kg. In the aerospace and military applications, aluminum casting components can be hundreds of kilograms in weight. Aluminum cast applications in the automotive and aerospace industries do have print-based specifications for mechanical properties, porosity limits, X-ray radiography requirements (destructive and fluoroscopy), cast surface roughness, dimensional stability, and corrosion. In most cases, ASTM standards for these tests are specified. While metallographic quality is important, it is difficult to establish specific parameters within limits to provide intended mechanical targets. However, available ASTM standards will be cited in order to indicate cast aluminum compliance methodologies in production. Table 7.1 summarizes many of the ASTM standards that can be found on automotive and aerospace casting prints.

During development, many prototype castings are made and extensively tested to evaluate the eventual process being established. After PPAP (Production Part Approval Process) stage, quality assessments using these established ASTM standards are made on an audit-type base. This means that a specific number of castings made on a particular shift, week, or month will be destroyed to make these checks. This process is carried to confirm that the casting components made are identical to what was PPAP approved.

In addition to material specifications, many automotive and aerospace cast parts are tested for cast wall leaks to meet specific requirement usually measured in cm^3/min and are indicated on a print. Manufacturers for these components must plan for the capital investment that allows leak testing to occur accurately with high cycle times. Normally leak testing is conducted after the casting component machining process.

Aerospace castings tend to be made using processes that are amenable to lower volume capacity but achieve complex architectures and specific mechanical, corrosion, and dimensional tolerances. In addition, aerospace castings have a service life

Table 7.1 Commonly listed ASTM standards for the testing of Al-Si castings

	Standard	Description
Corrosion	ASTM G31	Standard Practice for Laboratory Immersion Corrosion Testing of Metals
	ASTM G111	Standard Guide for Corrosion Tests in High Temperature or High Pressure Environment, or Both
	ASTM B117	Standard Practice for Operating Salt Spray (Fog) Apparatus
Mechanical	ASTM E466	Standard Practice for Conducting Force Controlled Constant Amplitude Axial Fatigue Tests of Metallic Materials
	ASTM B557	Standard Test Methods for Tension Testing Wrought and Cast Aluminum- and Magnesium-Alloy Products
	ASTM E8/ E8M	Standard Test Methods for Tension Testing of Metallic Materials
	ASTM B769	Standard Test Method for Shear Testing of Aluminum Alloys
	ASTM E9	Standard Test Methods of Compression Testing of Metallic Materials at Room Temperature
Hardness	ASTM E10	Standard Test Method for Brinell Hardness of Metallic Materials
	ASTM E384	Standard Test Method for Knoop and Vickers Hardness of Materials
Radiography	ASTM E155	Standard Reference Radiographs for Inspection of Aluminum and Magnesium Castings
	ASTM E94	Standard Guide for Radiographic Examination
	ASTM E2422	Standard Digital Reference Images for Inspection of Aluminum Castings
	ASTM E2973	Standard Digital Reference Images for Inspection of Aluminum and Magnesium Die Castings
	ASTM E1030	Standard Test Method for Radiographic Examination of Metallic Castings
Metallography	ASTM E3–11	Standard Guide for Preparation of Metallographic Specimens
	ASTM E112	Standard Test Methods for Determining Average Grain Size
	ASTM E340	Standard Test Method for Macroetching Metals and Alloys
	ASTM E407	Standard Practice for Microetching Metals and Alloys
	ASTM E883	Standard Guide for Reflected–Light Photomicrography

in most cases longer than that for automotive applications. This can be particularly reflected in high fatigue durability and corrosion resistance specified on aerospace casting prints.

Many of the quality checks done for the automotive industry are nearly the same for aerospace industry prior to PPAP. The required testing after production is conducted for all castings made for aerospace applications before submitted to the customer. For example, mechanical and corrosion testing would be conducted using what is called Integral Test Blanks, usually integrated in the runner system of the casting. Once the gating is removed from the casting, the saw operator would remove these Integral Test Blanks and have them sent to the foundries lab for testing.

During development or pre-PPAP stage, mechanical samples are tested from both casting and integrity to ensure adequate predictive value of properties post-PPAP.

X-ray radiography for aerospace applications would be done using X-ray film packed around the casting part with multiple angle shots to ensure no defects are present [ASTM E155, see Table 7.1]. Zyglo fluorescence testing is a nondestructive testing to ensure no stress cracking is present particularly after heat treatment. Finally, all quality information associated with a specific casting would be sent to the end user to ensure documented quality and traceability, but will be referenced again in the rare event of a field failure or air crash investigation.

The more complex casting process, along with the quality checks required, tends to result in a cost that can be as high as 20K USD per unit. Nearly half of this unit cost is due to the extensive and highly sophisticated testing conducted for each casting.

7.3 Applications of Aluminum Castings in Automotive and Aerospace Industries

Applications of aluminum casting components have grown steadily over the years to become the second most abundant alloy after steel and irons (gray cast iron, compacted graphite iron, etc.) [2]. Al-Si cast alloys have numerous applications in the automotive and aerospace industries. The use of aluminum cast alloys is expected to increase in these industries in the near future. The reason for this is that cast aluminum has many desirable qualities such as a high strength-to-weight ratio, good castability, excellent corrosion resistance, cosmetic surface quality, resistance to hot tearing, relatively good thermal conductivity, relatively lower melting temperatures, good machinability, and good weldability [2, 3]. Another important consideration, other than being lightweight, is that aluminum alloys are comparatively easy to recycle, which is becoming critical in current times when green energy, environmental, and recycling awareness are of paramount importance [2–4].

Cast aluminum components are allowing the industry to meet the current as well as future customer and legislative expectations (such as reduced weight and increased fuel economy) while retaining current vehicle size and performance. The legislative driver behind this endeavor was the introduction of the Corporate Average Fuel Economy (CAFE) in 1975. Since the National Highway Traffic Safety Administration (NHTSA) is required by the US Congress to set CAFE standards for no more than 5 years at a time [5], this gives a framework for automotive manufactures to adequately plan vehicle designs and material selection to meet CAFE standards for 10 years forward.

In the automotive industry, powertrain components are commonly designed using Al-Si castings. As an example, a V8 engine block, cast in Al, weighs approximately 32 kg, compared to 68 kg in gray cast iron [4]. The North American automotive industry has been in the process of converting cast production from iron cylinder heads and engine blocks to components made of Al cast alloys for many years.

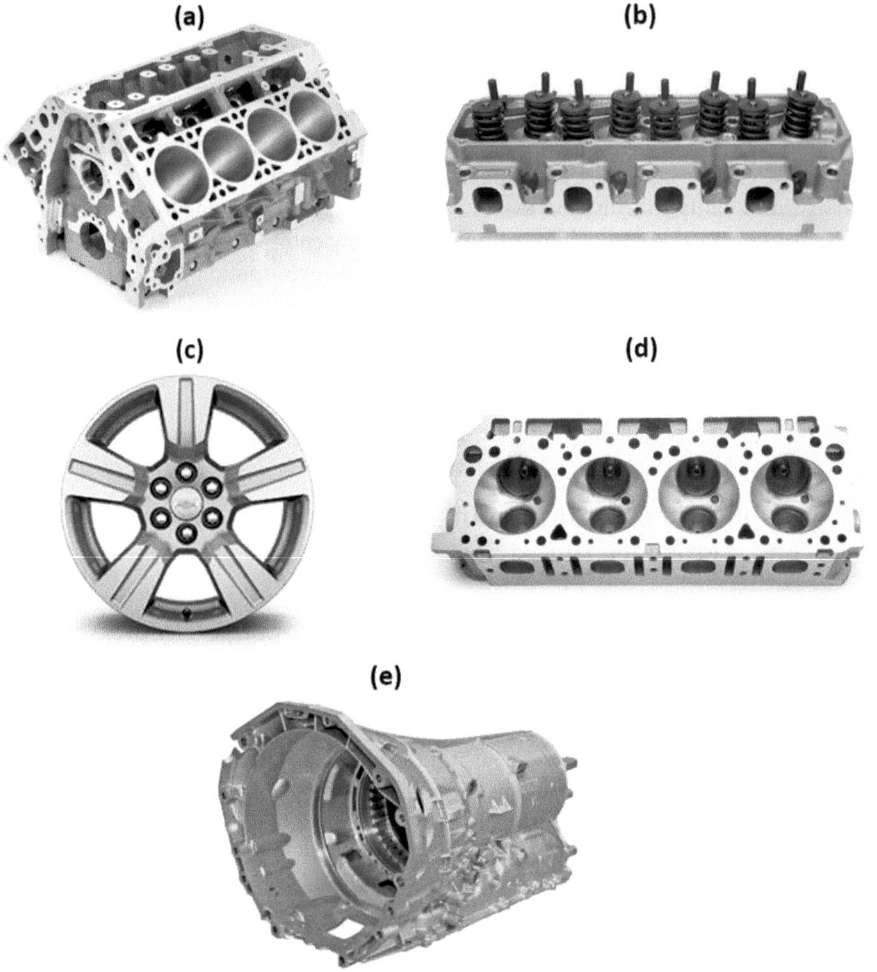

Fig. 7.1 (**a**) LT1 corvette engine block – sand casting process. (**b**) Ford C351 cylinder head – semipermanent casting process. (**c**) GM cast wheel rim – permanent casting process. (**d**) Chrysler cylinder head – semipermanent casting process. (**e**) 8 Speed transmission housing by GF Automotive – high-pressure die cast

Currently, about 80% of the heads and 25% of the blocks are made of aluminum cast alloys [4]. Examples of aluminum blocks, heads, wheels, and transmission housing are shown in Fig. 7.1. Recent advancements in process and microstructural control have allowed better properties that can stand up to the riggers encountered in suspension components and space-frame members in cars and light trucks.

Al-Si castings have been a significant contributor in aerospace applications since the inception of the airplane by the Wright Brothers. As mentioned in the Preface, the Kitty Hawk had a 69 kg cast aluminum crankcase – the heaviest component involved. During both World Wars, fuel supplies and mission ranges needed the

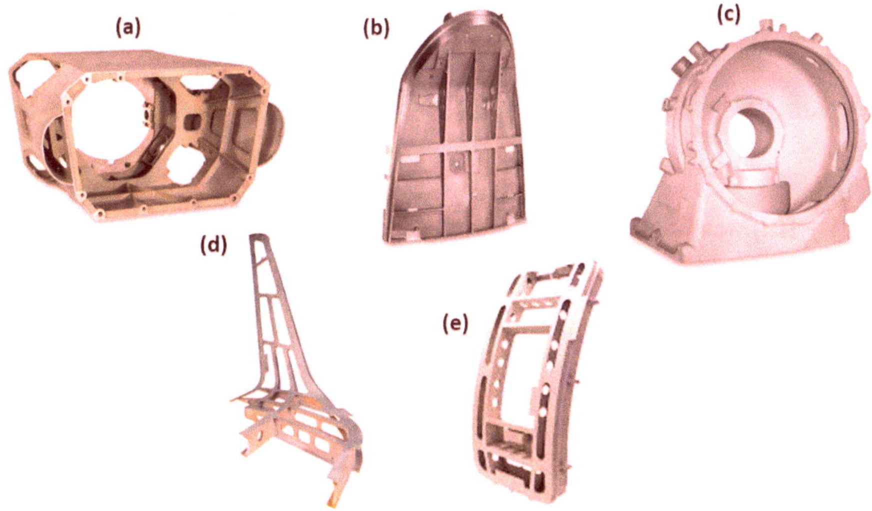

Fig. 7.2 (**a**) Rear frame avionics bracket in D357 alloy in the sand process – TITAL castings (now acquired by Alcoa) – 23.5 kg. (**b**) AFT canopy bracket in D357 alloy – TITAL castings (now acquired by Alcoa) – 21 kg. (**c**) Tail rotor gearbox in D357 alloy in the sand process – TITAL castings (now acquired by Alcoa) – 26 kg. (**d**) Winglet substructure for Embraer Phenom in F357 alloy in the investment process – Aeromet. (**e**) Overwing emergency door exit in A357 alloy in the investment process (Sofia) – Aeromet

strongest efforts at aircraft weight reductions feasible at the time. Aluminum was one of the most important factors in achieving this effort. Today, environmental issues (e.g., Advisory Council for Aeronautics Research in Europe [ACARE] targets of 50% reduction in CO_2 and a 50% reduction in perceived noise by 2020 [Source: Aluminum Association]) and fuel costs for both military and commercial airlines have been strong drivers for weight reductions of aircraft.

Examples of aerospace aluminum castings are shown in Fig. 7.2. The most significant recent increase in Al-Si castings in the aerospace industry has been the use of the investment casting process to make complex substructure components, replacing aluminum wrought substructures which have to be welded together and then formed into the desired shape. One example, shown in Fig. 7.2d, is the Embraer Phenom Winglet, which was previously made from machined two stock material aluminum blocks and then welded together; today it is cast from a single investment casting. Due to the lack of welds, a 50% weight reduction is realized, along with a lower component cost.

Another common casting process used in aerospace Al-Si castings is the precision sand casting process and it has been used almost since the aviation industry was born. Examples of aerospace aluminum castings made in sand are shown in Fig. 7.2a, c. More details of this process will also be described in this chapter.

Recently composite materials have made significant inroads toward aircraft weight reductions; however, Al-Si, and to a lesser extent Al-Li alloys, will continue to grow in use owing to their relative cost-effectiveness by comparison. In the competitive airline industry, a balance between the unit cost of the aircraft and the long-term operating cost in terms of fuel use and maintenance has to be struck. Composites can complement weight savings further, but add to the unit cost per aircraft.

7.4 Impact of Aluminum Cost on Casting Processes

It is not unusual that the cost of the aluminum used can be the single largest cost involved with the overall or final cast unit cost. However, both the automotive and aerospace industries incorporate the cost considerations of aluminum differently.

7.5 Aerospace

In the aerospace industry, mainly primary aluminum is used and this is generally sold from a major raw aluminum supplier directly to the foundry (e.g., D357 alloy from Alcoa), where the sale price is set by the primary supplier (and this is generally close to the major aluminum indexes at the time). The aluminum aerospace foundry will receive a heat and will blend this heat along with back scrap in the melt batch process (e.g., rigging, which is composed of the risers, ingates, and runners). The yield in aerospace can be as low as 30–40% so that the overall cost of the aluminum is really set as a purchase price averaged over several months and then is fixed in the unit cost. This process continues until the customer and aerospace foundry renegotiate the commercial agreement.

For example, if the going price of D357 alloy was 2.6 USD/kg and the back-scrap processing cost as determined by the foundry was 0.44 USD/kg, the commercially agreed upon cost of the aluminum for a 45 kg casting would be 60 USD. Finally, because of the investment required to assemble a single sand mold, or an investment cast mold, the portion of the aerospace castings unit cost that is composed of the aluminum itself can be as low as 5% (weight of the casting multiplied by the foundry/kg cost). This is why it is easier to have a fixed unit price that is agreed for a longer period. Large fluctuations in aluminum prices may not necessarily adversely affect profitability or loss to the aerospace foundry's balance sheet.

7.6 Automotive

Aluminum alloy cost in the automotive industry needs to have special considerations that are different from those of the aerospace industry. Aluminum costs of the unit price per casting can be the main portion of the unit casting cost, and the alloy is generally made from recycled scrap or secondary alloys. Thus, fluctuations in aluminum prices as tracked on the Platts 319 or the London Metal Exchange Aluminium Index could adversely affect the loss or profit of the automotive foundry if a commercial unit price is treated like in the aerospace comparative. To put this into further perspective, a 34 kg cast aluminum engine block would typically sell to the customer for 180–220 USD. If we assume a 319 alloy price of 2.25 USD/kg, and we use the same back-scrap cost of 0.44 USD/kg, as well as a 70% casting yield, then we are looking at 57 USD in metal cost of the block, or nearly 25% of the unit selling price.

It is now more common that the unit price of the cast aluminum component will comprise two separate costs. The first is of manufacturing the castings and the second is the cost of the metal (per kg weight). Basically, the unit cost commercially agreed between the supplier and customer is the manufacturing cost plus the aluminum cost as reflected on an aluminum index (e.g., Platts 319 Index). This method is reasonable as the foundry generally purchases aluminum alloy for use within days. This helps reducing networking capital and the yields for automotive castings are generally higher (e.g., 70% or higher). The balance which is made of back scrap most likely was removed from another alloy with the same composition during the previous days.

References

1. Verne, J. 1970. *From the Earth to the Moon, 1865*. Trans. J. and R. Baldick. New York: Dutton.
2. Bourcier, G.F., et al. 1985. *Aluminum recycling casebook*, 56. Washington, D.C.: The Aluminum Association, Inc.
3. Gruzleski, J.E., and B.M. Closset. 1990. *The treatment of liquid aluminum-silicon alloys*, 256. Des Plaines: American Foundry Society, Inc.
4. Zalensas, D.L. 1993. *Aluminum casting technology*, 201. Des Plaines: American Foundrymen's Society.
5. Thoma, T., and F. Vaca. 2004. National Highway Traffic Safety Administration (NHTSA).

Chapter 8
Principles of Solidification

8.1 Aluminum Alloy Systems

The Al-Si family of alloys is extremely important in the aluminum casting industry since it makes up 85% or more of the total aluminum cast parts produced [1]. These alloys usually contain between 5.5% and 12% silicon, and the most common alloying elements are copper and magnesium [1, 2]. The microstructure of an Al-Si alloy is usually composed of an α-Al dendritic aluminum matrix with the subsequent formation of the Al-Si eutectic structure. The Al-Si eutectic is usually found in the remaining space, and their characteristics can be controlled mainly via cooling rate. Chemical alternatives include the addition of certain master alloys to the melt prior to casting. The best way to view the Al-Si casting structure is from polished sections using light optical microscopy and scanning electron microscopy (backscattered electron mode). Examples of the microstructure are found in Fig. 8.1. The Al-Si alloys illustrated have two Fe levels that result in different characteristics of the Fe phase precipitates. The Fe phase precipitates can be detrimental to the properties and, hence, the integrity of the casting components. Alternatively, Mn additions modify Fe-rich phases into a lesser detrimental structure known as Chinese script [1–3, 5].

During solidification, boron and titanium boride compounds act in the aluminum melt as heterogeneous nucleants that refine the grain size [1–3]. They are also known as grain refiners. Sodium and strontium are added to the melt to modify the silicon of the eutectic structure from an acicular to a fibrous morphology [4–11]. Additions of strontium reduce solution heat treatment times [8, 11]. An increase in cooling rate can result in similar alterations to the microstructure. This variability of microstructure for the same alloy type via cooling rate or master alloy additions allows aluminum-silicon alloys to achieve a myriad of physical properties. This allows tuning of the properties of those alloys and makes them suitable for a wide range of applications in the automotive, aerospace, and other industry sectors.

© Springer International Publishing AG 2017

F.C. Robles Hernandez et al., *Al-Si Alloys*, DOI 10.1007/978-3-319-58380-8_8

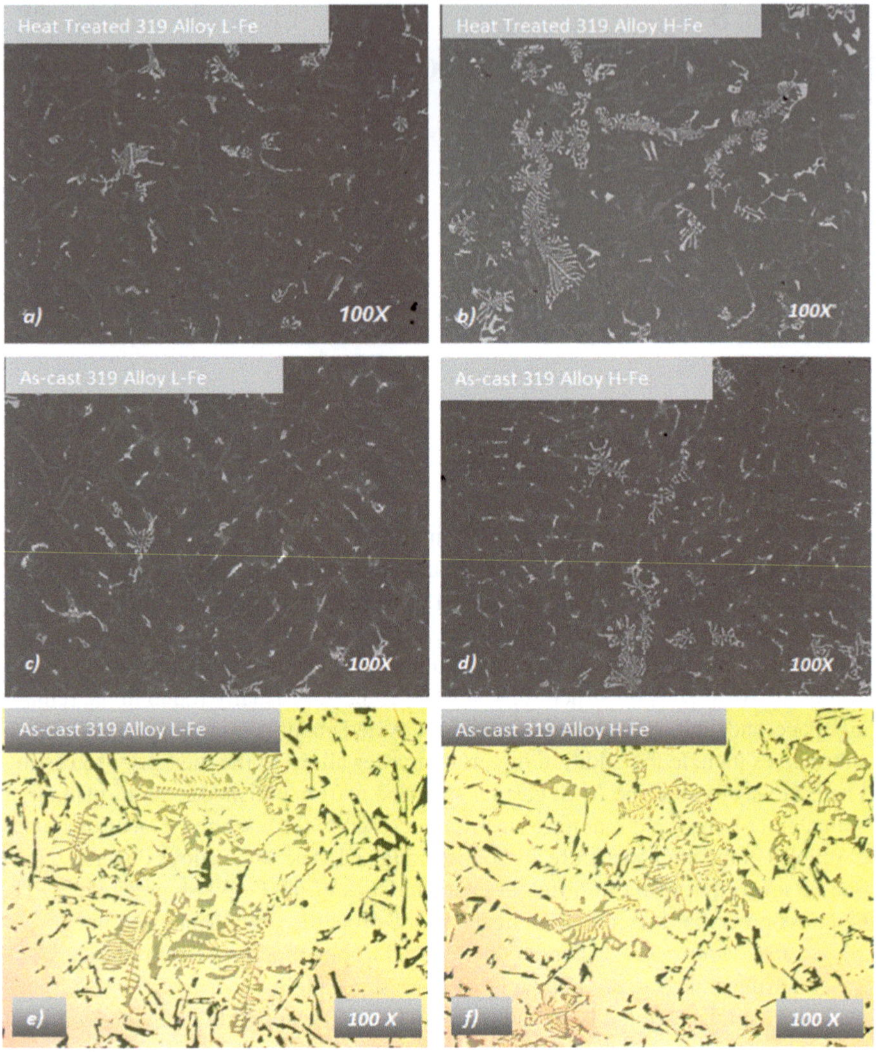

Fig. 8.1 SEM/BSE micrographs and LOM micrographs for the 319 alloy with low Fe (**a**, **c**, **e**) and for the 319 alloy with high Fe (**b**, **d**, **f**), both at low magnification

The most deleterious microstructural constituent in Al-Si cast alloys is the presence of voids known as porosity. Porosity usually occurs in Al-Si alloys through the evolution of segregated hydrogen gas (gas porosity) and/or liquid feeding difficulties (shrinkage) in the mushy zone during solidification. Another factor that can control cast structure is the relative shape of Si phases (part of the main Al-Si eutectic); the size and shape of Fe, Cu, and Mg phases; the condition of the dendritic matrix; and finally the relative cleanliness of the melt [1, 4, 9–13].

There exist two basic categories of aluminum casting alloy types: primary (new stocked alloy) and secondary (recycle stocked alloy). Secondary aluminum is generally cheaper than primary aluminum due to the cost of manufacturing and current availability of the scrap aluminum supply. The aerospace industry uses primary stock cast materials and has also driven a large part of this new knowledge on achieving lower porosity in cast parts. Research in aluminum cast aerospace parts has been ongoing since the Second World War. For the automotive industry, the increase in cast aluminum use has occurred in the last two decades, and secondary stock aluminum is used to cast a variety of these parts.

It should be pointed out that secondary stock materials have technical issues that do not face primary casting materials such as the effect of residual impurities on the role of pore development during solidification [2, 5, 10, 13–20]. Currently the automotive industries are eager to develop the most effective methods of assessing the propensity for stable pore nucleation and growth, as-cast soundness, and the porosity effect on cast durability. Additionally, the effects of those defects are a continuous topic of research.

8.2 Alloy Designations

There does not exist a globally recognized protocol for alloy designations used for the aerospace and automotive industries. Instead there are continent-based designations that exist in North America, Europe, and Asia. For instance, The Aluminum Association (AA), based in Arlington, Virginia, is one of the most commonly known advocacy organizations that promotes alloy designation and heat treat temper protocols that allow easier communication between customers and suppliers. In 1954, AA had 75 alloy designations identified, and today it sees approximately 530 designations. The most common Al-Si alloys are covered in the 300 and 400 series, which are the primary subject of this chapter.

The cast alloy designation system within AA is based on a three digit followed by a decimal designation xxx.x (i.e., 413.0). The first digit of the alloy designation indicates the principal alloying element, which has been added to the aluminum alloy. The second and third digits are arbitrary numbers given to identify a specific alloy in the series. The number following the decimal point indicates whether the alloy is a casting (.0) or an ingot (.1 or .2). A capital letter prefix indicates a modification to a specific alloy. If the capital letter prefix is absent, then it represents the original version of that alloy, where A and B versions would represent the second and third versions of that same alloy respectively. The capital letter prefix is used in cases where small modifications of an existing alloy are made. This leaves more unused designations for future alloy designation requirements.

The designation system as it stands does not distinguish between primary and secondary sources. However, the general rule to decide if an existing alloy designation would have to be made from primary sources reviews the Fe concentration of that alloy. If it is below 0.20 wt%, it is nearly impossible to manufacture an alloy cost-effectively from secondary sourced materials due to the repeated dilution of impurities (e.g., Fe) that would be required. Table 8.1 shows the cast aluminum designation system as outlined by AA. A brief description and examples can be found in Tables 8.2 and 8.3 for automotive and aerospace applications, respectively.

Table 8.1 Cast aluminum alloy designation system – Aluminum Association (AA)

Alloy series	Principal alloying element
1xx.x	99.000% minimum aluminum
2xx.x	Copper
3xx.x	Silicon plus copper and/or magnesium
4xx.x	Silicon
5xx.x	Magnesium
6xx.x	Unused series
7xx.x	Zinc
8xx.x	Tin
9xx.x	Other elements

Table 8.2 Examples of automotive-grade alloys identified using AA designations

Alloy (AA)	Si	Cu	Mg	Fe	Mn	Ti	Zn	Ni	Applications
356.0	6.5–7.5	0.20	0.20–0.45	0.6	0.10	0.25	0.35	–	Permanent mold – wheel rims
319.0	5.5–6.5	3.0–4.0	0.10	1.0	0.50	0.25	1.0	0.35	Sand – blocks, heads
A319.1	5.5–6.5	3.0–4.0	0.10	1.0	0.50	0.25	3.0	0.35	
B319.1	5.5–6.5	3.0–4.0	0.10	1.2	0.80	0.25	1.0	0.50	
380	7.5–9.5	3.0–4.0	0.10	1.3	0.50	–	3.0	0.50	HPDC – blocks, transmission housing
383	7.5–9.5	3.0–4.0	0.10	1.3	0.50	–	3.0	0.50	
390	16.0–18.0	4.0–5.0	0.10	1.3	0.10	0.20	0.10	–	
A390	16.0–18.0	4.0–5.0	0.30	0.50	0.10	0.20	0.10	–	
B390	16.0–18.0	4.0–5.0	0.45–0.65	1.3	0.50	0.20	1.5	010	

Applications taken from ASM Handbook Vol. 2. Composition balances with Al

Table 8.3 Examples of aerospace-grade alloys identified using AA designations

Alloy	Si	Cu	Mg	Fe	Mn	Ti	Zn	Example
D357	6.5–7.5	–	0.45–0.70	0.20	0.10	0.04–0.20	0.10	Investment cast fuselage substructures, winglets substructures, access and passenger door substructures, avionics chases
A355	4.5–5.5	1.0–1.5	0.45–0.6	0.09	0.05	0.04–0.20	0.05	Aircraft supercharger covers, fuel pump bodies, liquid-cooled aircraft engine crankcases
A356	6.5–7.5	0.20	0.25–0.45	0.20	0.10	0.20	0.10	Aircraft fittings and control parts, aircraft pump components
A201	0.10	4.0–5.2	0.15–0.55	0.15	0.20–0.50	0.15–0.35	–	Aerospace housings, missile fins, landing gear castings

Applications taken from the ASM Handbook Vol. 2. The composition balance is Al

Example: Alloy - A356.0 the capital A (Axxx.x) indicates the first modification of the original 356.0 alloy. The number 3 (A3xx.x) indicates that it is of the silicon plus copper and/or magnesium series. The 56 (Ax56.0) identifies the alloy within the 3xx.x series, and the .0 (Axxx.0) indicates that it is a final shape casting and not an ingot.

One of the main differences that can be noticed between automotive and aerospace Al-Si casting alloys is the limits for Fe and Zn. The Fe level contributes to brittle secondary-phase constituents which can compromise mechanical strength, particularly in aerospace castings that have slow solidification rates. The Mn level is generally tied to Fe concentrations as it can promote a more favorable Fe-bearing phase formation. The Zn levels are also generally lower as elevated Zn can promote a high rate of corrosion.

Examples of Al-Si alloys used in the automotive industries are listed in Table 8.2 and for aerospace industries in Table 8.3. As mentioned, 85% of all aluminum castings are in the Al-Si family, which falls under the 3xx and 4xx series. The primary phase that forms is the α-Al dendrite structure and is studied extensively to quantify the solidification rate of the casting structure. Eventually, the solute enrichment of the interdendritic liquid would be sufficient for the formation of the Al-Si eutectics and complex Al-Si-Ci and Al-Si-Mg complex eutectics.

8.3 Solidification Sequence of Al-Si Alloys

Before the discussion of subsequent sections, it would be prudent to list the typical phase formation in the casting structure and the sequence in their formation. This will allow a better perception of the casting quality checks and testing encountered with casting manufacturing.

The solidification sequence has been widely studied via the use of the thermal analysis method, which monitors the solidification of the alloy using a thermocouple [20, 21]. While not within the scope of the current chapter, one example of a thermal analysis curve for a 319-type alloy is shown in Fig. 8.2, along with the identification of the arrest points and the associated phase evolution. The arrest points are identified by changes in the calculated dT/dt curve, which is determined from the original cooling curve [7, 9, 19–21]. The dT/dt curve is also included in Fig. 8.2. The first arrest point is associated with the nucleation of α-Al phase (liquidus), followed by the point at which the growing α-Al phase comes in contact with neighboring α-Al crystals (known as the dendrite coherency point, DCP). As solute enrichment progresses, the main Al-Si eutectic structure forms. The remaining liquid becomes further enriched in Cu, Mg, Fe, Mn, and Si, which eventually leads to the formation of the Al-Cu-Si and Al-Cu-Mg-Fe-Si reactions. More of the phase evolution will be discussed in detail in the next sections.

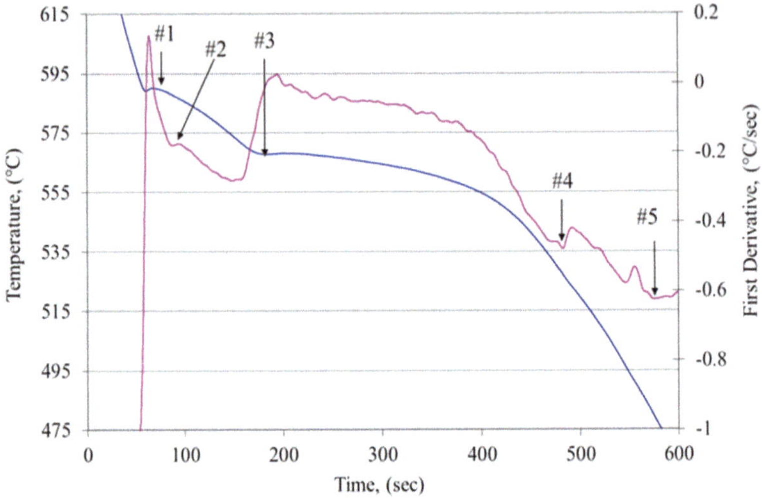

Fig. 8.2 Thermal analysis test sample cooling curve and calculated dT/dt curve of the 319-type casting alloy. The test sample solidification rate was 0.15 °C/s. Identified is *#1* α-Al primary dendritic, *#2* DCP, *#3* Al-Si eutectic, *#4* Al-Fe-Mg-Si-Cu reactions, and *#5* Solidus (T_{SOL})

8.4 Primary-Phase Evolution: α-Al Dendritic Structure

At the liquidus, the primary-phase nucleation of α-Al dendrites eventually achieve stable growth forming what is known as the primary dendrite stem [2, 5, 6, 20, 21]. Figure 8.3 sketches the FCC lattice or crystalline structure that is typical of the α-Al. The morphology evolves with perturbations on the surface of the primary stem that becomes stable secondary dendritic cells which grow at <100> crystallographic directions, orthogonally from the primary cell structure. It is illustrated in Fig. 8.7.

As solidification progresses, the α-Al cells which have now formed, referred to as secondary dendrite arms, undergo a coarsening process that affects both the size of the cells and the distance between neighboring cells. The dendrite coarsening (Oswald Ripening) process continues until it is encased in solidification casting structure [21]. In the remainder of the chapter, multiple references to an important microstructure parameter related to α-Al dendrites will be made and will be introduced before further discussion of the α-Al phase development and its impact on casting structure.

The coarsening and orthogonality of secondary dendrites can provide a measurable parameter known as the secondary dendrite arm spacing (λ_2), also labeled as SDAS in many texts and journal articles. The secondary dendrite arm spacing is determined using a polished 2-D cross-section of the cast structure under light optical microscopy with an image analyzer software. Several parallel secondary dendrite cells are identified, and a line is drawn over these dendrites with a final length

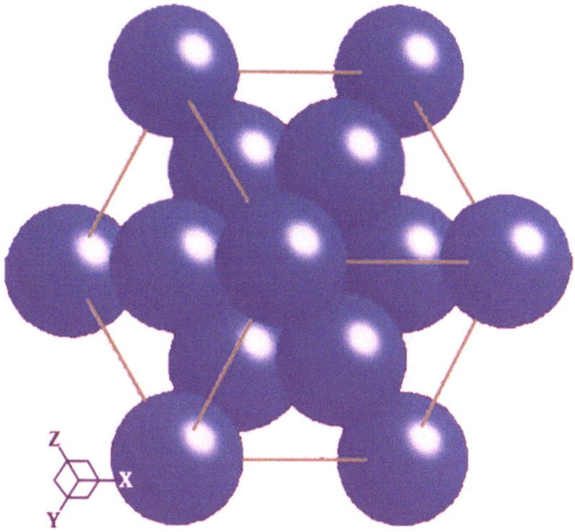

Fig. 8.3 Crystallography and parameters of Al

in microns determined. The number of secondary dendrite cells that should be incorporated is generally five to ten, although casting structures can vary in what a metallurgical lab technician can identify when conducting this type of analysis. This final length is then divided by the number of secondary dendrite cells intercepted. An example of the value of λ_2 is given in Figs. 8.4 and 8.5 [21].

The factors controlling dendrite coarsening stem are several thermodynamic driving forces. There is the effect of free energy associated with the surface of the dendrite (G_s), the bulk volume energy of the dendrite (G_v), and the partitioning of solute (also known as the equilibrium partitioning coefficient, $k = C_s/C_l$) [2]. To better visualize the role of each of the thermodynamic factors, it helps to consider a dendrite as a separate entity from the adjacent secondary dendrites and from the primary stock dendrite from which it grew. The schematics of dendrite ripening and the thermodynamic-driven diffusion process are shown in Fig. 8.4 for the initial stage of solidification and in Fig. 8.5 for the fully solidified cast structure.

The dendrites try to attain a larger portion of their free energy through bulk volume energy (G_v) at the expense of the dendrite surface energy (G_s). This thermodynamic driving factor exists for any phase such as the α-Al embryos and GPZ (Guinier-Preston zones). In the context of Fig. 8.4, this means that dendrite a (thicker dendrite) is more stable than dendrite b (thinner dendrite) because $G_{va} > G_{vb}$, $G_{va} > G_{sa}$, and $G_{sb} \geq G_{sa}$ (i.e., subscript va refers to the volume of dendrite a, see Fig. 8.4). Eventually, as solutes (Si and Cu) diffuse from dendrite a to dendrite b and solvent (Al) diffuses from dendrite b into dendrite a, dendrite b should eventually, with time, dissolve altogether. For the same given solidification time, a change in the alloy composition could change the rate of this dendrite ripening. Presumably, a larger quantity of solute present would dissolve thinner dendrites sooner [21].

Now alloy composition can be a factor in the rate of coarsening when a partitioning of solute elements is considered. For a given solidification rate, highly partitioning solute elements will drive the need to provide a larger liquid-solid interfacial

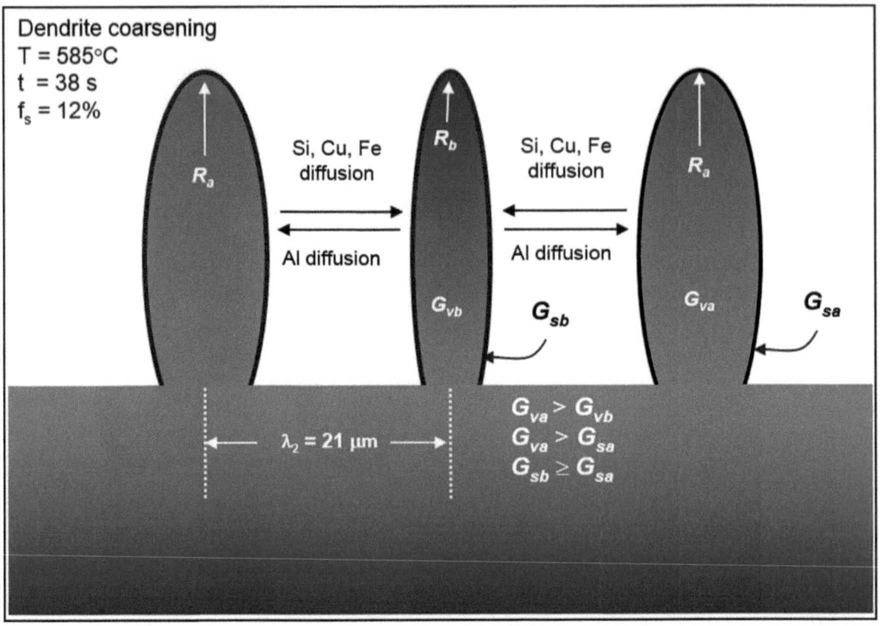

Fig. 8.4 In the first quenched stage, the dendrite size distribution results in dendrites with a radii of curvature at the tips. In the above case, $R_a > R_b$. This drives the diffusion paths of solute elements (Si, Cu, and Fe) along the diffusion paths indicated, while the solvent element (Al) diffuses in the reverse direction [21]

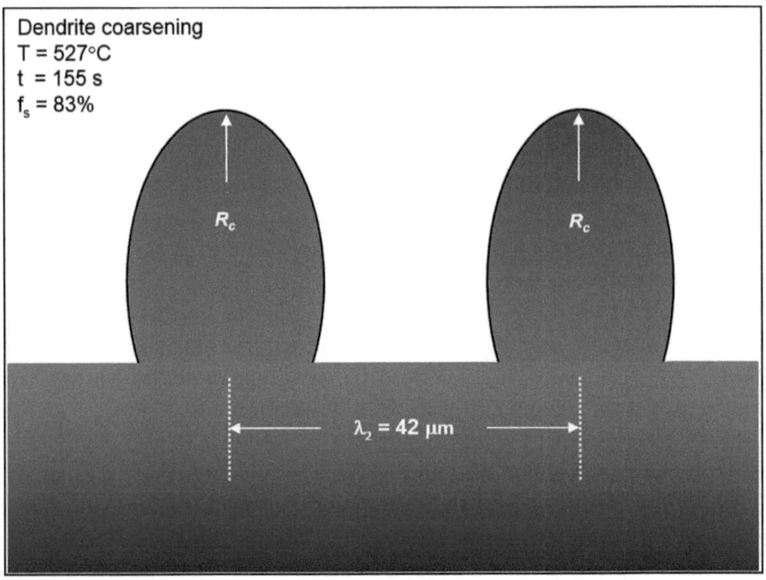

Fig. 8.5 At the end of solidification for the WA328 alloy, the dendrites were free to ripen (no quenching). In other words, thinner dendrites have dissolved at the expense of the larger dendrites. The cells are now larger yielding a radius of curvature of R_c, which is larger than R_a and R_b seen in Fig. 8.4 [21]

Table 8.4 The calculated equilibrium partition coefficient, *k*, and the slope of the liquidus line for Si, Cu, Fe, Mg, and Mn [21]

Element	*k*	Slope of the binary liquidus line – M (°C/wt%)	Dendrite coarsening kinetics
Si	0.13	−6.6	Contributes to larger dendrite cells
Cu	0.17	−3.43	
Fe	0.03	−2.96	
Ni	0.007	−3.3	
Mg	0.51	−6.2	Contributes to smaller dendrite cells
Zn	0.88	−2.9	
Cr	2.0	3.5	
Mn	0.82	−1.21	

area to eject that solute. Table 8.4 lists the most common alloying elements found in many Al castings and their effect on dendrite coarsening. Those elements having low partitioning, such as Si and Cu, will need to develop a larger surface area are of the solid-liquid interface, which more effectively ejects solute into the interdendritic liquid. Conversely, Zn has higher partitioning, meaning that more of the Zn is retained in the FCC structure and less to being ejected into the interdendritic liquid, and thus consequently leads to a lower surface area for the solid-liquid interface of forming dendrites.

The aforementioned discussion is important as dendrite formation not only reflects casting structure, but it can be used as an effective measure of the alloy solidification rate. This becomes important in the subsequent description of not only the aluminum casting processes, but how the understanding and measurement of the dendrite structure can be used to understand defect formation. This will be reviewed in a later section of this chapter.

The literature has discussed the relationship between dendrite coarsening and the overall solidification time at some length. Flemings et al. [22], among others [23–26], have introduced an empirical relationship that relates the effect of the final solidification time (t_f) on λ_2. The solidification time is usually established in solidifying alloys monitored with an embedded thermocouple where the liquidus and the solidus are determined. The mathematical trend that has been used to describe the relationship between both is given by [22]:

$$\lambda_2 = a t_f^{\,n}$$

where a is a constant, which depends on the alloy, and n is an exponential value reported to be between 0.3 and 0.4 [22–26]. The above relationship only reveals part of the phenomenon when it comes to understanding conditions of the mushy state, where factors like dendrite coherency point (DCP), permeability, and microsegregation are also critical [2–4]. Additionally, the above relationship relates to the fully solidified structure, but does not extensively relate to the coarsening of a partially solidified structure.

8.4.1 *Effect of Silicon Concentrations*

Si is the main alloying element in the vast majority of Al casting alloys used for automotive and aerospace applications, and thus its effect in cast structure performance is discussed here. Si is used mainly due to the effect it has on the solidification soundness that is possible and helps provide some wear characteristics. Most of the Si found in the cast aluminum structure will be part of an Al-Si eutectic phase, where a smaller portion of Si contributes to secondary-phase formation. As shown in Tables 8.2 and 8.3, it can be seen that Si concentrations will range from 5.0 wt% (hypoeutectic) up to 14 wt% (hypereutectic). The Al-Si alloys are divided into hypoeutectic, eutectic, and hypereutectic. The eutectic composition is located at approximately Al-12.6 wt% and at a temperature of 577 °C. The volume fraction of the Al-Si eutectic that can comprise of the total cast structure can be as low as 40% in 319 and 356 alloys, and be as high as 100% for eutectic alloys such as 413 alloy [22].

The eutectic Si that solidifies in these alloys forms a pure Si crystal in platelet morphology, but seen as needle like in 2-D metallographic examination. The crystal-type Si is called the diamond crystal structure, the same structure as C in diamond form. Figure 8.6 shows an example of the Si structure. Si is considered to be a semi-metal element, otherwise known as a metalloid [17]. The solidification kinetics of Si is described in Fig. 8.7a, b and shown in a metallographic image in Fig. 8.7c.

In Fig. 8.7a, the undercooling effect of Si nucleation is shown, which is higher than that for the primary α-Al dendrites. This is because Si is a highly faceted crystal phase and will not begin to grow until several degrees below the Al-Si eutectic

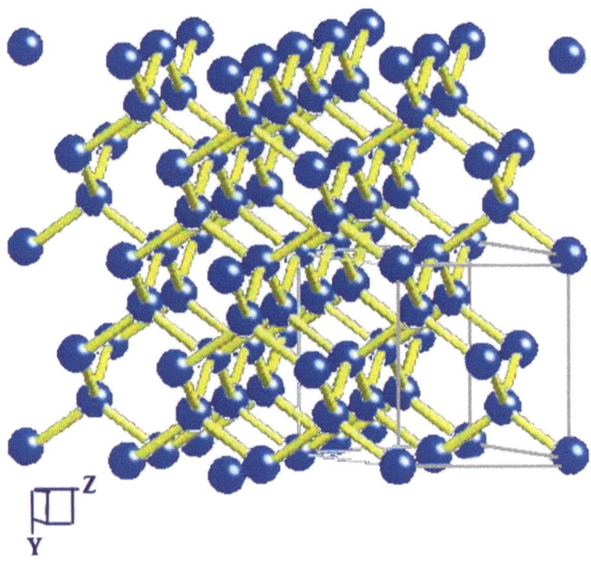

Fig. 8.6 Crystallography and parameters of Si

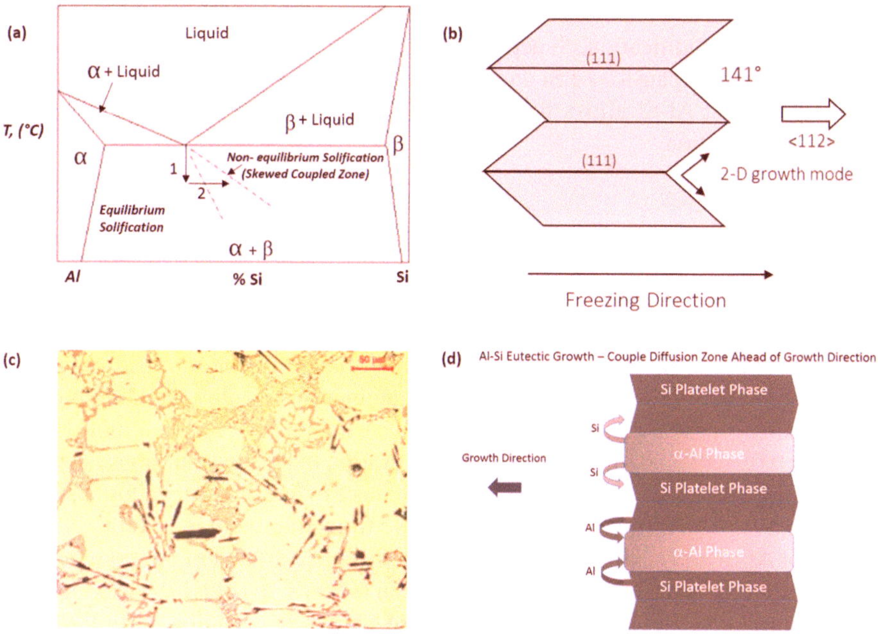

Fig. 8.7 (**a**) A binary Al-Si alloy of eutectic composition will undercool (*arrow 1*) before allowing Al-Si eutectic growth to occur (*arrow 2*). (**b**) Growth mode of platelet Si crystals: twin plane reentrant edge (TPRE) growth mechanism. (**c**) Light optical micrograph of Al-Si partially solidified structure showing the evolution of the Si platelets. (**d**) Illustration of the growth of the Al-Si eutectic phase

point. This is illustrated by the skewed coupled zone located below and off to the right of the Al-Si eutectic point. Si, when it crystallizes, gives off five times the latent heat of Al (398 kJ/kg for Al, 1926 kJ/kg for Si) and this in turn provides an enhancement of interdendritic feeding capacity. Si is ideal to increase the fluidity of partially solidified aluminum castings which not only provides hardness (due to the fact that the Si that forms are Si crystals) but the latent heat generation reduces porosity formation because of the enhanced interdendritic feeding.

As seen in Fig. 8.7b, the growth kinetics of platelet Si follows the twin plane reentrant edge (TPRE) growth mode [22]. Branching is possible at twin boundaries and can promote a fibrous structure. Both enhanced solidification and chemical modification have been reported to promote this fibrous structure, and more about this is discussed in a subsequent section of the chapter. In this zone is where Al-Si eutectic phase can begin its polyhedral growth (or coupled growth, where Al lamellae feed Si to adjacent Si lamellae and vice versa). This growth is illustrated in Fig. 8.7d, where the coupled Al and Si liquid diffusion feeding platelet Si and α-Al occurs.

Referring back to Fig. 8.7c, the Si seen growing in the micrograph of the evolving Si structure in a 319 alloy sample was captured via a method of quenching during a partially solidified state. In the microstructure, coarse acicular Si is seen, and the ultrafine Si was the result of Al-Si eutectic, which had yet to solidify normally and was quenched (Fig. 8.7c) [23].

Si has a profound effect on the soundness of the cast structure. This is due to the fact that upon solidification, Si actually expands 2.9%, while α-Al will contract by nearly 7%. This expansion of Si during crystallization then reduces the overall contraction of most Al-Si alloys to between 3% and 4% and reduces interdendritic shrinkage formation.

8.4.2 *Effect of Iron Concentrations*

Fe tends to be the most common impurity in Al-Si alloys and is also one of the most widely studied in the metal casting literature. Fe is always present and can be no lower than 0.005 wt% in 99.99 wt% pure aluminum. The alloy blending process, when making primary grades, will generally drive this Fe concentration higher, between 0.05 wt% and up to 0.3 wt%. This elevation in Fe concentrations due to the alloying process comes from the Fe concentrations existing in master alloys used [2].

Fe tends to be a difficult impurity to remove from aluminum in general. Primary grades of Al-Si alloys will tend to have a max Fe concentration of no more than 0.3 wt%. Secondary-grade sources may range from 0.5 wt% to over 1.0 wt%. The only aluminum casting application where elevated Fe can be desirable is in the high-pressure die casting (HPDC) industry. It has been shown in practice that the elevated Fe concentration in Al-Si melts deters die soldering, which can be a costly impairment in productivity and tool die surface repair. Having elevated Fe in HPDC alloys, however, does have the undesired impact on mechanical properties seen in all aerospace and automotive applications.

While the level of Fe concentration in Al-Si melts cannot be eliminated completely, targeting its upper and lower ranges can be important in terms of understanding cost impacts over the production life span of a casting made. Aluminum prices do fluctuate due to global demand, and the Fe concentration negatively impacts the overall cost of the alloy by 22 cent/kg for every 0.20 wt% Fe. The key question is what level of Fe concentration ranges should be listed on a casting component print. The below factors should be considered first:

1. What will the concentration of other impurity elements such as Sr, Mn, and Cr be, and will these be listed on the print specifications?
2. What is the solidification rate of the most critical location of the casting in terms of mechanical and thermomechanical properties? Most likely, the question is answered about what casting process is being considered.
3. What cost impact can be tolerated over the life span of the component being cast?

The two most important factors as indicated in Table 8.5 are the interfacial bonding with the α-Al FCC matrix and the stress concentration effect. The more deleterious β-bearing Fe phase is the platelet form having the stoichiometry of β-Al_5FeSi with another less commonly seen version as δ-Al_4FeSi_6 [1, 18, 19]. In metallographic 2-D view, this phase is seen as a needle and has been called the Fe-needle

Table 8.5 Fe-bearing phase characteristics

Characteristics	α-Phase	β-Phase
Metallography	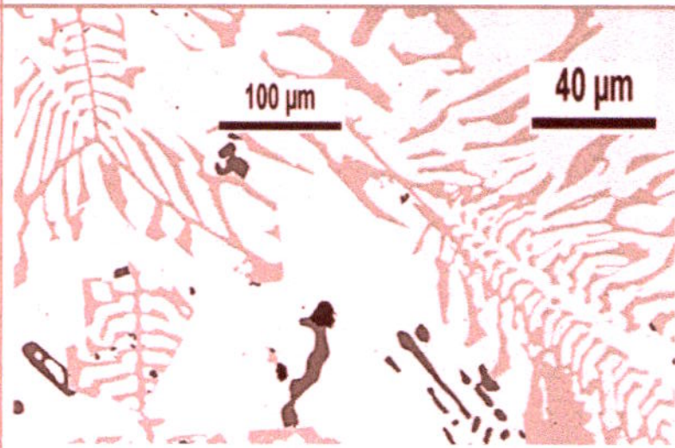	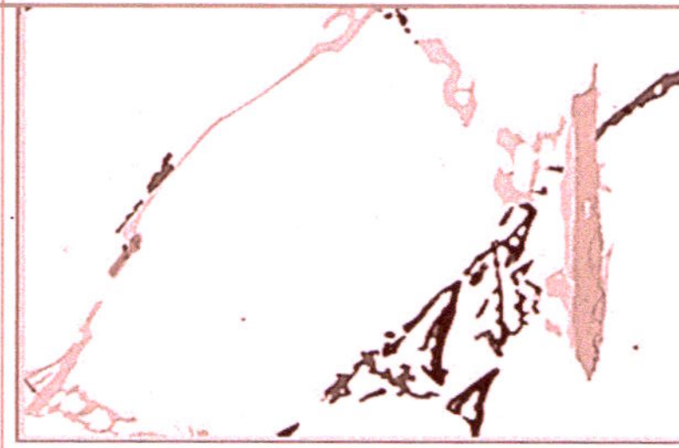
Morphology	Script	Platelet
Stoichiometry	$Al_{15}(Fe,Mn,Cr)_3Si_2$,	β-Al_5FeSi
Growth Kinetics	Normal	Twin plane reentrant edge
Bonding with α-Al FFC matrix	High	Low
Crystallography	Cubic (BCC)	Monoclinic
Density	3.3–3.6 g/cm^3	3.3–3.6 g/cm^3

phase. The β-bearing Fe phase growth shares the same kinetics of platelet Si crystals – twin plane reentrant edge. Growth is along the tips of the platelet, while bonding strength along the surface of the platelets with the α-Al FCC matrix is poor.

The α-bearing Fe phase has better interfacial bonding with the α-Al FCC matrix and also has a lower stress concentration effect. Both favorable factors tend to promote a better mechanical strength for the aluminum casting component. The most common stoichiometry cited is $\alpha\text{-}Al_{15}(Fe,Mn,Cr)_3Si_2$, but other stoichiometries also cited are $\alpha\text{-}Al_8Fe_2Si$, $\alpha\text{-}Al_{12}Fe_3Si_2$, and $\alpha\text{-}Al_{17}(Fe,Mn)_4Si_2$ [18–20, 22–26]. The importance of Fe concentration on item 1 has to do with the impact that impurities have on Fe phase formation. The most critical is the level of Mn in the melt, as its presence will favor more α-phase formation at the expense of β-phase formation. This is illustrated in the simplified Al-Fe-Si phase diagram with fixed Mn concentration in Fig. 8.8.

This is why many print specifications for casting components in the automotive industry tend to list both Fe and Mn targets. Mn is specified in some cases as a ratio to the given Fe concentration in the melt used at any one point to cast the components. Typical ratios that suppress β-phase and favor mainly α-phase formation can range from 0.5% to 0.7% of the Fe concentration [22, 25, 26]. Typically in the foundry, melt operators will have to get optical emission spectrometer (OES) feedback on Fe levels and then make the required Al-Mn master alloy addition to the melt. This follows a secondary OES reading to confirm target ratio of Mn to Fe concentration, before casting can commence.

The Fe-bearing phases listed in Table 8.5 are strongly affected by solidification rate, with the β-phase being more faceted and requires higher undercooling. Thus in higher freezing rate conditions, β-phase will require a larger undercooling, which then allows existing Fe concentrations to be consumed in α-phase formation. However, in slow freezing rate conditions, proper Mn/Fe ratios will force α-phase formation to occur as indicated in Fig. 8.8. For example, in high-pressure die casting (solidification times of 10–60 s), Fe can be high, but some Fe is retained in the α-Al matrix, and the resulting β-phase particles are very small. The Mn concentration can support the 1:2 ratio, and thus result in small α-phase formation instead. In both cases, the impact on mechanical properties is minimal.

In precision sand aluminum casting processes (solidification times of 5–15 min), Fe has to not only be as low as possible, but the Mn concentration has to maintain the 1:2 ratio to promote α-phase formation; if not, then β-phase formation will be large and thus will drastically impact on mechanical strength.

A wrong balance of high Fe concentration and low Mn in the Al-Si alloy and low solidification results in the worse possible scenario in terms of mechanical properties and alloy impairment. The stress concentration effect of the β-phase platelet morphology is magnified due to its size, coupled with the liquid feeding impairment giving rise to an associated pore formation, when β-phase nucleation occurs before the onset of the Al-Si eutectic [19].

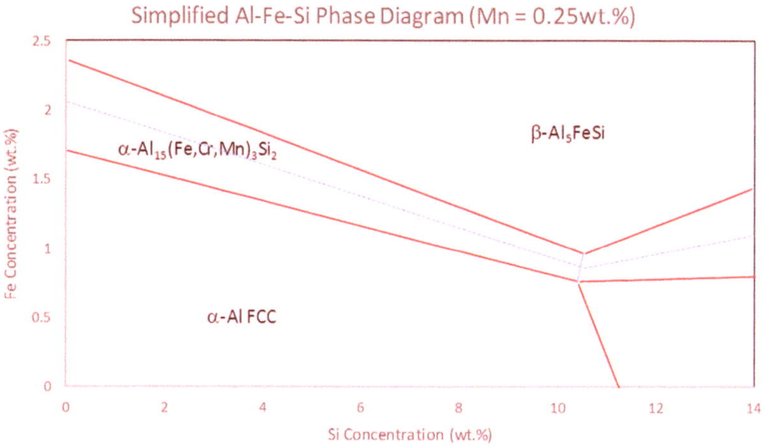

Fig. 8.8 Phase relations for α-phase and β-phase formation due to Mn concentrations

8.4.3 Effect of Copper Concentrations

Cu was the original alloying element used for liquid Al to cast parts. The 69 kg cast crankcase made for the Kitty Hawk was made from an alloy that was Al-Cu [ASM Handbook, Vol 15]. However, despite its susceptibility to hot cracking and poor corrosion resistance, its elevated temperature strength is still superior to most Al-Si alloys, and as a result, there are still applications for 2xx alloys.

After the Second World War, significant scrap supply of Al-Cu-based alloys existed, and this gave way to a new generation of Al-Si alloys containing Cu. This scrap made the Al-Si alloys cost effective for its emerging use in the automotive and heavy truck industries. Al-Si-Cu alloys generally are not used in aerospace applications due to high porosity present in the final cast structure. Cu has the effect of extending the overall freezing range and thus provides a greater interdendritic feed demand. However, in most automotive casting applications, using the high-pressure die casting, precision sand, or semisolid gravity processes, the resulting porosity levels can be acceptable.

The most common phase that forms in Al-Si-Cu casting is the Al_2Cu phase and can be in two forms – blocky and eutectic [26]. The crystal structure of Al_2Cu is body-centered tetragonal (BCT). Figure 8.9 shows an example of a 319-type alloy which contains the eutectic form of Al_2Cu. The Al_2Cu phase usually forms during the final stages of solidification and is the phase which partially or fully dissolves during the solution stage of heat treatment, where the Cu atom diffuses into the α-Al FCC matrix, and then Cu will follow a matrix precipitation process during artificial aging stage. After aging, the matrix will begin to increase in hardness and yield strength goes up [2].

One comment should be made about another Cu-bearing phase having the stoichiometry as the π-$Al_5Mg_8Cu_2Si_6$. Its formation is affected by Mg concentrations, and that is why most 319 alloys have their Mg limits cap at trace levels (e.g., 0.10 wt%, see Table 8.2).

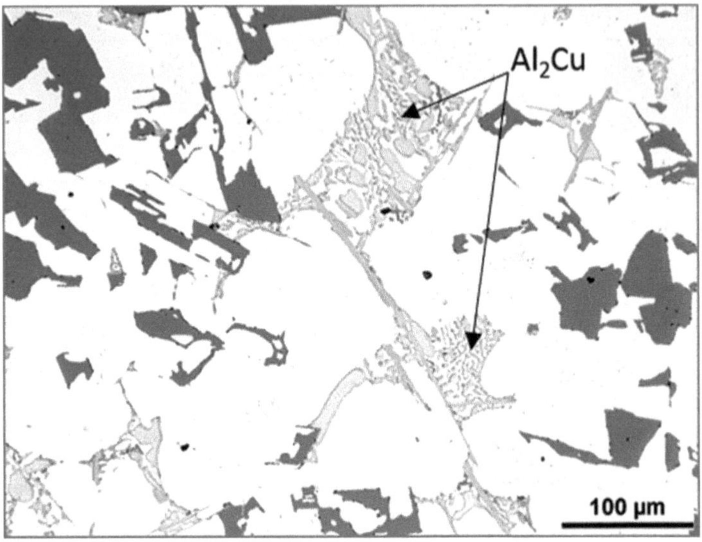

Fig. 8.9 Typical cast microstructure containing Cu-based phase. The most common stoichiometry is Al_2Cu

8.4.4 Effect of Magnesium Concentration

Mg is added to alloys that are used in both automotive and aerospace applications. Mg has benefits to improve the performance of casting components in terms of mechanical properties. Firstly, when compared to Al-Si-Cu, Mg actually shortens the solidification range and thus reduces feed demand, yielding a lower level of microshrinkage development. Finally, Mg-bearing phases solidify as a complex eutectic that is higher in temperature than for the Cu-bearing reactions of 319 alloy. This means that a higher solution temperature can be used which also allows for some Si-platelet spheroidization to occur. The main stoichiometry Mg has in Al-Si alloys is the Mg_2Si phase [26] and is seen as a black script phase as shown in Fig. 8.10. The crystallography for Mg_2Si is the cubic anti-fluorite crystal structure (compounds based on FCC packing of ions).

The possible stoichiometries of Mg-containing phases that can form in Al-Si cast structures are p-$Al_8Mg_3FeSi_6$ and Mg_2Sn. Their formation depends on the trace amounts of Fe, Sn, and Cu. For example, increasing Fe concentrations in a 356 or 357 alloy will promote a greater formation of the p-$Al_8Mg_3FeSi_6$ over the Mg_2Si phase [19], which improve yield strength and plastic deformation. This has the effect of reducing the amount of Mg_2Si phase, which can contribute to solutionizing the α-Al FCC structure. Sn in sufficient quantities can replace Si in Mg phase formation to crystallize Mg_2Sn, which also does not dissolve into the α-Al FCC matrix during solution treatment [27]. It is clear why aerospace versions of 356 and 357 alloys typically are made from primary sources where trace elements are considerably low.

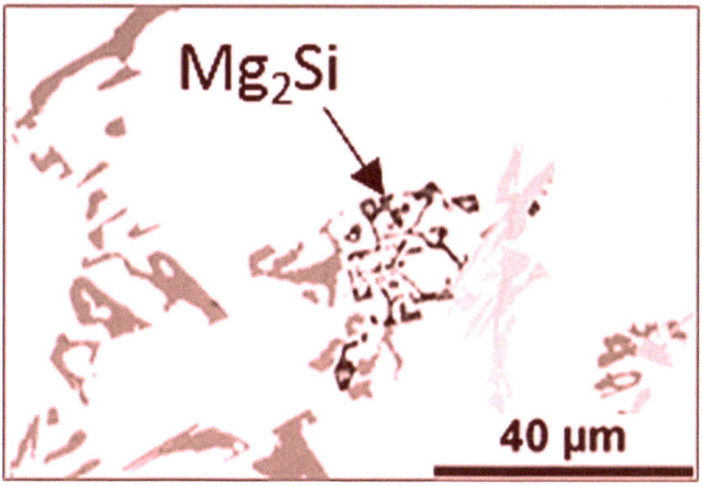

Fig. 8.10 Typical cast microstructure containing Mg-based phase. The most common stoichiometry is Mg_2Si

8.4.5 *Si Eutectic Modifiers: Strontium, Antimony, and Sodium*

Si eutectic modifiers are widely used as elemental additive to many aluminum casting applications. Sr is usually introduced as a master alloy to the alloy melt prior to producing the cast product. Sr modifies Si through the twin plane poisoning technique leading to a fibrous Si structure [1–5, 8–11, 14, 28–30]. Sr however has a fairly deleterious effect on mechanical properties. As Sr content increases, the level of porosity increases [1, 3–5, 30]. This has resulted in aluminum castings made for aerospace applications without Sr in order to achieve the highest possible cast mechanical properties [1, 30]. Examples of the effect of Si phase modification is shown in Fig. 8.11. The preferred etchant to remove Al in the SEM image of Fig. 8.11 is HF [31]; however, special precautions must be taken due to its corrosiveness and toxicity.

The main purpose of using Sr has been to take advantage of the fact that fibrous Si leads to a reduced solution duration (to achieve Si spheroidization) and to reduce machining costs [1, 2, 4, 5, 14]. Unlike Sr, Sb refines the size of Si particles as opposed to modifying Si particle structure [1, 4, 5, 14, 26, 27]. Sb is very popular in Japan and is used to refine the Si structure of aluminum engine blocks and cylinder heads. As these cars are recycled in North America, the Sb levels in some secondary batches may be higher than usual. Once a melt contains excess Sb and Sr, the Sr-Sb-based compound forms. The Sb-Sr that has been identified is the Mg_2Sb_2Sr intermetallic, which has a density of 4.2 g/cm^3 and sinks to the bottom of the melt upon formation [1, 11]. Thus, Sr becomes depleted from the melt, and modification becomes difficult to achieve. The only way to get a desired modification from these

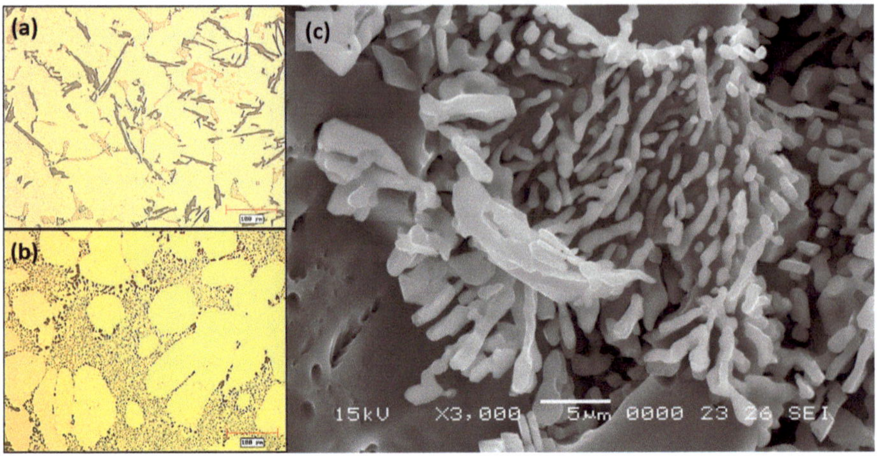

Fig. 8.11 (**a**) Nonmodified, (**b**) Modified, and (**c**) Modified Si SEM image of Sr-modified Si with aluminum etched away

melts is to add a sufficient level of Sr to react with most of the Sb, so the remainder can modify the Si particles.

Na has been used in the past as a Si particle modifier. Na is used less today due to its relatively rapid fading in comparison to Sr [1, 2, 9, 11, 30, 32]. Secondary stock materials can have elevated levels of Na in melts. Limited research into Sr-Na reactions found that no compounds containing both modifying elements exist in cast Al-Si melts [32]. In fact it has been found that a background level of Na can enhance the added level of Sr in addition to the melt since both Na and Sr induce twin planes in Si crystals [27].

8.4.6 Grain Refinement

Due to the growing popularity of grain refining, many remelted cast materials have higher Ti and B levels. Master alloys containing Ti and B will introduce particles such as Al_3Ti and AlB_2, which are believed to act as heterogeneous nuclei for α-Al nucleation [1–5, 32–37]. The effect of elevated Ti in the ingot material has been found to affect the grain refining potency [2, 4, 10, 32]. The initial Ti level has recently been referred to as "background Ti." Many publications have indicated that Ti levels need to be as low as possible so the full grain refining effect can be met with added grain refiners [2, 34]. Recent work, however, has challenged this concept and has indicated that a high background Ti level in the cast melt can reduce the level of further additions of grain refiners to get a desired grain size [3, 34].

Figure 8.12 shows the effect of grain refining on the cast structure of a precision sand component for an automotive application where a cast iron chill is used to help

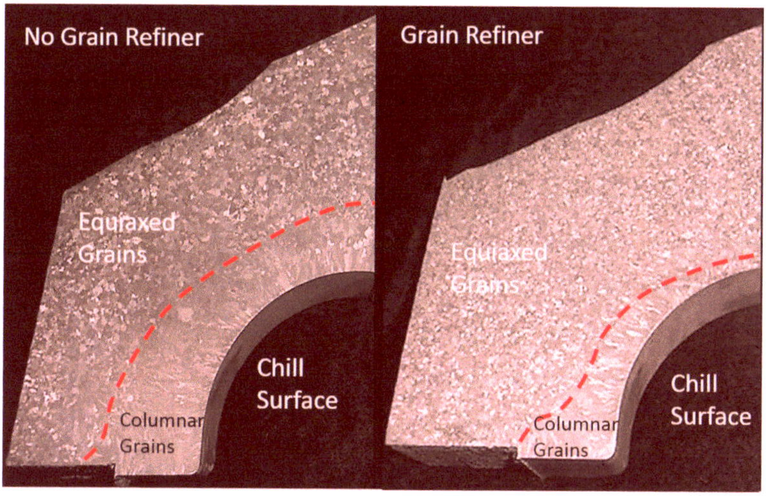

Fig. 8.12 Effect of grain refiner on cast Al-Si structure

extract heat during solidification. In the case where grain refiner is added (right image), not only the grains are smaller, but the length of the columnar grain is also shorter. Columnar grain tends to affect fatigue performance. An example of columnar grain structure resulting in fatigue failure will be shown in a later section of this chapter.

8.5 Mechanism for Pore Development

After reviewing the secondary-phase evolution in Al-Si alloys in the previous section, this section will review the voids that form and contribute as greatly, or greater toward overall mechanical cast section performance. An elemental constituent which is not desired in liquid aluminum is hydrogen (H^+) and a compound of Al and O that makes oxides (mainly in the form of Al_2O_3). Both of these are ultimately the main culprits that work together toward the pore development during solidification. There currently does not exist a single system that measures both, but there are separate instruments commercially available that help assess either hydrogen content or oxides.

The most common method to mathematically describe stable pore development does focus on hydrogen in terms of internal gas pressure. This pressure partially facilitates pore growth, along with aspects of hydrostatic stress buildup due to dendritic volume contraction and temperature drop. The decrease in temperature is between 600 and 575 °C for primary α-Al dendritic growth and 575–500 °C for Al-Si eutectic growth [20]. Mitrasinovich et al. [38] proposed a methodology to accomplish this analysis by means of thermal analysis.

As the hydrogen concentration grows in the liquid, the hydrostatic stress grows and the conditions for pore nucleation become favorable. The heterogeneous nucleation of porosity in aluminum alloys is generally an oxide film or bi-film containing trapped air [13]. The mathematical description for pore growth is generally represented by the homogeneous nucleation process shown in Eq. 8.1. The chemical activity of hydrogen is higher in a gas void than in the liquid state, and thus hydrogen readily difuses into an already nucleated gas void (or in the heterogeneous nucleation case, the oxide enveloped void). The threshold conditions that allow for pore expansion or growth is described in Eq. 8.1.

$$P_G + P_S \geq P_{atm} + P_H + \gamma\left(1/r_2 + 1/r_1\right) \tag{8.1}$$

- $P_G = (f_H C_o/S)^2/[1 - f_s(1 - k)]^2$, also internal gas pressure of hydrogen
- P_S = Shrinkage pressure (hydrostatic stress eluded to earlier)
- P_{atm} = Atmospheric pressure
- P_H = Metallostatic head pressure

Now the description must be evolved further from just describing internal gas pressure. Atmospheric pressure is an important component in pore growth in solidifying aluminum, and this will be discussed further in relation to the reduced pressure test (RPT). The reduced pressure test is one of the most important melt quality tests used in automotive and aerospace foundries. Metallostatic head pressure related to the amount of liquid aluminum that exists above a nucleating pore, where the larger this height is, the more pressure acting on that pore and thus potentially impeding its growth. The internal gas pressure P_G is due to the diffusion of H^+ and the transformation to H_2 gas. Surface pressure P_S is related to the surface tension at the gas-liquid interface.

The relationship expressed in Eq. 8.1 and its description is an oversimplification for the conditions of stable pore growth. Usually $P_G + P_S$ is insufficient to cause pore nucleation, and it is well known that for heterogeneous nucleation to occur, a foreign particle is required. Oxides are again the most common heterogeneous nucleant. Homogeneous nucleation is rather difficult due to the fact that a small gas pore of only several atomic diameters would be required to obtain an internal pressure of 30,000 atm (30,400 Bar) [12, 13].

Campbell [13] has introduced a concept of the gas-shrinkage map to show the relationship between internal gas pressure and the hydrostatic stress on the liquid. The gas-shrinkage map shown in Fig. 8.13 illustrates the conditions by which the interdendritic liquid may be forced to nucleate a pore through heterogeneous nucleant. The hydrostatic stress and the internal gas pressure required need to be high enough to overcome all the other parameters to form a pore. However, if feeding becomes retarded, the level of hydrostatic stress on the interdendritic liquid should be higher and thus activate more heterogeneous nucleants, such as nucleants 1 (highest potency), 2 (medium potency), and possibly 3 (lowest potency).

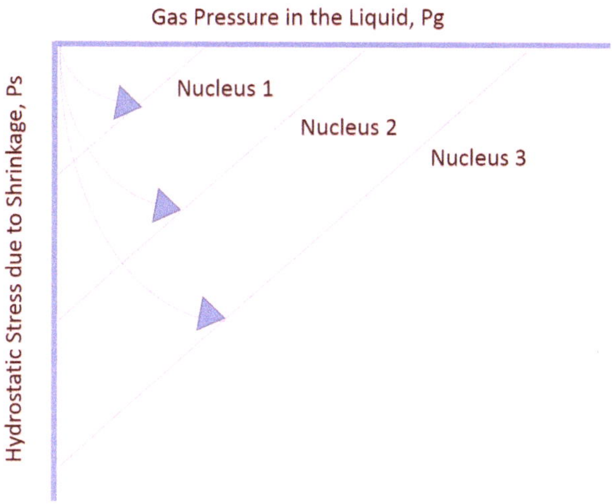

Fig. 8.13 The gas-shrinkage map illustrates the combination of shrinkage stress and internal gas pressure that can be relieved by heterogeneous nucleation from particles present within interdendritic regions of the mushy zone [13]

8.6 Hydrogen Concentrations: Measurement and Established Thresholds

The level of hydrogen cannot be determined by the common online chemical methods used in foundries (i.e., optical emission spectroscopy (OES)). Foundries have used the Alscan, Telegas, or the Alspek H. All these methods are especially designed to determine the presence of hydrogen. Unfortunately, they lack continuous monitoring capabilities and robustness to be widely used, and thus have been used to spot-check the efficiencies of hydrogen removal practices. The most common method to gauge pore development as part of a foundry testing protocol is the reduced pressure test (RPT), and more will be discussed about this testing method later on in this section. This method reflects pore development susceptibilities due to collaboration between hydrogen (H^+) and oxides (Al_2O_3).

For any given alloy and specific solidification conditions, there is a "threshold hydrogen content" below which no observable gas-assisted porosity is formed [28, 39]. If the area fraction of porosity is independent of the hydrogen level, this would indicate that only shrinkage plays a primary role in pore growth. Shrinkage porosity is due to the effect of hydrostatic stress development in the interdendritic liquid. Above the threshold; however, the percent area fraction of porosity increases linearly with respect to hydrogen content. This indicates that the growth in final porosity has an increasingly larger contributory factor by hydrogen.

Before further discussion, the units used to quantify dissolved hydrogen in liquid aluminum should be briefly reviewed, as it is used by chemists but almost never seen in engineering disciplines. The units for dissolved hydrogen in liquid aluminum are given as cc (or cm^3) H_2/100 g Al, and its temperature dependence is shown in Fig. 8.14. The unit basically implies a quantified volume of hydrogen gas will evolve from 100 g of Al. Now in liquid state, the dissolved hydrogen is in the form of H^+. Therefore, the solidification process is reflected in the nature of the unit.

For castings that solidify at low cooling rates, the accepted threshold level is approximately 0.10 cm^3 H_2/100 g Al [2, 11, 16, 28]. However, recent work by Anson et al. [29] has indicated that this threshold is actually slightly higher, at about 0.18 cm^3 H_2/100 g Al when the cooling rate is approximately 0.30 °C/s. Below this level of hydrogen in the melt, the resulting porosity is negligible. Hydrogen content, when above the threshold level (e.g., 0.2 cm^3 H_2/100 g Al), also affects the fraction solid at which new pores begin to grow in the solidifying alloy. In the 356 alloy quenched sample experiments performed by Anson et al. [29, 40], it was found that a baseline number of pores existed within the solidifying structure. The pores originated either by turbulent filling or from gas bubbles enveloped by oxide bi-films in suspension in the melt. The number of overall baseline pores did not change during solidification until about 50–60% fraction solid (f_s), where the nucleation of new pores occurred. The expansion of the baseline pores occurred because hydrogen diffuses into the pores and thus achieves the threshold pressure required for pore expansion. Figure 8.14 illustrates the above observations by Anson et al. [29].

As far as hydrogen solubility is concerned, various aluminum alloys have different hydrogen absorption tendencies due to differing solubility capacities. For exam-

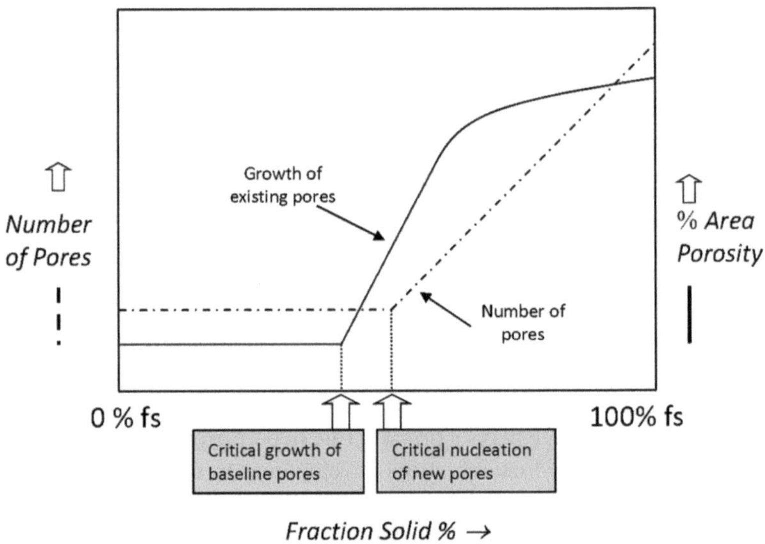

Fig. 8.14 This graph illustrates the variation between the final percent porosity and the final pore density with a given solid fraction. Since there is a change in pore density at the H threshold, it is assumed that the nucleation of the pores is partly dependent on the H content (Reproduced from [29])

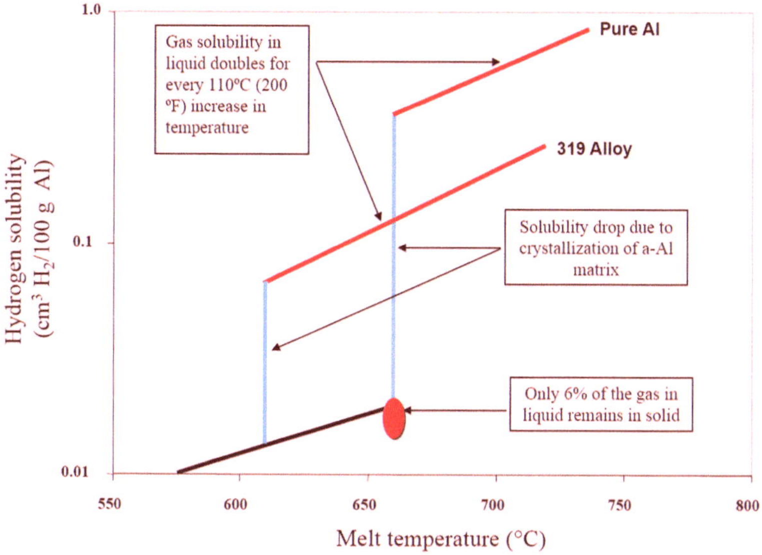

Fig. 8.15 Solubility of hydrogen in aluminum, with comparison between pure aluminum and the 319 alloy

ple, under the same conditions (i.e., melt temperature), soluble hydrogen content in the 356 alloy is 18% higher than in the 319 alloy, but 22% lower than in pure Al. This is illustrated for 319 alloy and pure aluminum in Fig. 8.15. The hydrogen uptake can be higher at higher melt temperatures due to thermal kinetics and also due to the fact that the hydrogen solubility in liquid aluminum doubles for every 110 °C above the liquidus, superheat, temperature.

Upon the nucleation of α-Al dendrites, the solubility of H^+ drops dramatically, and therefore, H^+ must defuse into the remaining interdendritic liquid. At the same time, the interdendritic liquid gradually becomes hydrostatically stressed as volumetric contraction of the forming dendrite structure continues. As temperature decreases and solidification progresses, the alloy will eventually have only 6 wt% or less of the original hydrogen that was dissolved from higher temperature. The rest of the hydrogen is not soluble in the solidified α-Al, which results in precipitation, forming bubbles, and if it is not removed by degassing, the bubbles will turn into pores. This is illustrated also in Fig. 8.15.

8.7 Methods of Hydrogen Absorption

Hydrogen is the only gas that is soluble in aluminum and its alloys [1, 2, 4–13, 15, 16, 29, 30, 40–49]. Gases such as oxygen and nitrogen may form stable compounds with aluminum or other elements, which are usually considered insoluble impurities. Those gases have almost no gaseous solubility in liquid aluminum.

Water vapor is mainly responsible for the presence of hydrogen in the aluminum melt. It may originate from the atmosphere or in any form of contamination of tools, flux tubes, grain refiners, and other master alloys [1, 33]. Upon exposure to the molten metal, the water vapor ($H_2O(v)$) reacts to form magnesium oxide (Eq. 8.2) and aluminum oxide (Eq. 8.3) and gaseous hydrogen [2, 29, 40]:

$$H_2O(v) + Mg \rightarrow 2H + MgO \tag{8.2}$$

$$3H_2O(v) + 2Al \rightarrow 6H + Al_2O_3 \tag{8.3}$$

Under the melt conditions, these reactions are considered irreversible in the sense that a large negative free energy change would be required (e.g., the melting point of Al_2O_3 is 2072 °C) to break those molecules. It should be noted that the initially formed oxide films are amorphous for milliseconds and then they crystallize [41]. During the short amorphous stage of the oxide layer sitting on the melt surface, hydrogen absorption does not readily take place. However, the crystallized oxide loses its protectiveness against hydrogen absorption due to the fact that diffusion of Al^{3+} cations from the liquid side to the O_2-rich side of the oxide can now occur. Once Al^{3+} cations react with H_2O to form Al_2O_3 (called dross in the foundry), the dissociated H^+ cations (protons in this case) diffuse in the reverse direction through the oxide into the melt. This is illustrated in Fig. 8.16.

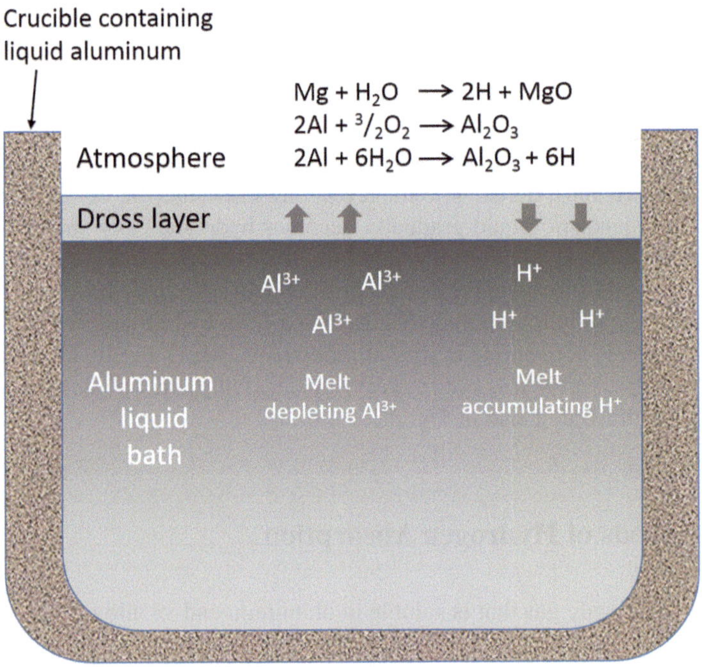

Fig. 8.16 Schematic of the oxide layer that separates liquid aluminum and atmosphere contain H_2O vapor

Previously it was believed that O_2 anions diffuse from the surface of the oxide into the melt. However, the concept of the Al^{3+} species diffusion through the oxide as opposed to O_2 anions is supported by the size of the diffusing species [48]. Al^{3+} cations are 0.51 Å while O_2 is 1.32 Å. The smaller the cation size, the easier it is for that species to diffuse through the crystallized Al_2O_3.

The H^+ cations that are generated at the oxide surface diffuse through to the melt quite readily. The rate at which this occurs is governed by the solubility of hydrogen in the melt (see Fig. 8.15) and the thickness of the oxide layer. As time passes, the oxide layer thickens, and the diffusion distances for H^+ become greater. The consequence of the greater diffusion distances is that the supply of Al^{3+} cations becomes less and the oxide thickening rate slows down.

8.8 Kinetics for Oxide Generation

The most deleterious oxides observed in aluminum alloy castings are Al_2O_3 (alumina) and $MgAl_2O_4$ (spinel) [1, 16]. Other slightly less deleterious compounds found in Al melts are MgO and Si-C particles. The general reactions by which these oxides form are shown as follows [1, 2, 16]:

$$2Al + 3/2O_2 \rightarrow Al_2O_3$$
$$Mg + 1/2O_2 \rightarrow MgO$$
$$Mg + 2Al + 2O_2 \rightarrow \mathrm{MgAl_2O_4}$$
$$MgO + Al_2O_3 \rightarrow \mathrm{MgAl_2O_4}$$
$$3/2SiO_2 + 2Al \rightarrow Al_2O_3 + 3/2Si$$

However, Al_2O_3 is the predominant oxide. The general classification is normally based on the types of alumina present, as follows [1, 12, 16, 26, 41]:

- Amorphous alumina film inclusions ("new oxide films") are the product of direct oxidation of aluminum melts which have no time to crystallize. New oxide films are a by-product of turbulent pouring, bubble trails, or confluent fronts. Their formation usually takes between 10 and 60 s [16, 25]. It has a continuous aspect and low thickness and exhibits a wrinkled appearance. Thickness ranges from 0.05 μm up to 10 μm [41].
- Gamma alumina film inclusions ("old oxide films") are the product of the crystallization of amorphous alumina. Their formation require a certain incubation; therefore, these types of inclusions are associated with turbulent operations before mold pouring. The formation usually requires from minutes to several hours. The thickness can be much larger than new oxides, ranging from 0.1 to 1 mm [1, 25].

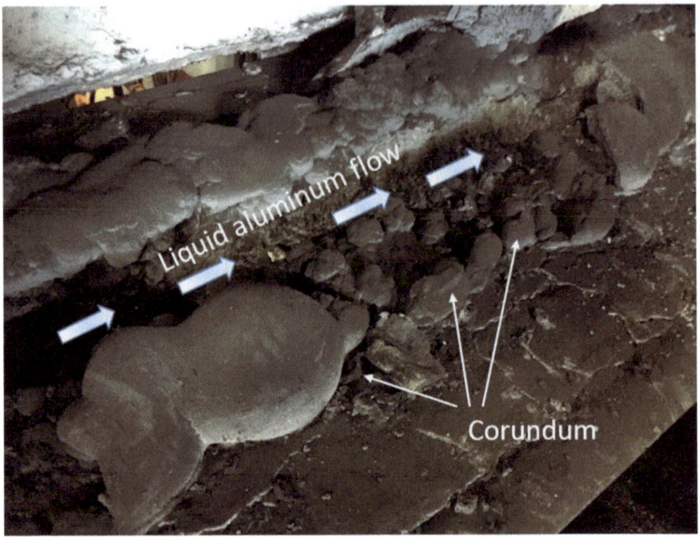

Fig. 8.17 Example of corundum (α-Al_2O_3) growth that has occurred in a launder melt system. The hard black growth, sometimes in spherical shape, will continue to grow after it has started. Prudent cleaning practices (dross removal – γ-Al_2O_3) is required to prevent its nucleation and growth

- Alpha alumina inclusions ("hard spots") are the product of recrystallization of γ-Al_2O_3 after long exposure to high temperatures. As the incubation conditions are severe, α-Al_2O_3 is normally formed due to adherence and accumulation of γ-Al_2O_3 films at the crucible walls, mainly in a holding furnace that contains "dirty metal" (high content of γ-Al_2O_3 films). Its main characteristics are the inclusion size, typically more than 1 mm, the black color, and the high hardness (corundum) [1, 25]. The example shown in Fig. 8.17 reflects actual long-term buildup of corundum which has grown into spherical hard spots, which can be several centimeters in diameter in a melting launder system. The melting launder system actually is covered by an inert gas blanket (nitrogen), but once younger oxides become incorporated in the flowing liquid metal stream, they will collect and undergo the time and temperatures required for continuous corundum growth.

8.9 Titanium Boride (TiB_2) Inclusions

In a study, Mohanty et al. [16] used TiB_2 particles and demonstrated that they are capable of nucleating pores. However, TiB_2 becomes active at pore nucleation if the excess Ti in the melt is kept below 0.001 wt% of what is needed to achieve TiB_2 particles for α-Al phase nucleation. The basic reason why background Ti in the Al melt is beneficial toward reducing porosity is due to the observation that TiB_2

particles are engulfed by the solidifying interface. Low-Ti-containing melts (below 0.001 wt%) develop TiB_2 crystals and push them out in front of the solid-liquid interface and into the interdendritic regions.

8.10 Strontium-Based Inclusions

Additions of Sr are common in foundry melts, and as a consequence, Sr-based inclusions are found in some cast structures. However, SrO and $Sr(OH)_2$ inclusions reduce in the melt to form complex compounds (Sr-40Al-40Si) that become engulfed by the advancing α-Al dendritic solid-liquid interface. Thus, it was proposed that Sr-based compounds do not contribute to the nucleation of pores within the solidifying melt. However, there are a growing number of publications which indicate that Sr increases the porosity when added to melts. Sr-based compounds are not engulfed and they actively contribute to pore nucleation [1, 3, 15, 16, 49–52].

8.11 Reduced Pressure Test Methodology

Alscan, Telegas, and Alspek H units are direct methods to assess hydrogen concentrations of the aluminum bath. However, they lack the robustness and simplicity of the reduced pressure test (RPT). The RPT is by far the most widely used test to gauge the porosity tendency of a liquid bath. Its only drawback is that this test may not be able to distinguish between higher porosity formation due to oxides and hydrogen concentration [38].

The basic fundamental aspect of the RPT is that a sample of liquid aluminum is allowed to solidify in a reduced atmospheric condition (Fig. 8.18). As pores nucleate, the effect of low atmosphere is that the internal gas pressure dominates the growth kinetics to the point where pores nucleate and grow, which enhances one's ability to assess pore formation tendencies [42].

There are several ways in which the solidified RPT sample can be analyzed to gauge porosity formation from a melt. Basic visual methods have been used in the past but are subject to judgment variability from operator to operator.

Figure 8.19 shows a typical chart rating for RPT samples, which have been sectioned in half to reveal the severity of the porosity. There is an associated density and a rating system from 1 to 12, which ranges from no porosity to the highest level of porosity seen. It should be noted that it is difficult to have such a chart as seen in Fig. 8.15, which is functional for the range of Al-Si alloys used in the casting industry. This is due to both the differences in density that exist for the alloys (porosity free state) and the differences in alloy freezing range (which contributes to the possible severity of porosity). This is why typically foundries tend to employ their metallurgists to develop an internal charting system for production use. In those

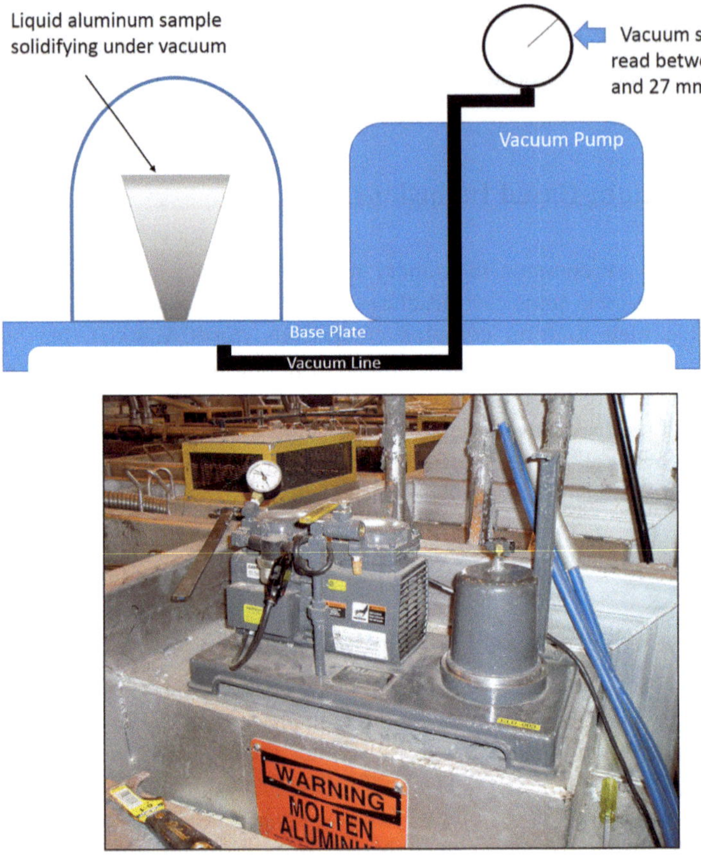

Fig. 8.18 Representation of the reduced pressure test

custom charts, the expected threshold of either rating number or its corresponding density is proposed. The plant needs to develop a corresponding level to which a casting has acceptable porosity levels for customer casting compliance. While the example shown in Fig. 8.19 is from a company which supplies the reduced pressure testing apparatus commercially, it may be prudent to construct a version most suited to the foundry's processes. More details are described in the next two sections.

Most metal casters today need a stronger determiner to gauge if more degassing will be required to make metal more acceptable. The operator referencing to visual aids like what is seen in Fig. 8.19 can lead to castings made with high porosity rejects. This arbitrary measure is usually because an operator believed a rating of 3 was more appropriate than 5. In foundries, this increases the loss in scrap. Conversely, an operator may want to play it safe by degassing more (production stoppage) until he/she is confident they are truly getting a 2.

The density measurement technique using the Archimedes method was the solution to improving the RPT conducted in foundries. No sample cutting/sectioning is

Fig. 8.19 Scale rating for the severity of pores as seen on a sectioned RPT test sample (Stahl Specialty Company, 1990)

required. The RPT sample is placed on a weight scale to measure the dry weight, followed by the determination of the wet weight. Through a basic calculation seen in Eq. 8.4, the relative density can be determined, giving a metric the operator can use to make a decision to continue degassing or not, and is free from any visual interpretation.

$$\ell = \frac{W_d}{W_d - \left[(W_d - \left(W_w - W_{wi}\right)\right]} \tag{8.4}$$

- W_d = "dry" weight in g
- W_w = "wet" weight in g
- W_{wi} = wire weight in g

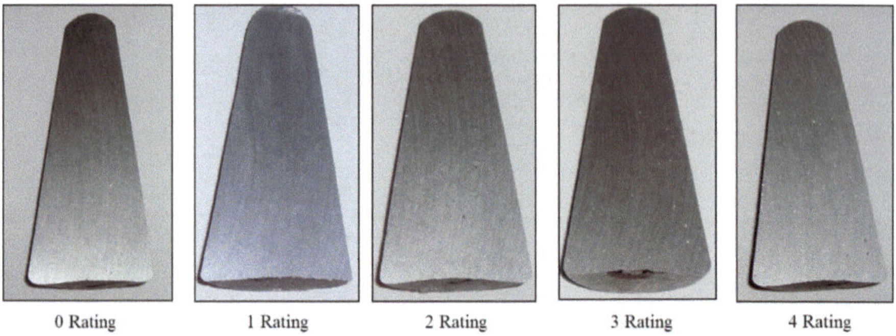

Date	Dry Weight [g]	Wet Weight [g]	Factor	T H_2O [°C]	ρ H_2O	ρ [g/cm³]	RPT Rating
9/1/2005(7:30AM)	120.65	43.86	76.79	22.0	0.9978	2.74	0
9/6/2005(7:30AM)	121.39	44.44	76.95	22.0	0.9978	2.73	1
9/7/2005(7:30AM)	106.58	39.11	67.47	22.0	0.9978	2.72	2
8/29/2005(12:30PM)	122.31	45.10	77.21	22.0	0.9978	2.71	3
10/14/2005(7:20PM)	116.85	43.64	73.21	21.0	0.9980	2.67	4

Fig. 8.20 Density measurements and visual ratings of 5 RPT samples

The term $W_d - [W_d - (W_w - W_{wi})]$ represents the mass of the displaced water. Since the density of water is 1 g/cm^3, then mass $_{H2O}$ = vol $_{H2O}$. The weight of the wire used to suspend the sample is normally included to minimize the noise. Figure 8.20 shows the density measurements and chart ratings together of five RPT samples taken from a 319 alloy liquid bath.

There are several other considerations that need to be taken into place to ensure that a repeatable RPT process is conducted at the foundry. One is the alloy composition; a common mistake is to have a threshold density used from one alloy and to have this used for another alloy. Figure 8.21 shows the D357 alloy and 319 alloy side by side, where the copper content does show the increase in density.

Another factor to consider is the impact of where the porosity resides in the final solidification structure. Long-freezing-range alloys like 319 show a more dispersed porosity in the center of the solidified cross-section. In the case of a short-freezing-range alloy such as 413, dispersed shrinkage does not happen, but cavity shrinkage, and at times this shrinkage can be exposed to the sample surface. As can be seen in Fig. 8.22, the density of the same quality 413 alloy can range widely if the resulting cavity shrinkage is in contact with the surface of the sample or is contained in the center cross-section. The position of the cavity shrinkage is most likely a function of how uniform the sample cup preheat was, a sensitivity not found for the longer-freezing-range alloys.

Other considerations that could lead to erroneous density values are if the start temperature of the poured RPT sample is too low or if the preheat of the sample cup is inadequate. Either or both conditions can promote a more rapid solidification, which in turn can result in a higher density (potential false positive). To get around this problem, one must establish a consistant preheat time for the cup, a consistant transfer time from bath to cup, and it would be prudent to also baseline the process

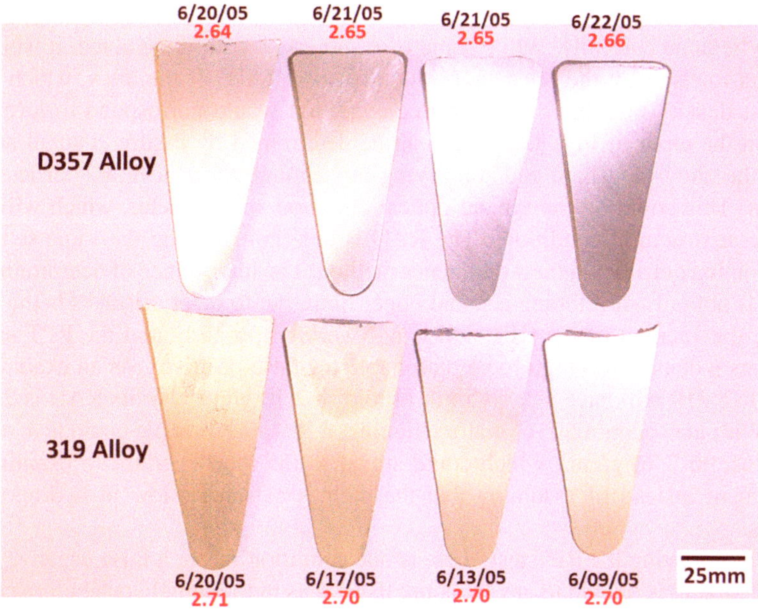

Fig. 8.21 Effect of alloy composition on density measured. All density measurements are in g/cm^3

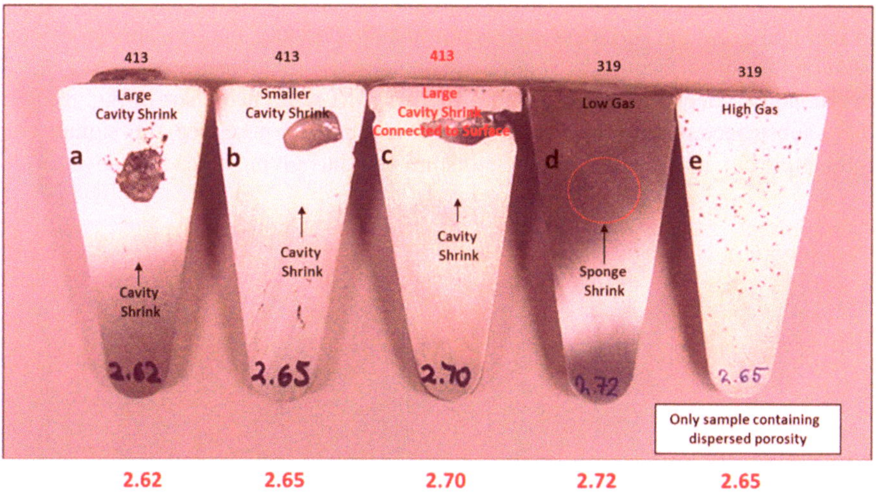

Fig. 8.22 Effect of solidification structure on density result. All density measurements are in g/cm^3

by quantifying the value of the secondary dendrite arm spacing λ_2 in the cast cross-section of the RPT test sample. An example of RPT test samples whose solidification rates are identical, and thus do not have the confluence effects of a slightly different solidification just mentioned, is shown in Fig. 8.23.

Finally, setting your threshold density for the determination of melt acceptability has to be done carefully. While some trial and error work is involved, it would be prudent to establish the relative absolute density (RAD) of the alloy, which is the highest density possible. The pore-free condition would correspond to level 1 as seen in the chart of Fig. 8.19. This can be determined by casting a small sample from the alloy melt bath and solidifying fast enough to achieve a λ_2 of less than 25 μm. This could be the typical optical emission spectrometer, which will have such cast structural conditions. The RAD is determined using the same setup for determining density via the Archimedes method. The importance of determining the RAD is not only developing a sound upper limit, but to deter mistakes being made where the vacuum of the RPT unit is not working properly and the RPT sample registers a density too high to be reasonable (or false positive). As an example, an alloy like 319 will have a pass/fail threshold of 2.70 g/cm^3, but its RAD is 2.76 g/cm^3. With the vacuum off, or malfunctioning, a 319 alloy sample could give a reading close to 2.76 g/cm^3, which could signal to the melt operator to examine the equipment instead of assuming that the melt is extremely low in hydrogen and oxides.

Before leaving the RPT test, there is one condition where a false negative result can occur and is related to any moisture that exists in the sample cup before casting the RPT sample. Figure 8.24 shows two 319 alloy samples (the threshold density of 2.70 g/cm^3 is required for a pass), but the first sample shows a 2.67 g/cm^3 and is therefore a negative result indicating that either oxides or dissolved hydrogen, or both, are elevated in the melt bath. It would be prudent to section the failed sample in half and grind the surface to assess not just the porosity, but to inspect for pores seen on the subsurface of the sample as shown in the failed test result in Fig. 8.19. A secondary check can be carried out by using the reference chart in Fig. 8.15. If these subsurface pores exist, but the interior is relatively pore-free, this signals that the sample cup may have had some moisture and thus giving a false negative result.

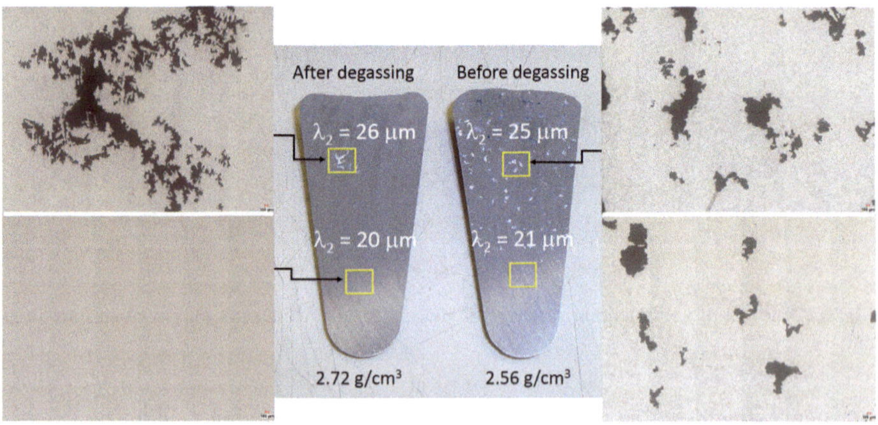

Fig. 8.23 RPT samples before and after degassing. The solidification rate shows that both the *upper* and *lower portions* of the solidified RPT sample are identical, and thus the true reflection of oxides and dissolved hydrogen can be made

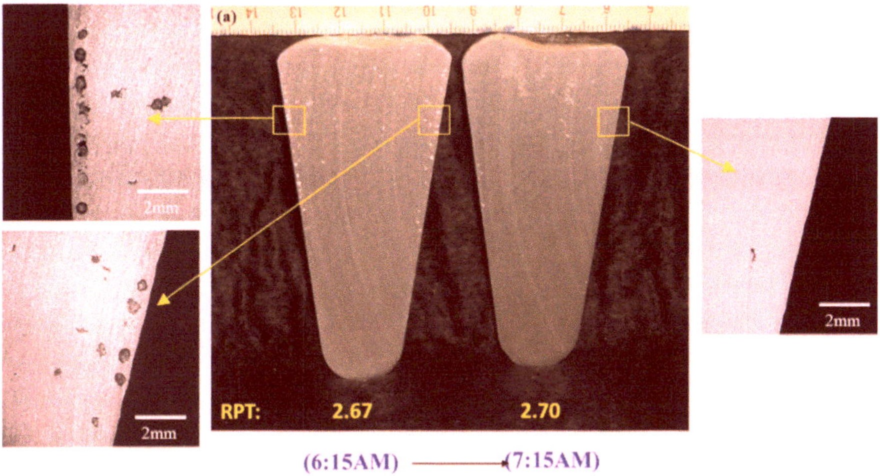

Fig. 8.24 Effect of RPT sample cup moisture on false negative readings. No changes were made to degas the melt when the recheck was conducted and finally measured 1 h later

8.12 Methods for Degassing

A significant discussion was made concerning assessing the dissolved hydrogen and oxides of an aluminum melt prepared for making a casting. Now we must review the most widely used foundry methods for reducing both dissolved hydrogen and oxides so that acceptable castings are made.

The principal method for reducing dissolved hydrogen is to disperse inert gas into an aluminum bath. The process allows for dissolved H^+ to diffuse into the circulating bubbles to form H_2 gas, and under the force of bouyancy, the bubble with hydrogen is carried away to the surface.

There have been several types of lances and probes used to deliver or bubble inert gas into the liquid aluminum bath, but the different methods will have their own efficiencies. The highest efficiency will have the best combination of a shorter time with minimal disturbance of the melt surface. Shorter time reduces melt heating cost involved with the degassing operation, and the minimal disruption of the melt surface reduces the amount of dross formation (melt loss) and hydrogen reabsorption that occurs with the breaking of the surface dross (refer back to the schematic in Fig. 8.16). Basically, if one runs a large foundry casting, 319 alloy, and you have a target density of 2.7 g/cm^3 to achieve. Before casting. it will be mandatory to get the most efficient degassing method to yield significant cost avoidance for its overall melt operations. In simple terms, smaller bubble size, the number of the bubbles, and their ability to circulate fully within the treated aluminum bath will be key toward achieving this goal.

The earliest lances used were with holes near the bottom, and they inevitably produced large bubbles, which, under the force of bouncy, go straight to the top,

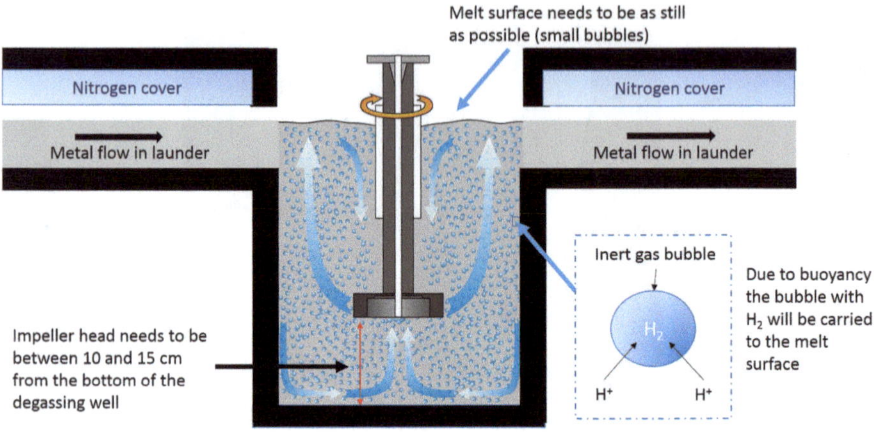

Fig. 8.25 Rotary degassing process with degassing times indicated. In the example indicated here, the degassing well has a capacity of 816 kg of liquid aluminum alloy and requires 18 min residence time to achieve 2.70 g/cm^3 threshold. This means that there is a 2720 kg/h throughput capacity with the above degassing system. If the bubble size increases (controlled by RPM) or the impeller head has erosion, throughput weight will drop

leaving portions of the aluminum bath untreated. This generates added dross generation at the top. Eventually, rotary head degassing systems were implemented that produced much smaller and a large number of bubbles, but also provided an outward centrifugal force that allowed bubble circulation away for the rotating impeller head. This ultimately provides the fastest way to degas liquid aluminum baths with minimal dross generation and thus is the most widely used degassing method in foundries. Figure 8.25 shows an example of a rotary degassing system used in an 816 kg capacity degassing well. It will take approximately 18 min for the hydrogen level to come down to acceptable levels. In this case, initial hydrogen was 0.25 cm^3 H_2/100 g Al and drops to 0.10 cm^3 H_2/100 g Al. Inert gas pressures and rotary speed (usually measured from PLC controls as rpm) have to be set for the optimum bubble size and dispersion for this to work.

Once optimal settings fail or if sufficient erosion of the impeller head occurs, hydrogen removal becomes less efficient. In cases where a continuous melt throughput occurs, the RPT readings will gradually begin to drop in density. In batch processes where rotary degassing is conducted, the RPT test is usually conducted, and if it fails, more degassing is required.

Oxides may be removed with the degassing as long as sufficient wetting can occur with the purging bubbles and then carry the captured oxides to the surface of the treated metal and then be skimmed off. This usually means using SF_6 (sulfur hexafluoride) or an inert gas with SF_6 mixture as the carrier gas. The SF_6 is however legislated as not permissible for use in many municipalities due to its toxicity to workers and the environment. Alternatives which can work will depend if you are treating your aluminum melts in a continuous process or a batch process. For continuous melting processes, filtration may be the easiest and most cost-effective

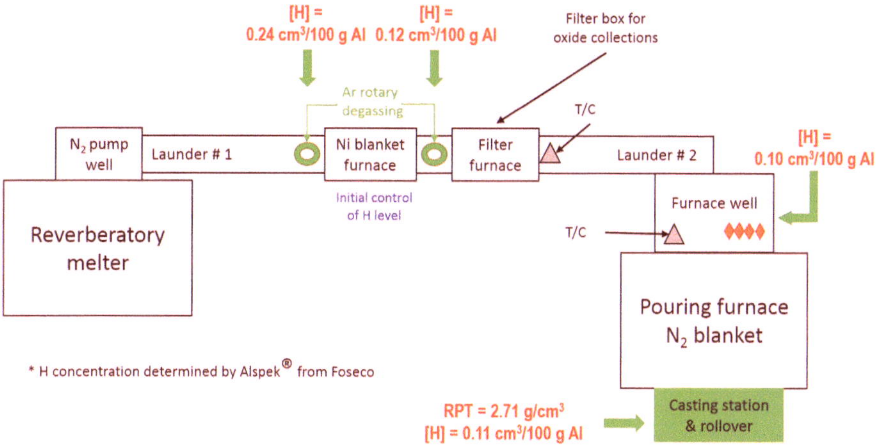

Fig. 8.26 Layout of continuous melt operations of a foundry not using fluxing or SF_6/Ar or SF_6/N gas mixtures for oxide removal. Hydrogen levels are controlled by a multiple strategy of rotary degassing system

Table 8.6 Correlation between RPT and direct hydrogen measurements using the Alspek probe technology

RPT (g/cm³)	Alspek (cm³ H_2/100 g Al)
2.64	0.33
2.69	0.13
2.69	0.13
2.69	0.13
2.70	0.13
2.70	0.12
2.70	0.12
2.71	0.12
2.71	0.12
2.71	0.12
2.71	0.11

approach for suspended oxides in the metal. A schematic of a continuous melting process is shown in Fig. 8.26, and it shows that with sufficient degassing strategy, an integrated filter box achieves a threshold of 2.7 g/cm³. In the batch process, clearly a continuous flow of liquid aluminum through a filter does not happen; therefore, the fluxing tablets would ultimately be the option to achieving a target RPT density which can reduce both dissolved hydrogen and suspended oxides.

The filter box system in Fig. 8.26 will remove oxides in suspension while the rotary degassing system removes hydrogen. As mentioned, the RPT is not capable of separating both hydrogen and oxide contributions to pore formation. However, in this continuous process setup, since oxides are removed by the filter box, comparative analysis can be done between a direct hydrogen concentration measurement system (example here was the Alspek H by Foseco) and the RPT unit. Mackay et al. [51] showed this comparative analysis in Table 8.6, which shows that the lower densities from the RPT were due to dissolved hydrogen in the melt.

References

1. Gruzleski, J.E., and B.M. Closset. 1990. *The treatment of liquid aluminum-silicon alloys*. Des Plaines: American Foundry Society, Inc.
2. Zalensas, D.L. 1993. *Aluminum casting technology*. Des Plaines: American Foundrymen's Society.
3. Hamed, Q.S., M. Dogan, and R. Elliott. 1993. The dependence of secondary dendrite arm spacing on solidification conditions of Al-7 Si-0.5 Mg alloys treated with TiBAl and TiBAl/Sr additions. *Cast Metals* 6: 47–53.
4. Paray, F., and J.E. Gruzleski. 1994. Microstructure-mechanical property relationships in a 356 alloy. Part I: Microstructure. *Cast Metals* 7: 29–40.
5. ———. 1994. Microstructure-mechanical property relationships in a 356 alloy. Part II: Mechanical properties. *Cast Metals* 7: 153–163. https://doi.org/10.1080/09534962.1994.11819174.
6. Djurdjevic, M., J. Sokolowski, and T. Stockwell. 1998. Control of the aluminum-silicon alloy solidification process using thermal analysis. *Meta* 4: 237–248.
7. Djurdjevic, M., T. Stockwell, and J. Sokolowski. 1997. The effect of strontium on the microstructure of the aluminum-silicon and aluminum-copper eutectics in the 319 aluminum alloy. *International Journal of Cast Metals Research* 12 (2): 67–73.
8. Djurdjevic, M., H. Jiang, and J. Sokolowski. 2001. On-line prediction of aluminum–silicon eutectic modification level using thermal analysis. *Materials Characterization* 46 (1): 31–38. https://doi.org/10.1016/S1044-5803(00)00090-5.
9. Paray, F., and J.E. Gruzleski. 1994. Factors to consider in modification. *Transactions of the American Foundrymen's Society* 102: 833–842.
10. Backerud, L., G. Chai, and J. Tamminen. 1990. *Solidification characteristics of aluminum alloys. Vol. 2. Foundry alloys*, 266. American Foundrymen's Society, Inc.
11. Paray, F., and J.E. Gruzleski. 1993. Modification-A parameter to consider in the heat treatment of Al-Si alloys. *Cast Metals* 5: 187–197.
12. Campbell, J. 1997. 10 rules for good castings. *Modern Casting* 87: 36–39.
13. Campbell, J. 2003. *Castings*. 2nd ed. Oxford, UK: Butterworth-Heinemann.
14. Bourcier, G.F., J. Dickinson, J. Tessandori, and D. Schiffer. 1985. *Aluminum recycling casebook*, 56. The Aluminum Association, Inc.
15. Rooy, E.L. 1992. Mechanisms of porosity formation in aluminum. *Modern Casting* 82 (9): 34–36.
16. Mohanty, P.S., F.H. Samuel, and J.E. Gruzleski. 1995. Experimental study on pore nucleation by inclusions in aluminum castings. *Transactions of the American Foundrymen's Society* 103: 555–564.
17. Callister, W.D. 1994. *Materials science and engineering*. 3rd ed. New York: Wiley Publishing.
18. Crepeau, P.N. 1995. Effect of iron in Al-Si casting alloys: A critical review. *Transactions of the American Foundrymen's Society* 103: 361–366.
19. Mackay, R.I., and J.E. Gruzleski. 1998. Quantification of magnesium in 356 alloy via thermal analysis. *International Journal of Cast Metals Research* 10: 255–266.
20. MacKay, R., and J. Sokolowski. 2010. Comparison between wedge test castings and component engine block casting properties. *International Journal of Metalcasting* 4 (4): 33–50. https://doi.org/10.1007/bf03355501.
21. ———. 2008. Experimental observations of dendrite coarsening & Al-Si eutectic growth in progressively quenched structures of Al-Si-Cu casting alloys. *International Journal of Metalcasting* 2 (2): 57–75. https://doi.org/10.1007/bf03355428.
22. ———. 2010. Effect of silicon & copper concentrations, and cooling rate on soundness in casting structure. *International Journal of Cast Metals Research* 23: 7–22. https://doi.org/10.1179/174313309X449282.
23. Mackay, R., J. Sokolowski, R. Hasenbush, and W. Evans. 2002. Effect of Cu and Si on the solidification kinetics of 3XX. X Series alloys during Mushy Zone Developemet. *Transactions of the American Foundry Society*.

24. Caton, M.J., J.Wayne Jones, J.M. Boileau, and J.E. Allison. 1999. The effect of solidification rate on the growth of small fatigue cracks in a cast 319-type aluminum alloy. *Metallurgical and Materials Transactions A* 30 (12): 3055–3068.
25. Gall, K., N. Yang, M. Horstemeyer, D.L. McDowell, and J. Fan. 1999. The debonding and fracture of Si particles during the fatigue of a cast Al-Si alloy. *Metallurgical and Materials Transactions A* 30 (12): 3079–3088.
26. Backerud, L., E. Krol, and T. Tamminen. 1996. *Solidification characteristics of aluminum alloys. Vol. 1. Wrought alloys*. Vol. 1, 156. AFS/Skanaluminum.
27. Emadi, D., L.V. Whiting, V.Y. Gertsman, M. Sahoo, R. MacKay, and G.E. Byczynski. 2006. Effect of tin on the mechanical properties of aluminum 319 alloy. *AFS Transactions* 114: 263.
28. Gauthier, J., and F. Samuel. 1995. Tensile properties and fraction behaviour of solution heat treated 319.2 Al automotive alloy. *AFS Transactions* 103: 849–855.
29. Anson, J.P., and J.E. Gruzleski. 1999. A quantitative evaluation of the effect of hydrogen content on the relative amounts of shrinkage and gas microporosity in a cast Al–7% Si foundry alloy. *AFS Transactions* 107: 456–467.
30. Dahle, A.K. 1996. *Mushy zone properties and castability of aluminum alloys*. PhD Thesis. Norwegian University of Science and Technology.
31. Vander Voort, George F. 1984. *Metallography, principles and practice*. Materials Park: ASM International.
32. Sokolowski, J., C. Kierkus, B. Brosnan, and W. Evans. 2000. Formation of insoluble Ti (Al, Si) 3 crystals in 356 alloy castings and their sedimentation in foundry equipment: Causes, effects and solutions. *Transactions of the American Foundrymen's Society* 108: 491–496.
33. Davis, J.R. 1996. *ASM specialty handbook: Aluminum and aluminum alloys*. Materials Park: ASM International.
34. Gustafsson, G., T. Thorvaldsson, and G.L. Dunlop. 1986. The influence of Fe and Cr on the microstructure of cast Al-Si-Mg alloys. *Metallurgical Transactions A* 17 (1): 45–52.
35. Kanicki, D.P. 1994. Changing casting demands shape Ford's new foundry. *Modern Casting* 84 (9): 24–27.
36. Mocarski, S.J., G.V. Scarich, and K.C. Wu. 1991. Effect of hot isostatic pressure on cast aluminum airframe components. *American Foundrymen's Society, Inc.(USA)* 99: 77–81.
37. Reinhart, T. 1994. *Characterization of Alloy D357.0-T6*. Anaheim: AEROMAT.
38. Mitrasinovic, A., F.C. Robles Hernandez, M. Djurdjevic, and J.H. Sokolowski. 2006. On-line prediction of the melt hydrogen and casting porosity level in 319 aluminum alloy using thermal analysis. *Materials Science and Engineering A* 428 (1): 41–46.
39. Sigworth, G., S. Shivkumar, and D. Apelian. 1989. The influence of molten metal processing on mechanical properties of cast Al-Si-Mg alloys. *AFS Transactions* 98: 811–823.
40. Anson, J.P., M. Stucky, and J. Gruzleski. 2000. Effect of modification on the growth of microporosity during the solidification of aluminum–7% silicon foundry alloy. *AFS Transactions* 108: 611–623.
41. Fuoco, R., and E.R. Correa. 1999. *Characterization of some types of oxide inclusions in aluminum alloy castings*. Philadelphia: AFS Castexpo.
42. La-Orchan, W., M.H. Mulazimoglu, X.G. Chen, and J.E. Gruzleski. 1995. Quantification of the reduced pressure test. *AFS Transactions* 103: 565–574.
43. Crepeau, P.N. 1995. Molten aluminium contamination: Gas, inclusions and dross. In *4th international conference on molten aluminium processing. Orlando.*
44. Caceres, C.H., M.B. Djurdjevic, T.J. Stockwell, and J.H. Sokolowski. 1999. The effect of Cu content on the level of microporosity in Al-Si-Cu-Mg casting alloys. *Scripta Materialia* 40 (5): 631–637.
45. Chai, G. 1994. *Dendrite coherency during equiaxed solidification in aluminium alloys*. PhD Thesis. Stockholm University.
46. Dupuis, C., Z. Wang, J.P. Martin, and A. Allard. 1992. An analysis of factors affecting the response of hydrogen determination techniques for aluminum alloys. *Light Metals* 27: 1055–1067.

47. Edwards, G.A., G.K. Sigworth, C.H. Cáceres, D.H. St John, and J. Barresi. 1997. Microporosity formation in Al-Si-Cu-Mg casting alloys. *Transactions of the American Foundrymen's Society* 105: 809–818.
48. Emadi, D. 1995. *Porosity formation in Sr-modified Al-Si alloys*. PhD, Department of Mining and Metallurgical Engineering. PhD Thesis, McGill University.
49. Tynelius, K.E. 1992. *A parametric study of the evolution of microporosity in Al-Si foundry alloys*. PhD, PhD Thesis. Drexel University.
50. Mackay, R.I. 2003. *Development of a new durable Al-Si alloy for the next generation of engine block casting*. Ph.D., PhD Thesis. University of Windsor.
51. Mackay, R., and G. Byczynski. 2011. The use of the Weibull statistical method to assess the reliability of cast aluminum engine blocks made from different casting processes. In *Shape casting: 4th international symposium*, 191–198. John Wiley & Sons, Inc.
52. Arnberg, L., L. Backerud, and G. Chai. 1996. *Solidification characteristics of aluminum alloys. Vol. 3. Dendrite coherency*. 247. AFS/Skanaluminum.

Chapter 9
Grain Refinement

9.1 Introduction to Grain Refinement

Al-Si alloys can be improved by increasing the homogeneity, quality, integrity, and thus the service characteristics of cast components. This is also possible by means of microstructure refinement and phase modification. The following are the four basic methods and their combinations are used for grain refinement and phase modification in the liquid and semisolid states: thermal, chemical, mechanical, and electromagnetic. All those processes are discussed in this chapter.

In summary, we can say that larger grains tend to solidify at higher temperatures, causing nucleation of a smaller number of nuclei that result in fewer, but larger, grains and larger dendrites. Larger dendrites are known for their detrimental effects on proper feeding [1, 2]. This problem is associated with a larger propensity for hot tearing, and therefore it does compromise the integrity of the castings, leading to a negative effect on the mechanical properties. Grain refinement is desirable due to its effects on improving feeding characteristics and reducing shrinkage porosity, which in turn increases the quality and the mechanical properties of the castings[3, 4]. Additionally, machinability is improved along with the machining tool life, thereby decreasing the cost for tool repair and replacement [5–17]. Primary Si modification is the main concern in Al-Si hypereutectic alloys to improve their tribomechanical applications such as engine blocks. The main problem in Al-Si hypoeutectic alloys is their low matrix hardness and hence wear resistance. Here it is important to clarify that Si has a high hardness; however, it is brittle and fractures relatively easily leaving expose the Al-matrix (mainly aluminum). Therefore, its refinement is key to improve tribological characteristics. The bulk hardness of Al-Si hypereutectic is comparable to that of cast iron.

Microstructure refinement of aluminum castings is highly desirable; fine and equiaxed grains are always the "ideal" microstructures [5]. Grain refinement is possible by increasing the number of heteronucleants which is a common mechanism

© Springer International Publishing AG 2017

F.C. Robles Hernandez et al., *Al-Si Alloys*, DOI 10.1007/978-3-319-58380-8_9

in grain refiner additions. The most common grain refiners contain Ti [18, 19]. Alternatively, Heteronucleation can be facilitated by mechanical or electromagnetic (e.g., stirring or vibration) methods where shear forces are induced during solidification. The shear forces induced mechanically or electromagnetically are effective in the semisolid state by breaking the newly developed dendrites and/or precipitating phases. In the Al-Si alloy family, the only alloys that can be susceptible to refinement in liquid state are the Al-Si hypereutectic [20–24].

9.2 Chemical Grain Refinement

9.2.1 *Principles and Grain Growth Restriction Factor*

In order to control grain growth, it is important to understand the nucleation and coarsening mechanisms that are explained by various theories. The solute theory, formalized by Johnsson and Bäckerud [25, 26], suggests that both nucleants and segregation influence grain growth. A mathematical model to predict the effects of grain growth was developed, and this model is known as the grain growth restriction factor. The growth restriction factor (Q) is used to determine the degree of segregation and can be translated as the measurement of the restricting effect of the solid–liquid interface to form new grains. Ti is the most common and effective grain refiner and has a grain grow index of (m (k-1)) of 220. Comparing the grain refiner effectiveness of Ti and Si, one can conclude that Ti is 38 times more effective as a grain refiner than dissolved Si. In other words, 100 ppm of Ti is equivalent to 3.8 wt% Si. Based on the binary Al-Xi system, Q_{Xi} is defined as:

$$Q_{Xi} = m_{Xi}\left(k_{Xi} - 1\right)C_0 \tag{9.1}$$

where m_{Xi} is the slope of the liquidus temperature line based on the phase diagram, C_0 is the concentration of the solute in the melt, and k_{Xi} is CS/CL, where CS and CL are the solute concentrations at the interface for the solid and liquid, respectively. k is also known as the equilibrium partitioning coefficient.

For dilute alloys, Q and the constitutional undercooling (P) can be determined as the sum of the individual Q and P for the particular elements in the alloy. The techniques to assess Q and P for nondiluted and multicomponent alloys are controversial and not well accepted by some authors [27–29]. On the other hand, this method is effective for alloys with no elements that can react with Ti or Zr. This method is recommended for binary Al-Si alloys and grain refiner additions; any further additions (secondary elements) can easily overestimate the values of Q and P. We present the equations that are used to determine the supercooling and the growth restriction parameters. The total Q for any Al-Si alloy, respectively, is calculated using Eqs. 9.2, 9.3, and 9.4 [30–32]:

$$P = -\frac{m\left(1-k\right)C_0}{k} \tag{9.2}$$

$$Q = m_i C_0 (k - 1) = kP \tag{9.3}$$

$$Q_{TOT} = \sum m C_0 (k - 1) = \sum m_1 C_1 (k_1 - 1) + \sum m_2 C_2 (k_2 - 1) + \ldots. \tag{9.4}$$

where m is the liquidus slope, k is the equilibrium partition coefficient, and C_0 is the alloy composition. P is determined based on the following assumptions: It is equal to the freezing range of the alloy assuming that both the liquidus and solidus are straight lines and the alloy solidifies entirely as a solid solution. Therefore, P is the degree of undercooling induced by growth restriction. Several researchers have proposed mathematical models that are limited to particular factors (temperature, composition, cooling rate, etc.) [33–35].

Easton et al. [29] proposed a method to calculate the effect of alloy composition on *grain size (GS)*. Bäckerud et al. [27] patented the methodology to determine the required amount of Ti for a desired grain refinement effect. Such method is expressed in Eq. 9.5.

$$\begin{aligned} \text{Amount of } Ti(wt\%) &= \frac{Q_\mathrm{d} - Q_\mathrm{b}}{(k_{Ti} - 1) m_{Ti}} \\ Q_\mathrm{b} &= Q_\mathrm{d} - \left[\text{Amount of } Ti(wt\%) \times (k_{Ti} - 1) m_{Ti}\right] \end{aligned} \tag{9.5}$$

The "*Amount of Ti*" is the percentage in weight of Ti to be added to the melt, Q_d is the grain growth index resulting in aluminum castings having a minimal *GS*. Q_b is the grain growth index of the original aluminum base material, m_Ti is the slope of the liquidus in the binary (Al-Ti) system, and k_Ti is the distribution coefficient of Ti between solid and liquid.

According to Johnson and Bäckerud [36–38], a low Q slows the dendritic growth due to diffusion of solute in front of the interface. But at higher concentrations, the dendrites develop lancet-like tips which grow into the liquid and reject solute orthogonally to the growth direction. This is due to a change from diffusion control growth to dendrite tip radius-controlled growth, where capillarity effects dominate [39].

9.2.2 Grain Refinement of Aluminum-Rich Phases

The peritectic theory is well accepted to explain the effects of Ti additions as grain refiners for Al-Si alloys. In this theory, enough additions of Ti (>0.15 wt%) in the form of $TiAl_3$ intermetallic serve as a heteronucleus. The grain refinement efficiency for Al depends on the number and size of the $TiAl_3$ particles formed. Amounts between 0.1 and 0.15 wt% Ti result in a peritectic reaction with the liquid aluminum. This peritectic nucleation effect of the $TiAl_3$ particles is shown in Fig. 9.1.

The Hulk's theory formulates that all $(Al_7Ti)B_2$ particles usually dissolve in Al-Si melts [40–42]. On the other hand, $TiAl_3$ particles dissolve partially and establish a titanium diffusion profile around them. This increases the titanium concentration in the surroundings to the aluminides exceeding the solubility of

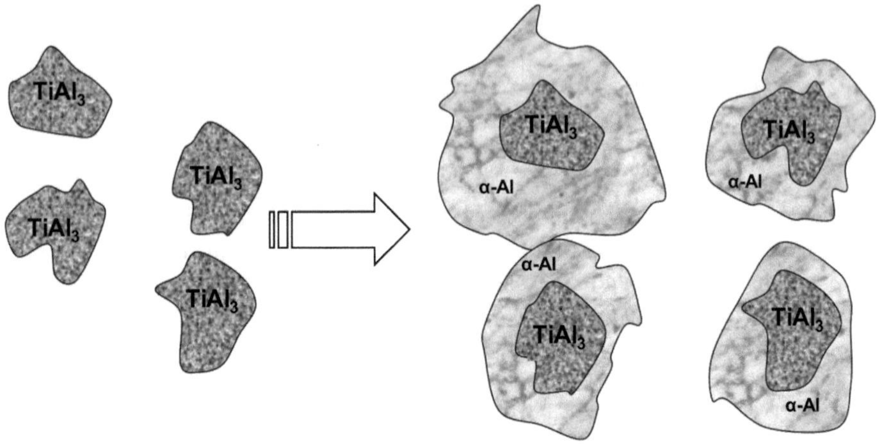

Fig. 9.1 Nucleation of α-Al on Al-Ti_3 by the peritectic reaction

TiB_2. This follows a precipitation on the surface of the aluminum-rich phases forming a shell that prevents the dissolution of the $TiAl_3$ particles. At the same time, a simultaneous diffusion of titanium (outward) and aluminum (inward) occurs. This diffusion mechanism takes place through the shell and forms pools of liquid aluminum that are rich or saturated in titanium. These are usually peritectic cells activating heterogeneous nucleation near the peritectic temperature (665 °C). When this occurs, α-Al nucleates inside the shell and the heterogeneous nucleation continues to grow into the surrounding liquid [43]. Hulks' theory is shown in Fig. 9.2.

The hypernucleation theory proposed by Jones and Pearson [28, 29, 43–45] establishes that when there is Ti in excess (above the ratioTi/B $TiA_3 = 2.21$) in molten aluminum, solute titanium segregates from the melt to the TiB-melt interface. This results in the formation of a thin layer of TiAl, which during solidification behaves in a peritectic manner sponsoring the nucleation of α-Al. According to this mechanism, the grain refinement effect is the result of the agglomeration and settling of bonded particles. Figure 9.3 shows the hypernucleation theory.

Excessive holding times after the addition of the grain refinement have undesirable effects. Usually longer holding times decompose $TiAl_3$. Once $TiAl_3$ is decomposed, it cannot be easily reactivated, and this reduces its refining effect and promotes sedimentation [46]. The presence of boron in the Al-Ti master alloys improves the refinement ability of the $TiAl_3$ particles. This results from a coupling effect between TiB_2 and the $TiAl_3$ compound. Unfortunately, the refinement effect using Al-Ti-B enhances pore formation on folded oxides due to pore nucleation and growth [47]. Other disadvantages observed in refiners include the increase in the hydrogen content of the melt, and they generate corrosive fumes. The recovery of the refining elements increases the level of inclusions in the melt [43]. Therefore, excessive amounts of Al-Ti-B master alloys are detrimental to the service and mechanical properties of the Al-Si components. This is in addition to the health hazard that they represent. An ideal master alloy (containing $TiAl_3$ and TiB_2) should react as fast as possible, and the melt needs to be used or cast as soon as possible.

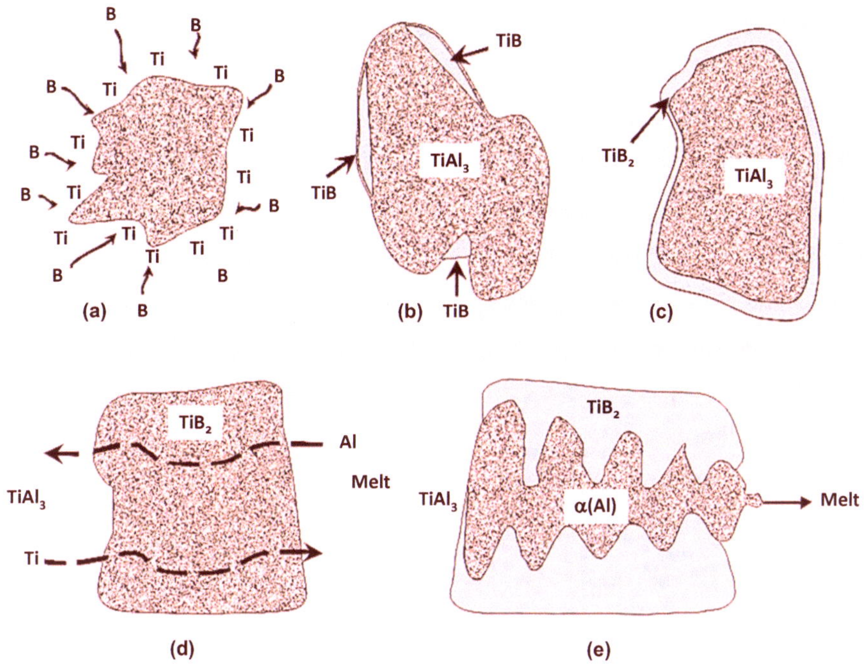

Fig. 9.2 Hulk's Theory (**a**) Partial dissolution of $TiAl_3$, and diffusion of B toward $TiAl_3$, (**b**) Solubility product of TiB_2 is exceeded, (**c**) Formation of the protective TiB_2 shell, (**d**) Simultaneous diffusion of Al and Ti through the shell, (**e**) Nucleation and growth of α-Al [43]

9.3 Silicon in Liquid and Semisolid States

9.3.1 Structure in the Liquid State

Up to this point, the presence of the "Si agglomerates" has not been described. The following section focuses on liquid state and a deep analysis of the Si agglomerates and their transformation to the well-known primary Si particles. It is expected that in the future a number of research documents will be published on this topic. The understanding of this topic is crucial to launch Al-Si eutectic and hypereutectic alloys for applications under cyclic loading, wear, fretting, among other environments.

9.3.2 Al-Si Hypereutectic Alloys

Traditionally, it has been reported that the precipitation sequence of the Al-Si hypereutectic alloys commences at the liquidus temperature by the precipitation of the primary Si followed by the Al-Si eutectic. However, several reports indicate that Si forms agglomerates with short- to medium-range order in the liquid state [25, 48–56], and these agglomerates form at temperatures significantly above liquidus. Presumably, the Si agglomerates are the same particles as the primary Si; however, they have

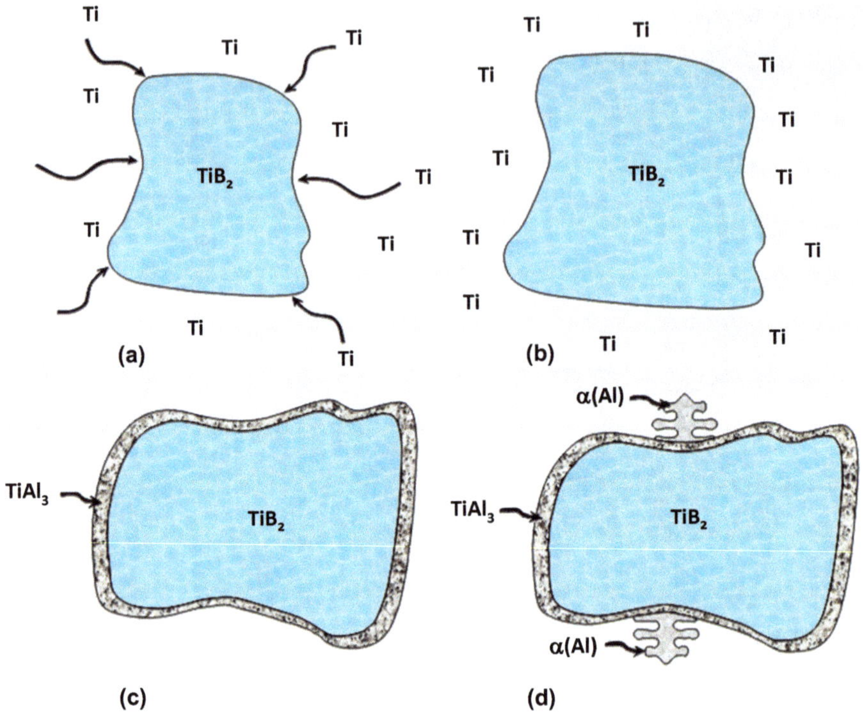

Fig. 9.3 Hypernucleation theory. (**a**) Excess Ti (Ti > 2.21) in solution, (**b**) Ti segregated to the TiB_2-melt interface, (**c**) formation of $TiAl_3$ layer on TiB_2, (**d**) Nucleation of α-Al by peritectic reaction [28, 29, 43–45]

different crystalline structures. The corresponding amount of Si in Si agglomerates for alloys containing between 13 and 25 wt% Si is almost negligible (0.0004–0.12 wt% Si respectively) [57].

It has been reported that during cooling the Si in the liquid melt starts agglomerating at temperatures above the respective liquidus of the Al-Si eutectic and hypereutectic alloys [11, 23, 25, 48–56, 58–63]. The identification of those agglomerates can be carried by a rapid solidification (e.g., quenching) test methods. During the rapid quenching, the Si agglomerates are frozen instantly, at cooling rates above 100 °C/s. The investigation of those samples using optical and electron microscopy as well as Raman spectroscopy allowed to determine the presence of those agglomerates. The so-called Si agglomerates hold atomic arrangements in the liquid state [23, 64, 65]. The atomic arrangements are in the short and medium atomic range order and they do not necessarily form crystals.

The Si agglomerates have been characterized using different methods such as X-ray diffraction (XRD), synchrotron radiation, electron and neutron beams, and magnetic and electrical resistance [11, 23, 25, 48–56, 58–63]. Using these methods it is demonstrated that the Si agglomerates held short and medium atomic range in liquid state. For instance, XRD permits to determine the structure factor ($S(Q)$) of melts as well as the radial atomic distribution function. The structure factor is defined by Eq. 9.6, which is an important relation in X-ray crystallography, since it permits the discrimination of the atomic planes that do not obey Braggs' law [66].

$$S(Q)_{hkl} = \sum_{N=1}^{N} f_n e^{2\pi i(hu_n + kv_n + lw_n)} \tag{9.6}$$

where: $S(Q)$ is the structure factor,
f is the atomic scattering factor,
hkl are the Miller indices and
uvw are the fraction coordinates (unit vectors).

Brodova et al. [11] conducted a series of liquid metal characterization and determined that electrical resistance and magnetic susceptibility have a good correlation with the size of the atomic agglomerations of Si. Such relation is used to predict the size of the agglomerates at different temperatures. In addition, gamma ray absorption spectroscopy showed the differences in the density of liquid elements, and was used to measure nanometric "agglomerations" that held short to medium atomic arrangements [11]. Using Raman spectroscopy (Fig. 9.4), it can be observed that the number of Si–Si bonds increases considerably as the quenching temperature of the molten and/or semisolid alloy decreases. It is particularly interesting to see the changes of the in quenched samples at temperatures below the Al-Si eutectic (liquidus). However, the most important result is that at temperatures of ~150 °C above liquidus (casting temperature), the Si–Si bonds are identifiable and hence the presence of the Si agglomerates. This result is shown in Fig. 9.4, which presents Raman spectra for various Al-Si hypereutectic samples that were quenched from various temperatures above and below liquidus. The characteristic Raman band for Si–Si bonds is observed at approximately 521 cm^{-1}, which is in agreement with the results in Fig. 9.4. The intensity of the Si band in the Raman spectra increases as the quenching temperature decreases. The samples analyzed using Raman spectroscopy were

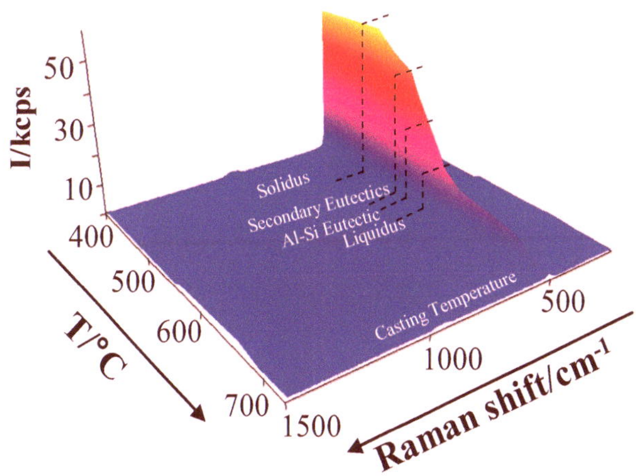

Fig. 9.4 Raman shift of Si particles present in 390.1 Al-Si hypereutectic alloy quenched at different temperatures

quenched following the method indicated in Chap. 2, Sects. 2.5, 2.6 and 2.7 and the respective Figs. 2.10–2.15 and references [23, 25, 67].

Inui et al. [44] reported the first neighbor parameter and its atomic radius using a geometrical model to explain the structure of a binary liquid of an amorphous alloy. The results of this analysis concluded that Si in Al-Si hypereutectic alloys tends to segregate. The segregation of the Si atoms at high temperatures takes place in the microscale due to the strong bonding interaction of the Si–Si atoms. A similar phenomenon was observed in the segregation of C (graphite) in cast iron ingots and in Sn in Na-Sn alloys in the liquid state [44, 68]. The agglomeration of atoms in liquid state is a common phenomenon presented by several alloys and is classified by their binary (eutectic, peritectic, intermetallic, etc.) nature [50]. In the liquid state, the amount of covalent Si–Si bonds in liquid Si is up to 25–30% [51].

Wang et al. [52] measured the nearest neighboring atomic distances and the average atomic densities of Al-Si hypo- and hypereutectic alloys at different temperatures in the liquid state. The nearest neighbor atomic distance of pure Al changes in the liquid state with Si additions, suggesting that the melt maintains the structure of pure Al. The density of Al-Si hypoeutectic alloys in a liquid state increases with the concentration of Si, implying that some of the free volume in pure Al (interstitial sites) is taken by Si atoms. For Al-Si concentrations close or above the eutectic composition, agglomerations of Si_4 covalent bonds (tetrahedral) are promoted and the Si agglomerates are observable. The presence of the Si agglomerates is a function of the Si content in the Al-Si hypereutectic alloys. The Si agglomerates are insoluble in liquid Al. Lowering the alloy's temperature close to the respective liquidus reduces the density of Al-Si hypereutectic melts. Conversely, it increases the superheat temperature and the density of the Al-Si hypereutectic compositions because the Si–Si bonds decompose into Si atoms that diffuse into the free volume of the Al-rich melt [52]. According to Gabalthular et al. [51, 69], the total coordination number of Si atoms can be determined as follows:

$$N^{I} = N^{f}\left(C_{Al}^{f} + C_{Si}^{f}\right) + N_{Si}^{t} C_{Si}^{t} \tag{9.7}$$

where N_f is the coordination number of free Al and Si atoms in the Al-Si melt (equal to 10.2), N_{Sit} is the coordination number of Si atoms in a tetrahedral structure (equal to 4), and C_{Alf}, C_{Sif}, and C_{Sit} are the respective Si and Al concentrations and the Si atoms in tetrahedral structure. Therefore:

$$C_{Al}^{f} + C_{Si}^{f} + C_{Si}^{t} = 1 \tag{9.8}$$

And the atom's nearest distance is given by:

$$r^{I} = r^{f}\left(C_{Al}^{f} + C_{Si}^{f}\right) + r_{Si}^{t} C_{Si}^{t} \tag{9.9}$$

where r_f is the atomic neighbor distance for the Al and Si atoms (equal to 0.282 nm) and r_{Sit} is the distance between Si atoms in a tetrahedral structure in the melt (equal to 0.252 nm). Using the above model, the distance among Si atoms in liquid melts can be determined.

Figure 9.5 shows the XRD results where the Si agglomerates were characterized using as a function of composition, temperature, and in the presence of additions of Sb. The Sb additions are used because they promote the Si–Si bonding. The XRD diffractograms present the main peak (reflection) with $S(Q)$ similar to pure Al

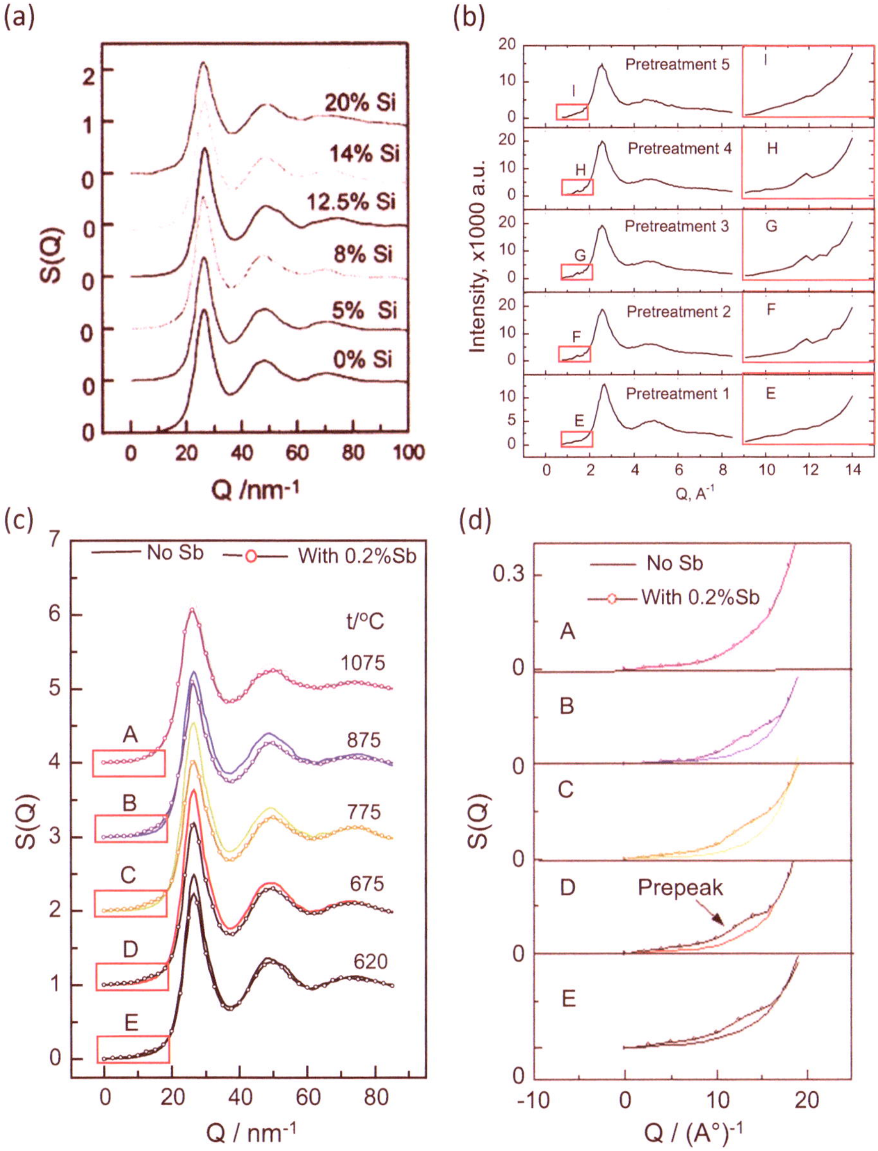

Fig. 9.5 (**a**) Structure factor (S(Q)) of the Al-Si alloys with different compositions at 670 °C (**b**) radial distribution function of the Al-12.5 wt% Si alloys (**c**) effect of Sb additions and (**d**) its respective radial distributtion at various temperatures [46, 50–52] (Figures courtesy of Prof. X. Bian, Shandong University of Technology, China)

(26.8 nm^{-1}) [51] and prereflection of Si for different Si contents and at temperatures as high as 1075 °C. The prereflection had an $S(Q)$ range from 1.4 to 1.6 nm^{-1} [56], which was identified as Si agglomerations with medium-range order confirming the Raman spectroscopy findings (Fig. 9.4) [23]. XRD at room temperature did not present the Si prereflection. This indicates that Si agglomerations have a different crystalline structure than solid Si. The distance among the Si atoms in the Si–Si agglomerates can be determined using the superimposed principle developed by Cervink as reported in reference [56] that is carried out based on the cube diagonal method.

The study of the liquid state for Al-Si hypereutectic alloys is important since the Si agglomerates are susceptible to modification by means of thermal, chemical [26, 39, 49, 57, 58, 64, 70–79], and electromagnetic stirring and vibration (ESV) [25, 51, 80]. In other words, the primary Si agglomerates are key to effectively treat and modify Al-Si alloys (particularly eutectic and hypereutectic) at temperatures above liquidus. Similar effect has been observed in other elements such as Al, C, Cu, Fe, Ni, and Sn [50, 68].

The ESV is the only reported method that promotes full refinement of the primary Si when applied to Al-Si eutectic and hypereutectic alloys in the liquid and semisolid state. Most importantly, this may be the only reported method that resulted effective in liquid melts. This is important because this allows to refine primary Si particles, allowing widespread applications for the Al-Si eutectic and hypereutectic alloys. Their main applications are attributed to their relatively high hardness, and they are ideal for environments where wear resistance is critical. Some examples include automotive and aerospace industries. The refined Si particles could prevent the fracture of the coarse primary Si. Instead a small eutectic particles may fracture reducing the detrimental mechanical effects of broken primary Si particles. This result in improved service characteristics and quality, making them a suitable substitute for heavier or more costly materials [25].

9.3.3 Primary Si Growth

Silicon nucleates in the liquid metal as a spherical-like particle that grows forming ribbons that is the typical structure prior to the formation of the octahedral primary Si particles. Primary Si particles grow in three stages, the first stage is diffusion controlled, and therefore the resultant crystal is isotropic (0–15 μm) [81, 82]. In the second stage, the growth form branches along the [100] directions forming ribbons (15–50 μm). In this stage, the fastest growth occurs along the [100] directions and the slowest is the formation of the {111} facets that are also known as skeletal growth. In the third stage, octahedral particles are formed as shown in Fig. 9.6. At this stage, the particles become unstable due to the precipitation of the impurities along the {111} planes. This makes these locations poor in Si, promoting instability (>50 μm) such as branching or empty pockets. In this mechanism, the growth of the corners is still allowed, but the growth along the {111} planes forms macrosteps [81, 83] as seen in Fig. 9.7.

The primary Si particles grow along the <211> directions and form twin planes in families of parallel planes that are aligned with the {111} planes. The growth rate along

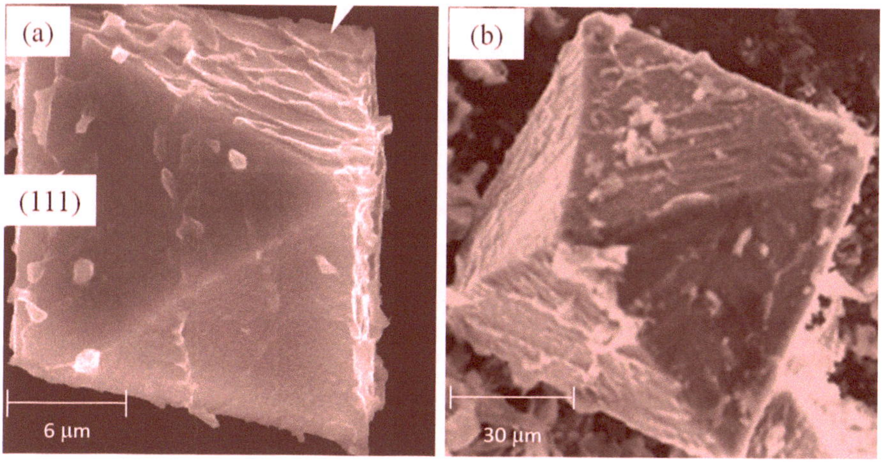

Fig. 9.6 (**a**, **b**) Pictures of perfect octahedral primary Si crystals characterized in Al-20 wt% Si alloys (Micrographs courtesy of Prof. Q.C. Jiang, Jilin University, China. Labels made by the author)

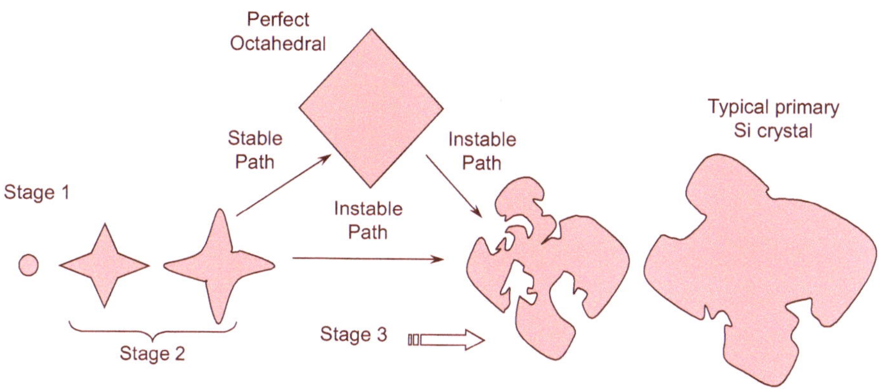

Fig. 9.7 Growth sequence of primary Si particles in unmodified Al-Si hypereutectic melt

the directions <211> can be up to five times faster than the <111> directions [84, 85]. Figure 9.8 shows the growth directions and planes as well as the angles formed along the twins on primary Si particles. The twin formation is usually the result of impurity elements having a relative, to Si, atomic radius close to 1.646 [84, 86]. The primary Si crystals growth is favored along the planes formed both sides of a twin, along the common groove. On the other side, the growth out of the grooves is slow because it is controlled by surface nucleation on the {111} planes [87]. This mechanism explains the typical imperfect and elongated (usually tetragonal) primary Si particles observed in Al-Si hypereutectic alloys with more than 20 wt% Si. High cooling rates during solidification forces the primary Si growth to hinder stage 3. This keeps the primary Si particles in sizes below 50 μm. These sizes are usually perfect octahedrons (Figs. 9.6, 9.8, and 9.9).

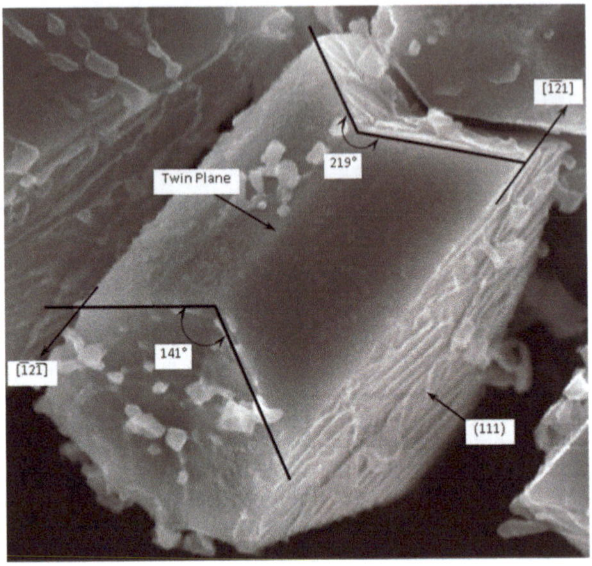

Fig. 9.8 Scanning Electron Micrograph (SEM) micrograph of a primary Si particle indicating their growth (planes and directions) (Micrograph courtesy of Prof. Q.C. Jiang, Jilin University, China. Labels made by the author)

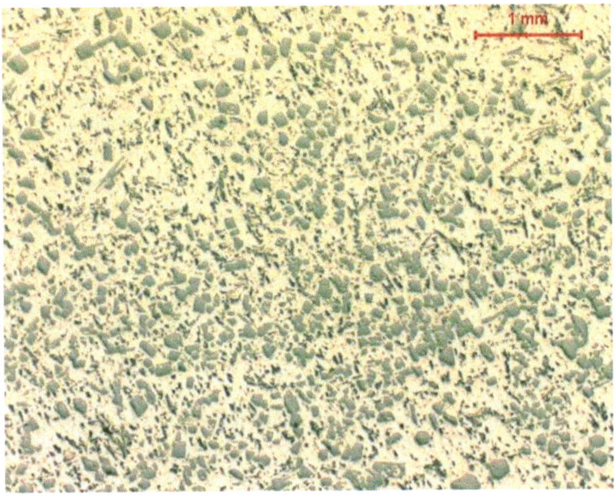

Fig. 9.9 Al-25 wt% Si rapid solidified (C.R. > 1.6 °C/s) showing perfect octahedrons

9.4 Silicon Modification Methods

9.4.1 Chemical Modification of Si Rich Phases

Phosphorous (P) is the most widely used primary Si modifier for Al-Si hypereutectic alloys. P is added in the form of Al-P, Al-Cu-P, Cu- 8 wt% P or Al-Fe-P master alloys [58, 88]. The Al-Cu-P [88, 89] and/or Cu- 8 wt% P [7] master alloys are known as the most effective primary Si modifiers. P additions transform the primary Si from the star-shaped into polyhedral, resulting in an increase in the number of particles in up to three times and hence reducing its size. The respective reduction in their distance is approximately 50% [88]. The refinement effect is accomplished due to the formation of the AlP compound that has a high melting point and acts as heterogeneous nuclei. The AlP compound forces the epitaxial growth of Si [64, 65, 88, 89]. In Fig. 9.10a–c is shown the effect of heteronucleation of the AlP particle in the center of a primary Si crystal. In order for the AlP particles to be effective the

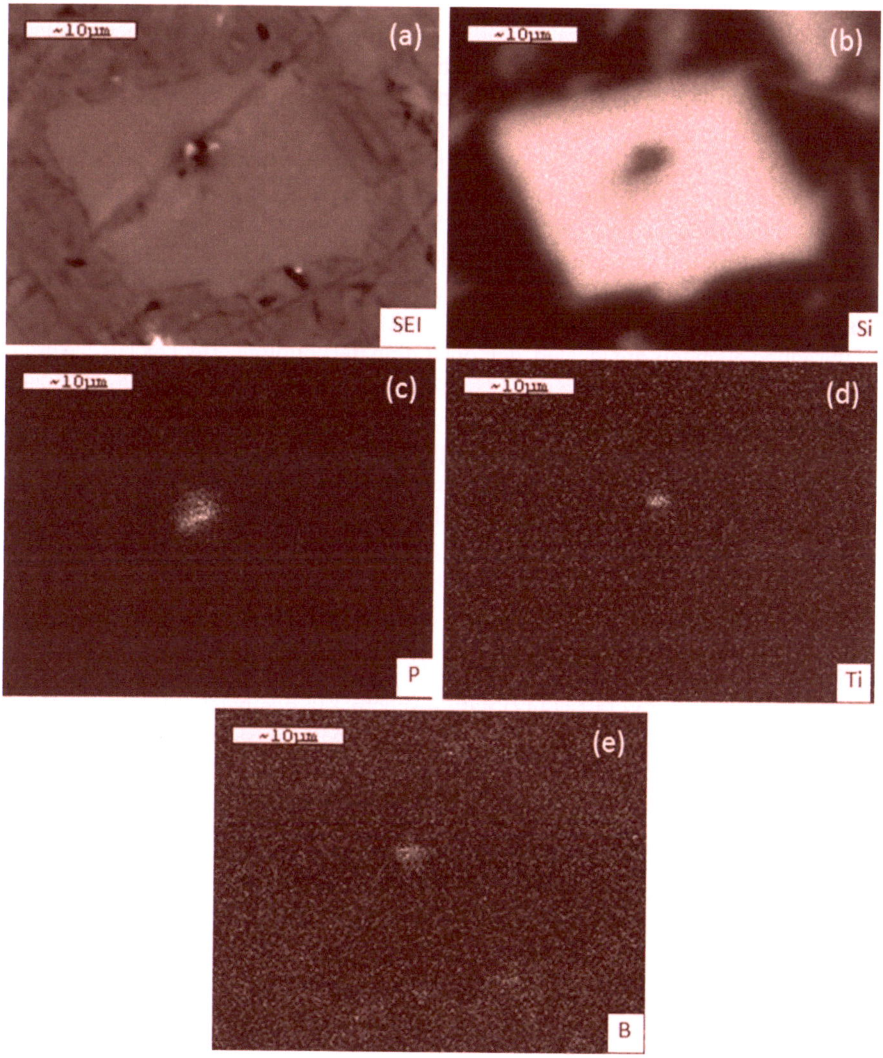

Fig. 9.10 Electron probe microanalysis (EPMA) of a chemically modified primary Si particle: (**a**) micrograpmh taken using secondary electrons mode of a modified primary Si particle taken from an Al-20 wt% Si alloy and EPMA images of (**b**) Si, (**c**) P, (**d**) Ti and (**e**) B (Micrographs courtesy of Prof. X. Liu, Shandong University of Technology, China)

superheat temperature must be kept below 980 °C [56]. Higher temperatures are detrimental due to the decomposition of AlP promoting its dissolution in liquid Al minimizes the refining effect for primary Si.

This compound forms at temperatures as high as 982 °C, which is well above the liquidus temperature of most commercial Al-Si hypereutectic alloys. This temperature is above the nucleation of the Si agglomerates of most commercial Al-Si hypereutectic alloys [48]. The AlP compound has a cubic crystalline structure with a lattice parameter of 5.45 Å similar to the diamond-like crystalline structure for Si ($a = 5.43$ Å). This similarity in crystalline structure promotes heterogeneous nucleation by seeding the primary Si particles that starts depositing in an epitaxial fashion on top of the AlP particles. Al-Ce compound plays similar role, but Al-Ce has an orthorhombic structure with lattice parameters $a = 9.27$ Å, $b = 7.68$ Å, and $c = 5.75$ Å. Therefore, the epitaxial growth of the primary Si particles is limited to the "c" direction of the Al-Ce lattice cell [64, 65], reducing its effectiveness when compared to AlP.

Yu et al. [90, 91] developed a methodology to refine primary Si for Al-Si near eutectic and hypereutectic alloys. This methodology consists of the coupling additions of AlP and TiB_2 or Al_4C_3 [90, 91]. Figure 9.10 shows the location where each of the added elements are located, demonstrating the coupling effect of TiB_2. Similar effect is observed with the use of Al_4C_3. The coupling effect of AlP and/or TiB_2 and Al_4C_3 particles form a peritectic-like reaction that results in their multiplication and refinement of the AlP particles. The coupled particles have the same effect as the pure AlP. But in this case their number increases and enhances their refining effect. This was previously demonstrated for grain refinement of Al-Si alloys in the Hulk's and hypernucleation theories [28, 29, 43].

The couple refining mechanism takes effect as a result of the high coherency among TiB_2 and AlP. Yu et al. [90, 91] proposed a modified Turnbull-Vonnegut equation to determine the atomic planes that maximize the effectiveness of coupling AlP as primary Si refiner. In the case of TiB_2 and AlP, the planes with the best coupling effectiveness are (112)||(442) with respective lattice parameters of 1.1049 Å and 1.1065 Å, and the (110)||(311) for Al_4C_3 and AlP, respectively, with the corresponding lattice parameters 1.6675 Å and 1.6341 Å.

Additions of P (as primary Si modifier) required incubation time. Ideally, the P master alloy is added after degassing, avoiding any chemical reactions of P with entrapped gasses. After adding P, it is recommended to mechanically stir the melt for approximately 10 min to ensure proper dissolution of the master alloy. Following the additions of P, the melt is left to rest for another 10 min. This process is known as activation time. During the activation time, the formation of the AlP compound is promoted.

Calcium is added as an inoculant and enhances the refinement effect of the P-rich master alloys. Calcium additions catalyze the formation of AlP, and therefore their incubation time is shorter [6, 92, 93]. Refinement of primary Si improves yield strength, tensile strength, and elongation, but the most important is the benefits on wear resistance [89, 94].

Combinations of Si modifiers (Sr + P) were investigated as an alternative method for chemical modification of Al-Si hypereutectic alloys [89]. The resultant microstructure in an Al-14 wt% Si hypereutectic alloy was similar to the eutectic composition. However, the Sr additions must be limited to 300 ppm or less. Further additions have no effects for the above-mentioned compositions. The effect of Sr for compositions containing more than 14 wt% Si is rather controversial since the main effect of Sr is the refinement of Si eutectic. Therefore, the use of strontium must be limited to near the eutectic composition or Al-Si hypoeutectic alloys. The refinement of the primary Si particles with P and Sr has a positive effect on yield strength, tensile strength, and elongation; however, a specific combination or ratio to maximize the three properties at the same time is not reported [89, 95].

Additions of rare earth elements such as Ce, La, Nd, and Pr have a refining effect on Al-Si hypereutectic alloys [65, 96, 97]. The morphology of the primary Si particles changes with the master alloy additions and the silicon content. Additions of up to 3 wt% of a master alloy containing La, Ce, Nd, Pr, and Pa showed successful results in the microstructure refinement for Al – 13–20 wt% Si alloys. Rare earth elements modify eutectic Si, changing the particles from a plate-like to a fibrous form. The effect of La on primary Si is limited to binary Al-Si compositions. In the presence of Cu, the $Al_7Si_7Cu_2La_{3.5}$ intermetallic is formed and it is detrimental to casting integrity by promoting the formation of two-dimensional defects (twins or stacking faults) [49].

Xu et al. [84, 98] developed primary Si modifiers to treat Al-Si hypereutectic alloys containing Al-P-Ti-TiC, Al-P-Ti-TiC-Y, and Al-P-Ti-TiC-Nd. These modifiers are effective for Al-Si hypereutectic alloys with 20–29 wt% Si. The master alloy Al-P-Ti-TiC with and without yttrium (Y) is also effective to modify primary Si. The morphology of the primary Si in alloys containing more than 20 wt% Si is considerably affected, and it is modified from an elongated or hollow to a blocky (octahedral-like) form. The master alloy with neodymium (Al-P-Ti-TiC-Nd) further modifies the primary Si particles to a rod-like shape. The presence of Nd modified not only the primary Si particles but also eutectic Si particles.

Master alloys having Al-P-Ti-TiC-Nd_2O_3 are known for their positive effect to refine the primary Si particles. The elimination of internal defects in coarser primary Si together with improved isotropy of the modified alloys after heat treatment promotes improvements of up to 20% in tensile strength [99].

9.4.2 Thermal Modification

Two kinds of thermal modification are the most commonly applied for Al-Si alloys that are superheating: variable cooling rates and their combination. Vivès et al. [100] used cooling rates from 0.01 to 20 °C/s to reduce the interdendritic spacing from 0.1 to 0.02 μm. High cooling rate promotes grain refinement, and enough super heating temperature combined with high cooling rates refine the primary Si. The combined effect of superheat temperatures of at least 500 °C and cooling rates of 20 °C/s is quite effective as a refiner, allowing a reduction in interdendritic spacing.

For Al-Si hypereutectic alloys an increase in the superheat temperatures has significant refinement effects of Si particles (primary and eutectic). This is in agreement with the fact that Si agglomerates form in Al-Si hypereutectic alloys at temperatures higher than their liquidus [23]. Therefore, higher temperatures reduce the nucleation potential for the Si agglomerates. Also, isothermal holdings at various superheat temperatures starting from 3 °C up to 300 °C above liquidus combined with rapid cooling are effective to refine primary Si, interdendritic spacing and intermetallic compound size [101, 102].

Isothermal holdings at temperatures slightly below liquidus and rapidly cooling rates transformed the primary Si particles from blocky to petal shapes [103]. Additionally, Bergsma et al. [104] reported that quenching using cooling rates between 15 and 30 °C/s on near eutectic compositions in liquid or semisolid states has similar refining effects as melt stirring. All these treatments, except for super heat temperature, can refine the microstructure, but are limited to the reduction in size of the primary Si particles. It means, primary Si particles are still present in the microstructure.

The XRD analysis of Al-13 wt% Si alloy treated using thermal rate treatments [48] shows the presence of Si agglomerates. The size of the agglomerates colonies forming Si–Si bonds is considerably reduced at temperatures above 875 °C. This is observed by the reduction in the coordination number of Si atoms, thus agglomerates as well as their atomic density. This suggests that rapid solidification from temperatures above 875 °C can have significant effects on primary Si refinement. This is demonstrated by pouring the alloy from 875 °C into a cast iron mold for rapid solidification resulting in a coral-like structure [48]. Figure 9.11a, b show the microstructure of the Al-Si hypereutectic microstructure when quenched from the liquid state. In these figures, it is observed that Si forms transport channels to diffuse toward a common center to contribute the coarsening of Si agglomerates or primary Si particles [23, 105]. In a eutectic-like fashion, another group of channels

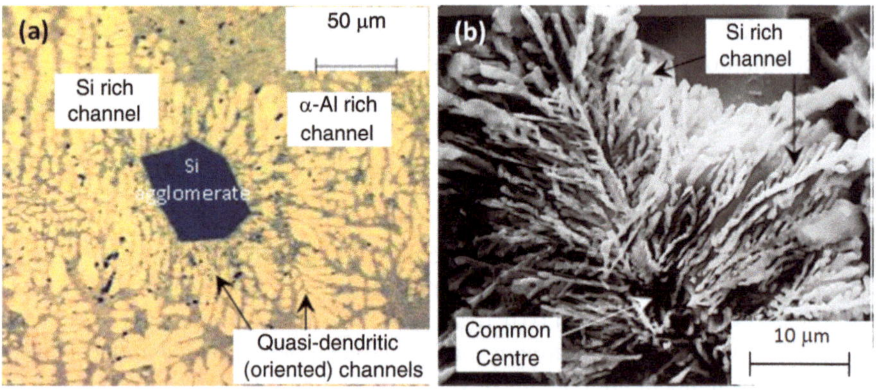

Fig. 9.11 Micrograph of thermal-treated Al-Si hypereutectic samples (**a**) 390.1 [25, 105] alloy quenched from 620 °C on a saline bath and (**b**) coral-like structure on a deep etched Al-13 wt% Si rapid solidified on a cast iron mold from a temperature of 875 °C (Micrograph (**b**) courtesy of Prof. X. Bian, Shandong University of Technology, China)

are formed where the Al and the other elements are rejected out of this common center. This transport mechanism is enhanced for higher cooling rates, promoting higher constitutional undercooling that results in dendrite formation. Figure 9.12 is included as a comparative example of the effects of a primary silicon particle that was allowed to grow in the absence of any refining process.

9.4.3 Mechanical Modification

Thixoforming methods have been used to treat Al-Si hypereutectic alloys with positive results [70, 106]. The use of thixoforming methods has provided significant advantages over the processing of Al-Si hypereutectic alloys that result in a potential substitute for cast iron. The improvement in mechanical properties by the use of thixoforming is presumably due to the improved quality and the effects in the modification of primary Si particles. Ni additions of up to 4 wt% play an important role in the modulus of elasticity and yield strength; however, Ni additionas may negatively affect elongation and the effects on hardness are rather negligible. Unfortunately, this technology is still expensive, but mass production can significantly reduce the cost [70, 106].

Chen et al. [107] combined thixocasting and powder metallurgy methods to produce Al-Si hypereutectic alloys containing up to 25 wt% Si. Using this method, the powders of the particular composition are compacted in a press surrounded by an induction coil. The powders are injected to mold(s) at variable consolidation temperatures. The thixocast powders have relatively high density that increases with the consolidation temperature. Conversely, this practice promotes grain coarsening.

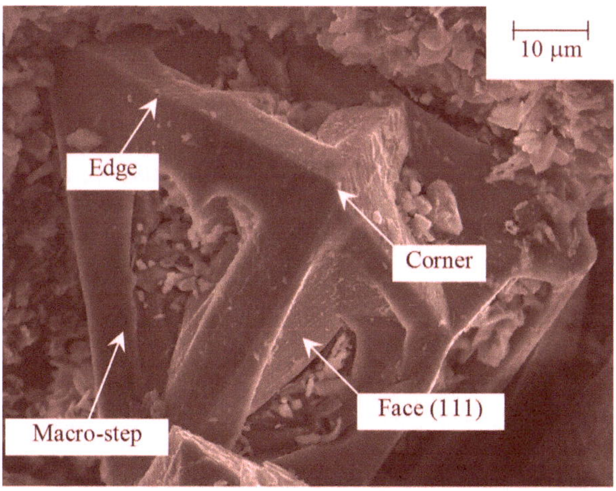

Fig. 9.12 SEM micrographs of an imperfect octahedral primary Si particle observed on an Al-Si hypereutectic alloy (Micrograph courtesy of Prof. Q.C. Jiang, Jilin University, China. Labels made by the author)

This method guarantees densities of up to 90% at sintering temperatures between 500 and 550 °C. The mechanical strengths show improvements from 193 to 224.7 MPa, depending on the processing temperatures and density. The grain growth (ripening) of the consolidating particles follows the classical coarsening described by the Lifshittz Slyzov Wagner (LSW) theory [107, 108].

Rotary-die equal-channel angular pressing is a novel method where Al-Si hypereutectic castings of any composition can be treated in a semisolid state [109]. The idea of using this method is to break connected eutectic Si particles, but most importantly to break and refine primary Si particles resulting in improvements of mechanical properties [110, 111]. The method consists of processing mechanically the Al-Si hypereutectic castings at relatively low temperatures (approximately 400 °C). In this process the alloys is mechanically processed by plastic deformations induced along different directions to refine the microsturcture while improving isotropy. This process occurs under uniaxial compressions to break the microstructure. The mechanical refinement helps in reducing the detrimental effects of brittle phases, such as the Si-rich ones and intermetallics. The number of uniaxial compressions in different directions promotes isotropy, fine structure, and pore reduction. The resultant components showed improvements in toughness of up to 18 times from 0.7 to 13 kJ/m^2 for Al-Si hypereutectic alloys. The same process applied for Al-Si alloys with near eutectic compositions shows improvements from 13.0 to 102 kJ/m^2 [112]. This process is highly recommended for research, and has potentia for future industrialization.

References

1. Fainstein-Pedraza, D., and G.F. Bolling. 1975. Superdendritic growth. *Journal of Crystal Growth* 28 (3): 311–318. https://doi.org/10.1016/0022-0248(75)90068-8.
2. Quaresma, J.M.V., C.A. Santos, and A. Garcia. 2000. Correlation between unsteady-state solidification conditions, dendrite spacings, and mechanical properties of Al-Cu alloys. *Metallurgical and Materials Transactions A* 31 (12): 3167–3178. https://doi.org/10.1007/s11661-000-0096-0.
3. Dahle, A.K., Y.C. Lee, M.D. Nave, P.L. Schaffer, and D.H. StJohn. 2001. Development of the as-cast microstructure in magnesium–aluminium alloys. *Journal of Light Metals* 1 (1): 61–72. https://doi.org/10.1016/S1471-5317(00)00007-9.
4. Efzan, M.N.E., H.J. Kong, and C.K. Kok. 2014. Review: Effect of alloying element on Al-Si alloys. *Materials, Industrial, and Manufacturing Engineering Research Advances* 845 (1.1): 355–359. https://doi.org/10.4028/www.scientific.net/AMR.845.355.
5. Gruzleski, J.E., and B.M. Closset. 1990. *The treatment of liquid aluminum-silicon alloys*, 256. Des Plaines: American Foundrymen's Society, Inc.
6. Campbell, J. 2003. *Castings*. 2nd ed. Oxford, UK: Butterworth Heinemann.
7. Bäckerud, L., G. Chai, and J. Tamminen. 1990. *Solidification characteristics of aluminum alloys: Foundry alloys*. Vol. 2. 1st ed., 256. Stockholm: AFS/Skan Aluminium.
8. Bäckerud, L., G. Chai, and J. Tamminen. 1990. *Solidification characteristics of aluminum alloys: Wrought alloys*. Vol. 1. 1st ed. AFS/Skan Aluminium.
9. Adams, J.H., et al. 1990. *ASM metals handbook: Properties and selection: Nonferrous alloys and special-purpose materials*, Vol. 2. Materials Park: ASM International.

10. Jorstad, J.L. 1971. The hypereutectic aluminum-silicon alloy used to cast the Vega engine block. *Modern Casting* 60 (4): 59–64.
11. Brodova, I.G., P.S. Popel, and G.I. Eskin. 2001. *Liquid metal processing: Applications to aluminium alloy production*. New York: Taylor & Francis.
12. Jorstad, J.L. 1970. *Paper no. 105*, 5–11. Detroit: SDCE.
13. Jostard, J.L. 1971. *AFS casting congress*, 71–76.
14. Green, R.E. 1970. Die casting the Vega engine block. *Die Casting Engineer* 14: 12–26.
15. Jorstad, J.L. 1970. *6th SDCE international die casting congress, Cleveland*, 1–6.
16. Norbye, J.P. 1970. Chevrolet vega 2300 engine. *Automobile Engineering*: 320–326.
17. *Development of power*. 2016. Available from: http://www.tpub.com/content/engine/14037/css/14037_91.htm.
18. Lee, Y.C., A.K. Dahle, D.H. StJohn, and J.E.C. Hutt. 1999. The effect of grain refinement and silicon content on grain formation in hypoeutectic Al-Si alloys. *Materials Science and Engineering A-Structural Materials Properties Microstructure and Processing* 259 (1): 43–52. https://doi.org/10.1016/S0921-5093(98)00884-3.
19. Kori, S.A., B.S. Murty, and M. Chakraborty. 2000. Development of an efficient grain refiner for Al-7Si alloy and its modification with strontium. *Materials Science and Engineering A-Structural Materials Properties Microstructure and Processing* 283 (1–2): 94–104. https://doi.org/10.1016/S0921-5093(99)00794-7.
20. Robles Hernandez, F.C., and J.H. Sokolowski. 2006. Comparison among chemical and electromagnetic stirring and vibration melt treatments for Al-Si hypereutectic alloys. *Journal of Alloys and Compounds* 426 (1–2): 205–212. https://doi.org/10.1016/j.jallcom.2006.09.039.
21. Robles Hernandez, F.C., and J.H. Sokolowski. 2006. Thermal analysis and microscopical characterization of Al-Si hypereutectic alloys. *Journal of Alloys and Compounds* 419 (1–2): 180–190. https://doi.org/10.1016/j.jallcom.2005.07.077.
22. Robles Hernandez, F.C., and J.H. Sokolowski. 2009. Effects and on-line prediction of electromagnetic stirring on microstructure refinement of the 319 Al-Si hypoeutectic alloy. *Journal of Alloys and Compounds* 480 (2): 416–421. https://doi.org/10.1016/j.jallcom.2009.02.109.
23. Robles Hernández, F.C., and J.H. Sokolowski. 2005. Identification of silicon agglomerates in quenched Al-Si hypereutectic alloys from liquid state. *Advanced Engineering Materials* 7 (11): 1037–1043.
24. Robles Hernandez, F.C., J.H. Sokolowski, and J.J. De Cruz Rivera. 2007. Micro-Raman analysis of the Si particles present in Al-Si hypereutectic alloys in liquid and semi-solid states. *Advanced Engineering Materials* 9 (1–2): 46–51.
25. Robles-Hernández, F.C. 2004. *Improvement in functional characteristics of Al-Si cast components through the utilization of a novel electromagnetic treatment of liquid melts*, 251. Windsor: Mechanical Engineering, University of Windsor.
26. Sadigh, B., M. Dzugutov, and S.R. Elliott. 1999. Vacancy ordering and medium-range structure in a simple monatomic liquid. *Physical Review B* 59 (1): 1–4.
27. Bäckerud, L. 1968. Kinetic aspects of the solidification of binary and ternary alloy systems. *Jernkontorets Annaler* 152 (3): 109–138.
28. Quested, T.E., A.T. Dinsdale, and A.L. Greer. 2005. Thermodynamic modelling of growth-restriction effects in aluminium alloys. *Acta Materialia* 53 (5): 1323–1334. https://doi.org/10.1016/j.actamat.2004.11.024.
29. Easton, M.A., and D.H. StJohn. 2001. A model of grain refinement incorporating alloy constitution and potency of heterogeneous nucleant particles. *Acta Materialia* 49 (10): 1867–1878. https://doi.org/10.1016/S1359-6454(00)00368-2.
30. Spittle, J.A., and S. Sadli. 1995. Effect of alloy variables on grain refinement of binary aluminium alloys with Al–Ti–B. *Materials Science and Technology* 11 (6): 533–537. https://doi.org/10.1179/mst.1995.11.6.533.
31. Tarshis, L.A., J.L. Walker, and J.W. Rutter. 1971. Experiments on the solidification structure of alloy castings. *Metallurgical Transactions* 2 (9): 2589–2597. https://doi.org/10.1007/BF02814899.

32. Greer, A.L., A.M. Bunn, A. Tronche, P.V. Evans, and D.J. Bristow. 2000. Modelling of inoculation of metallic melts: Application to grain refinement of aluminium by Al–Ti–B. *Acta Materialia* 48 (11): 2823–2835. https://doi.org/10.1016/S1359-6454(00)00094-X.
33. Easton, M., and D. StJohn. 2005. An analysis of the relationship between grain size, solute content, and the potency and number density of nucleant particles. *Metallurgical and Materials Transactions A* 36 (7): 1911–1920. https://doi.org/10.1007/s11661-005-0054-y.
34. Maxwell, I., and A. Hellawell. 1975. A simple model for grain refinement during solidification. *Acta Metallurgica* 23 (2): 229–237. https://doi.org/10.1016/0001-6160(75)90188-1.
35. Lee, Y.C., A.K. Dahle, D.H. StJohn, and J.E.C. Hutt. 1999. The effect of grain refinement and silicon content on grain formation in hypoeutectic Al–Si alloys. *Materials Science and Engineering A* 259 (1): 43–52. https://doi.org/10.1016/S0921-5093(98)00884-3.
36. Backerud, L., M. Johnsson, and G.K. Sigworth. 2000. Method for optimization of the grain refinement of aluminum alloys. 1, 6,073,677.
37. Johnsson, M. 1995. Grain refinement of aluminium studied by use of a thermal analytical technique. *Thermochimica Acta* 256 (1): 107–121. https://doi.org/10.1016/0040-6031(94)02167-M.
38. Backerud, L. 1983. How does a good grain refiner work. *Light Metal Age* 41 (9–10): 6–10.
39. Gruzleski, J.E. 2000. *Microstructure development during metal casting*, 238. Des Plaines: American Foundrymen's Society.
40. Iqbal, N., N.H. van Dijk, T. Hansen, L. Katgerman, and G.J. Kearley. 2004. The role of solute titanium and TiB2 particles in the liquid-solid phase transformation of aluminum alloys. *Materials Science and Engineering A-Structural Materials Properties Microstructure and Processing* 386 (1–2): 20–26. https://doi.org/10.1016/j.mesa.2004.06.068.
41. Murty, B.S., S.A. Kori, and M. Chakraborty. 2002. Grain refinement of aluminium and its alloys by heterogeneous nucleation and alloying. *International Materials Reviews* 47 (1): 3–29. https://doi.org/10.1179/095066001225001049.
42. Easton, Mark, and David StJohn. 1999. Grain refinement of aluminum alloys: Part I. The nucleant and solute paradigms—a review of the literature. *Metallurgical and Materials Transactions A* 30 (6): 1613–1623. https://doi.org/10.1007/s11661-999-0098-5.
43. Gloria Ibarra, D. 1999. *Control of grain refinement of Al-Si alloys by thermal analysis*, 123. Montreal: Department of Mining and Metallurgical Engineering, McGill University.
44. Inui, M., S.I. Takeda, Y. Shirakawa, S. Tamaki, Y. Waseda, and Y. Yamaguchi. 1991. Structural study of molten silver halides by neutron diffraction. *Journal of the Physical Society of Japan* 60 (9): 3025–3031. https://doi.org/10.1143/JPSJ.60.3025.
45. Jones, G.P. 1987. *Solidification processing*, 496–499. London: The Institute of Metals.
46. Lidman, W.G. 1984. Master alloys improve aluminum casting properties. *Foundry M&T* 112 (8): 46–47.
47. Mohanty, P.S., F.H. Samuel, and J.E. Gruzleski. 1995. Experimental study on pore nucleation by inclusions in aluminum castings. *AFS Transactions* 103: 555–564.
48. Bian, X., and W. Wang. 2000. Thermal-rate treatment and structure transformation of Al–13 wt.% Si alloy melt. *Materials Letters* 44 (1): 54–58. https://doi.org/10.1016/S0167-577X(00)00011-2.
49. Djurdjevic, M.B., W. Kasprzak, C.A. Kierkus, W.T. Kierkus, and J.H. Sokolowski. 2001. Quantification of Cu enriched phases in synthetic 3XX aluminum alloys using the thermal analysis technique. *AFS Transactions* 16: 1–12.
50. Xiufang, B., and Q. Jingyu. 2004. Aluminium alloys: Their physical and mechanical properties. In *9th International Conference on Aluminium Alloys (ICAA9)*. North Melbourne: Institute of Materials Engineering Australasia.
51. Xiufang, B., W. Weimin, and Q. Jingyu. 2001. Liquid structure of Al–12.5% Si alloy modified by antimony. *Materials Characterization* 46 (1): 25–29. https://doi.org/10.1016/S1044-5803(00)00089-9.
52. Wang, W., X. Bian, J. Qin, and T. Fan. 2000. Study on atomic density changes in the liquid Al-Si alloys by X-ray diffraction method. *Journal of Materials Science Letters* 19 (17): 1583–1585. https://doi.org/10.1023/A:1006741626728.

53. Eskin, G.I. 1998. *Ultrasonic treatment of light alloy melts*. CRC Press Taylor and Francis Group.
54. Robles Hernández, F.C., and J.H. Sokolowski. 2009. Effects and on-line prediction of electromagnetic stirring on microstructure refinement of the 319 Al–Si hypoeutectic alloy. *Journal of Alloys and Compounds* 480 (2): 416–421.
55. Jain, S.C., H.E. Maes, K. Pinardi, and I. De Wolf. 1996. Stresses and strains in lattice-mismatched stripes, quantum wires, quantum dots, and substrates in Si technology. *Journal of Applied Physics* 79 (11): 8145–8165. https://doi.org/10.1063/1.362678.
56. Wang, W., X. Bian, J. Qin, and S.I. Syliusarenko. 2000. The atomic-structure changes in Al-16 pct Si alloy above the liquidus. *Metallurgical and Materials Transactions A* 31 (9): 2163–2168. https://doi.org/10.1007/s11661-000-0134-y.
57. Robles Hernández, F.C., and J.H. Sokolowski. 2006. Thermal analysis and microscopical characterization of Al–Si hypereutectic alloys. *Journal of Alloys and Compounds* 419 (1–2): 180–190. https://doi.org/10.1016/j.jallcom.2005.07.077.
58. Qiao, J., X. Song, X. Bian, L. Zhu, and Q. Zhang. 2002. *Special Casting & Nonferrous Alloys*. Oct.–Dec: 43–45.
59. Abramov, V., O. Abramov, V. Bulgakov, and F. Sommer. 1998. Solidification of aluminium alloys under ultrasonic irradiation using water-cooled resonator. *Materials Letters* 37 (1–2): 27–34. https://doi.org/10.1016/S0167-577X(98)00064-0.
60. Vives, C. 1992. Elaboration of semisolid alloys by means of new electromagnetic rheocasting processes. *Metallurgical Transactions B* 23 (2): 189–206.
61. Zhang, W., Y. Yang, Q. Liu, Y. Zhu, and Z. Hu. 1997. Effects of forced flow on morphology of Al – CuAl2 eutectic solidified with electromagnetic stirring. *Journal of Materials Science Letters* 16 (23): 1955–1957. https://doi.org/10.1023/A:1018555121395.
62. Lim, S.-C., E.-P. Yoon, and J.-S. Kim. 1997. The effect of electromagnetic stirring on the microstructure of Al-7 wt% Si alloy. *Journal of Materials Science Letters* 16 (2): 104–109. https://doi.org/10.1023/A:1018525506838.
63. Desnain, P., Y. Fautrelle, J.L. Meyer, J.P. Riquet, and F. Durand. 1990. Prediction of equiaxed grain density in multicomponent alloys, stirred electromagnetically. *Acta Metallurgica et Materialia* 38 (8): 1513–1523. https://doi.org/10.1016/0956-7151(90)90119-2.
64. Nafisi, S., R. Ghomashchi, J. Hedjazi, and S.M.A. Boutorabi. 2004. Advances in lightweight automotive castings and wrought aluminum alloys. In *SAE world congress*. Detroit: Society of Automotive Engineers.
65. Chang, J., I. Moon, and C. Choi. 1998. Refinement of cast microstructure of hypereutectic Al-Si alloys through the addition of rare earth metals. *Journal of Materials Science* 33 (20): 5015–5023. https://doi.org/10.1023/A:1004463125340.
66. Cullity, B.D., and S.R. Stock. 2001. Elements of x-ray diffraction. In *Addison-Wesley series in metallurgy and materials*, 3rd ed., 680. Reading: Addison-Wesley Pub. Co.
67. Robles Hernandez, F.C., J.H. Sokolowski, and J. de J. Cruz Rivera. 2007. Micro-Raman analysis of the Si particles present in Al-Si hypereutectic alloys in liquid and semi-solid states. *Advanced Engineering Materials* 9 (1–2): 46–51. https://doi.org/10.1002/adem.200600173.
68. Radjai, A., and K. Miwa. 2002. Structural refinement of gray iron by electromagnetic vibrations. *Metallurgical and Materials Transactions A* 33 (9): 3025–3030. https://doi.org/10.1007/s11661-002-0287-y.
69. Gabalthular, J.P., S. Steeb, and P. Laparner. 1979. Über die Struktur von Aluminium-Silizium-Schmelzen. *Zeitschrift für Naturforschung* 34a: 1305–1315.
70. Kapranos, P., D.H. Kirkwood, H.V. Atkinson, J.T. Rheinlander, J.J. Bentzen, P.T. Toft, C.P. Debel, G. Laslaz, L. Maenner, S. Blais, J.M. Rodriguez-Ibabe, L. Lasa, P. Giordano, G. Chiarmetta, and A. Giese. 2003. Thixoforming of an automotive part in A390 hypereutectic Al–Si alloy. *Journal of Materials Processing Technology* 135 (2–3): 271–277. https://doi.org/10.1016/S0924-0136(02)00857-9.
71. McDonald, S.D., K. Nogita, and A.K. Dahle. 2004. Eutectic nucleation in Al–Si alloys. *Acta Materialia* 52 (14): 4273–4280. https://doi.org/10.1016/j.actamat.2004.05.043.
72. Callister, W.D. 2007. *Materials science and engineering: An introduction*. 7th ed., 832. Wiley.

73. Porter, D.A., K.E. Easterling, and M. Sherif. 2009. *Phase transformations in metals and alloys (Revised reprint)*, 520. Boca Raton: CRC press.
74. Evans, E.L., C.B. Alcock, and O. Kubaschewski, eds. 1967. *Metallurgical thermochemistry*. 4th ed. Oxford, UK/New York: Pergamon Press.
75. Wang, F., Z. Zhang, Y. Ma, and Y. Jin. 2004. Effect of Fe and Mn additions on microstructure and wear properties of spray-deposited Al–20Si alloy. *Materials Letters* 58 (19): 2442–2446. https://doi.org/10.1016/j.matlet.2004.02.027.
76. Narayanan, L.A., F.H. Samuel, and J.E. Gruzleski. 1992. Thermal analysis studies on the effect of cooling rate on the microstmcture of 319 aluminum alloy. *AFS Transactions* 100: 383–391.
77. Kierkus, W.T., and J.H. Sokolowski. 1999. Recent advances in cooling curve analysis: A new method of determining the 'Base Line' equation. *AFS Transactions* 66: 161–167.
78. Polishchuck, V.P., and N.R. Aranova. 1965. *Trudy Donetsk N-I Inst. Chern. Met.* 2: 146–152.
79. Barclay, R.S., P. Niessen, and H.W. Kerr. 1973. Halo formation during unidirectional solidification of off-eutectic binary alloys. *Journal of Crystal Growth* 20 (3): 175–182. https://doi.org/10.1016/0022-0248(73)90001-8.
80. Radjai, A., and K. Miwa. 2000. Effects of the intensity and frequency of electromagnetic vibrations on the microstructural refinement of hypoeutectic Al-Si alloys. *Metallurgical and Materials Transactions A* 31 (3): 755–762. https://doi.org/10.1007/s11661-000-0017-2.
81. Xu, C.L., H.Y. Wang, C. Liu, and Q.C. Jiang. 2006. Growth of octahedral primary silicon in cast hypereutectic Al–Si alloys. *Journal of Crystal Growth* 291 (2): 540–547. https://doi.org/10.1016/j.jcrysgro.2006.03.044.
82. Wang, R.-Y., W.-H. Lu, and L.M. Hogan. 1999. Growth morphology of primary silicon in cast Al–Si alloys and the mechanism of concentric growth. *Journal of Crystal Growth* 207 (1–2): 43–54. https://doi.org/10.1016/S0022-0248(99)00347-4.
83. Chernov, A.A. 1974. Stability of faceted shapes. *Journal of Crystal Growth* 24–25: 11–31. https://doi.org/10.1016/0022-0248(74)90277-2.
84. Xu, C.L., Q.C. Jiang, Y.F. Yang, H.Y. Wang, and J.G. Wang. 2006. Effect of Nd on primary silicon and eutectic silicon in hypereutectic Al–Si alloy. *Journal of Alloys and Compounds* 422 (1–2): L1–L4. https://doi.org/10.1016/j.jallcom.2005.03.128.
85. Kobayashi, K.F., and L.M. Hogan. 1985. The crystal growth of silicon in Al-Si alloys. *Journal of Materials Science* 20 (6): 1961–1975. https://doi.org/10.1007/BF01112278.
86. Lu, S.-Z., and A. Hellawell. 1987. The mechanism of silicon modification in aluminum-silicon alloys: Impurity induced twinning. *Metallurgical Transactions A* 18 (10): 1721–1733. https://doi.org/10.1007/BF02646204.
87. Wagner, R.S. 1960. On the growth of germanium dendrites. *Acta Metallurgica* 8 (1): 57–60. https://doi.org/10.1016/0001-6160(60)90145-0.
88. Kyffin, W.J., W.M. Rainforth, and H. Jones. 2001. Effect of phosphorus additions on the spacing between primary silicon particles in a Bridgman solidified hypereutectic Al-Si alloy. *Journal of Materials Science* 36 (11): 2667–2672. https://doi.org/10.1023/A:1017904627733.
89. Müller, K. 1997. Advanced light alloys and composites. In *NATO advanced study institute*. Zakopane: Kluwer Academic Publishers.
90. Yu, L., X. Liu, H. Ding, and X. Bian. 2007. A new nucleation mechanism of primary Si by peritectic-like coupling of AlP and TiB2 in near eutectic Al–Si alloy. *Journal of Alloys and Compounds* 432 (1–2): 156–162. https://doi.org/10.1016/j.jallcom.2006.06.005.
91. ———. 2007. A new nucleation mechanism of primary Si by like-peritectic coupling of AlP and Al4C3 in near eutectic Al–Si alloy. *Journal of Alloys and Compounds* 429 (1): 119–125. https://doi.org/10.1016/j.jallcom.2006.04.011.
92. Kim, H.J. 2003. Effect of calcium on primary silicon particle size in hypereutectic Al–Si alloys. *Materials Science and Technology* 19 (7): 915–918. https://doi.org/10.1179/026708303225002820.

93. Kyffin, W.J., W.M. Rainforth, and H. Jones. 2001. Effect of treatment variables on size refinement by phosphide inoculants of primary silicon in hypereutectic Al–Si alloys. *Materials Science and Technology* 17 (8): 901–905. https://doi.org/10.1179/026708301101510870.
94. Kattoh, H., A. Hashimoto, S. Kitaoka, M. Sayashi, and M. Shioda. 2002. Critical temperature for grain refining of primary Si in hyper-eutectic Al–Si alloy with phosphorus addition. *Journal of Japan Institute of Light Metals* 52 (1): 18–23. https://doi.org/10.2464/jilm.52.18.
95. Apelian, D., G.K. Sigworth, and K.R. Whaler. 1984. Assessment of grain refinement and modification of Al-Si foundry alloys by thermal analysis. *AFS Transactions* 92 (Paper 84–161): 297–307.
96. Yi, H., and D. Zhang. 2003. Morphologies of Si phase and La-rich phase in as-cast hypereutectic Al–Si–xLa alloys. *Materials Letters* 57 (16–17): 2523–2529. https://doi.org/10.1016/S0167-577X(02)01305-8.
97. Yi, H., D. Zhang, T. Sakata, and H. Mori. 2003. Microstructures and La-rich compounds in a Cu-containing hypereutectic Al–Si alloy. *Journal of Alloys and Compounds* 354 (1–2): 159–164. https://doi.org/10.1016/S0925-8388(03)00022-7.
98. Jiang, Q.C., C.L. Xu, M. Lu, and H.Y. Wang. 2005. Effect of new Al–P–Ti–TiC–Y modifier on primary silicon in hypereutectic Al–Si alloys. *Materials Letters* 59 (6): 624–628. https://doi.org/10.1016/j.matlet.2004.10.042.
99. Xu, C.L., H.Y. Wang, Y.F. Yang, and Q.C. Jiang. 2007. Effect of Al–P–Ti–TiC–Nd2O3 modifier on the microstructure and mechanical properties of hypereutectic Al–20 wt.%Si alloy. *Materials Science and Engineering A* 452–453: 341–346. https://doi.org/10.1016/j.msea.2006.10.114.
100. Vives, C., and C. Perry. 1986. Effects of electromagnetic stirring during the controlled solidification of tin. *International Journal of Heat and Mass Transfer* 29 (1): 21–33. https://doi.org/10.1016/0017-9310(86)90031-1.
101. Mondolfo, L.F., and J.G. Barlock. 1975. Effect of superheating on structure of some aluminum alloys. *Metallurgical Transactions B* 6 (4): 565–572. https://doi.org/10.1007/BF02913849.
102. Yamada, S.-I., K. Kobayashi, H.G. Ji, and S. Tsukahara. 2002. Morphology of primary silicon crystals during isothermal holding at solid-liquid co-existent temperature on hyper-eutectic Al–Si alloys. *Journal of Japan Institute of Light Metals* 52 (4): 174–178. https://doi.org/10.2464/jilm.52.174.
103. Lojen, G., A. Krizman, and I. Anzel. 1997. *Livarski Vestnik* 44: 1–8.
104. Bergsma, S.C., M.C. Tolle, M.E. Kassner, X. Li, and E. Evangelista. 1997. Semi-solid thermal transformations of Al–Si alloys and the resulting mechanical properties. *Materials Science and Engineering A* 237 (1): 24–34. https://doi.org/10.1016/S0921-5093(97)00112-3.
105. Robles Hernandez, F.C., M.B. Djurdjevic, W.T. Kierkus, and J.H. Sokolowski. 2005. Calculation of the liquidus temperature for hypo and hypereutectic aluminum silicon alloys. *Materials Science and Engineering A* 396 (1–2): 271–276. https://doi.org/10.1016/j.msea.2005.01.024.
106. Ward, P.J., H.V. Atkinson, P.R.G. Anderson, L.G. Elias, B. Garcia, L. Kahlen, and J.M. Rodriguez-ibabe. 1996. Semi-solid processing of novel MMCs based on hypereutectic aluminium-silicon alloys. *Acta Materialia* 44 (5): 1717–1727. https://doi.org/10.1016/1359-6454(95)00356-8.
107. Chen, C.M., C.C. Yang, and C.G. Chao. 2004. Thixocasting of hypereutectic Al–25Si–2.5Cu–1Mg–0.5Mn alloys using densified powder compacts. *Materials Science and Engineering A* 366 (1): 183–194. https://doi.org/10.1016/j.msea.2003.09.063.
108. Gonzalez-Reyes, L., I. Hernández-Pérez, F.C. Robles-Hernández, H. Dorantes-Rosales, and E.M. Arce-Estrada. 2008. Sonochemical synthesis of nanostructured anatase and study of the kinetics among phase transformation and coarsening as a function of heat treatment conditions. *Journal of the European Ceramic Society* 28 (8): 1585–1594. https://doi.org/10.1016/j.jeurceramsoc.2007.10.013.
109. Ma, A., K. Suzuki, N. Saito, Y. Nishida, M. Takagi, I. Shigematsu, and H. Iwata. 2005. Impact toughness of an ingot hypereutectic Al–23 mass% Si alloy improved by rotary-die equal-

channel angular pressing. *Materials Science and Engineering A* 399 (1–2): 181–189. https://doi.org/10.1016/j.msea.2005.03.009.

110. Dasgupta, R. 1997. Property improvement in Al–Si alloys through rapid solidification processing. *Journal of Materials Processing Technology* 72 (3): 380–384. https://doi.org/10.1016/S0924-0136(97)00198-2.
111. Satoh, T., K. Okimoto, S. Nishida, and K. Matsuki. 1995. Superplastic-like behavior of rapid-solidification-processed hyper-eutectic Al-Si P/M alloys. *Scripta Metallurgica et Materialia* 33 (5): 819–824. https://doi.org/10.1016/0956-716X(95)00291-3.
112. Ma, A., K. Suzuki, Y. Nishida, N. Saito, I. Shigematsu, M. Takagi, H. Iwata, A. Watazu, and T. Imura. 2005. Impact toughness of an ultrafine-grained Al–11mass%Si alloy processed by rotary-die equal-channel angular pressing. *Acta Materialia* 53 (1): 211–220. https://doi.org/10.1016/j.actamat.2004.09.017.

Erratum to: Al-Si Alloys: Automotive, Aeronautical, and Aerospace Applications

Francisco C. Robles Hernandez, Jose Martin Herrera Ramírez, and Robert Mackay

Erratum to:
F.C. Robles Hernandez et al., *Al-Si Alloys: Automotive, Aeronautical, and Aerospace Applications*, DOI 10.1007/978-3-319-58380-8

The original print and online versions of this book was inadvertently published with an uncorrected reference list and also the spelling of the (Dr. Robles Hernandez') name was incorrect. The name should read as follows:

Francisco C. Robles Hernandez

The updated online version of this book can be found at
https://doi.org/10.1007/978-3-319-58380-8

Francisco C. Robles Hernandez (✉)
College of Technology
University of Houston
Houston, Texas, USA
e-mail: fcrobles@central.uh.edu

Jose Martin Herrera Ramírez
Centro de Investigación en Materiales
Avanzados
Chihuahua, Mexico
e-mail: martin.herrera@cimav.edu.mx

Robert Mackay
Metallurgical & Heat Treatment
Nemak US/Canada Business Unit
Windsor, Ontario, Canada
e-mail: Robert.Mackay@Nemak.com

© Springer International Publishing AG 2017
F.C. Robles Hernandez et al., *Al-Si Alloys*, DOI 10.1007/978-3-319-58380-8_10

Index

© Springer International Publishing AG 2017

F.C. Robles Hernandez et al., *Al-Si Alloys*, DOI 10.1007/978-3-319-58380-8

MIX
Papier aus verantwortungsvollen Quellen
Paper from responsible sources
FSC® C105338
FSC
www.fsc.org

If you have any concerns about our products,
you can contact us on
ProductSafety@springernature.com

In case Publisher is established outside the EU,
the EU authorized representative is:
Springer Nature Customer Service Center GmbH
Europaplatz 3, 69115 Heidelberg, Germany

Printed by Libri Plureos GmbH
in Hamburg, Germany